Food Microbiology and Food Safety

Series Editor:
Michael P. Doyle

More information about this series at http://www.springer.com/series/7131

Food Microbiology and Food Safety Series

The Food Microbiology and Food Safety series is published in conjunction with the International Association for Food Protection, a non-profit association for food safety professionals. Dedicated to the life-long educational needs of its Members, IAFP provides an information network through its two scientific journals (Food Protection Trends and Journal of Food Protection), its educational Annual Meeting, international meetings and symposia, and interaction between food safety professionals.

Michele Jay-Russell • Michael P. Doyle
Editors

Food Safety Risks from Wildlife

Challenges in Agriculture, Conservation, and Public Health

Springer

Editors
Michele Jay-Russell
Western Center for Food Safety
University of California, Davis
Davis, CA, USA

Michael P. Doyle
Center of Food Safety
University of Georgia
Griffin, GA, USA

Food Microbiology and Food Safety
ISBN 978-3-319-79627-7 ISBN 978-3-319-24442-6 (eBook)
DOI 10.1007/978-3-319-24442-6

Springer Cham Heidelberg New York Dordrecht London

Softcover re-print of the Hardcover 1st edition 2016

Printed on acid-free paper

Springer International Publishing AG Switzerland is part of Springer Science+Business Media (www.springer.com)

To Dr. William E. Keene (1957–2013), Senior Epidemiologist, Oregon Public Health Division.

Preface

The purpose of this book is to share information and insights into emerging topics related to wildlife and food safety. While other texts have delved extensively into wildlife conservation, agriculture, and resource management, this book uniquely brings these subject areas together in a food safety context.

The first half of the book addresses the prevalence, epidemiology, and ecology of foodborne pathogens in wildlife hosts. Beginning with a review of the major bacterial, parasitic, and viral pathogens associated with wildlife, the following chapters cover a wide range of international wildlife and food safety topics. The chapters not only bring together the available science in this relatively new area of research but also highlight the societal and economic implications where public health, food production, and wildlife conservation priorities sometimes conflict.

In the second half of the book, strategies to mitigate microbial food safety risks from wildlife hosts are presented. Novel approaches in risk communication, co-management, and One Health are highlighted broadly by the authors through the description of real-world experiences. There is an emphasis on produce food safety because of the many recent foodborne disease outbreaks linked to contaminated fruit and vegetable crops, and the promulgation of new on-farm regulations by the US Food and Drug Administration (FDA) Food Safety Modernization Act (FSMA) Produce Safety Rule. Wildlife intrusions are one of the potential sources of microbial contamination addressed in the proposed FSMA rule.

Because of the unique role of wildlife in human societies, the book covers topics not usually addressed in scholarly discussions on food safety (e.g., human-wildlife conflict, wildlife-livestock interactions, habitat loss, illegal wildlife trade, endangered and invasive species). Balancing food safety, agriculture, and conservation goals are underlying themes throughout the book. Historically, wild animals have been used by humans for food, clothing, recreation, entertainment, and other utilitarian purposes. But, increasingly attitudes about wildlife are shifting toward more ecologistic, humanistic, and moralistic feelings and beliefs (as defined by Kellert and Westervelt (1983)). These influences are addressed throughout the book, with many examples provided by the authors.

Our goal for this book is to advance the understanding of wildlife and food safety and assist in the development of the best science and policy to protect the public health, support a robust agriculture industry, and promote environmental stewardship in a world shared by humans and wildlife.

We thank our esteemed authors for sharing their expertise, time, and passion to create this book. It was an honor to work with such a distinguished group who provided a variety of timely and well-informed perspectives. We are also grateful to Susan Safren for recognizing the importance of this topic and inviting us to publish this book, and to Michael Koy for his conscientious assistance with production. We have dedicated this book to the memory of Dr. William Keene, an incredible epidemiologist who had a passion for solving food-associated outbreaks in the interest of preventing future outbreaks. He never shied away from a public health challenge (or fecal sample).

Davis, CA, USA Michele Jay-Russell
Griffin, GA, USA Michael P. Doyle

Reference

Kellert SR, Westervelt MO (1983) Historical trends in American animal use and perception. Int J Study Anim Probl 4:133–146

Contents

1 **Overview: Foodborne Pathogens in Wildlife Populations** ... 1
Marilyn C. Erickson

2 **Emerging Viral Zoonoses from Wildlife Associated with Animal-Based Food Systems: Risks and Opportunities** ... 31
Kris A. Murray, Toph Allen, Elizabeth Loh, Catherine Machalaba, and Peter Daszak

3 **A European Perspective on the Transmission of Foodborne Pathogens at the Wildlife–Livestock–Human Interface** ... 59
Nora Navarro-Gonzalez, María Ugarte-Ruiz, Lucas Domínguez, and Francisco Ruiz-Fons

4 **Microbiological Hazards of Wild Birds and Free-Range Chickens** ... 89
Susan Sanchez, Monique França, and Nicole M. Nemeth

5 **Molecular Tools for Monitoring and Source-Tracking *Salmonella* in Wildlife and the Environment** ... 131
Anita Wright, Amber Ginn, and Zhiyao Luo

6 **Reducing the Risk of Foodborne Transmission of Nipah Virus** ... 151
Stephen P. Luby, Nazmun Nahar, and Emily S. Gurley

7 **A Survey of How Growers, Shippers, and Handlers Address Food Safety Risks from Wildlife in Leafy Greens** ... 169
Henry Giclas and Diane Wetherington

8 **Keeping Wildlife Out of Your Food: Mitigation and Control Strategies to Reduce the Transmission Risk of Food-Borne Pathogens** ... 183
Alan B. Franklin and Kurt C. VerCauteren

9 Co-management: Balancing Food Safety, the Environment, and the Bottom Line 201
Mary Bianchi and Karen Lowell

10 Recommendations to Regulations: Managing Wildlife and Produce Safety on the Farm 217
Gretchen L. Wall and Elizabeth A. Bihn

11 A One Health Approach to Wildlife and Food Safety 241
Amanda Arens, Cheryl Scott, and Bennie Osburn

Index 249

Contributors

Toph Allen EcoHealth Alliance, New York, NY, USA

Amanda Arens Western Institute for Food Safety and Security, School of Veterinary Medicine, University of California, Davis, Davis, CA, USA

Mary Bianchi University of California Cooperative Extension, San Luis Obispo, CA, USA

Elizabeth A. Bihn Department of Food Science, Cornell University, Geneva, NY, USA

Peter Daszak EcoHealth Alliance, New York, NY, USA

Lucas Domínguez VISAVET Health Surveillance Centre, Complutense University of Madrid, Madrid, Spain

Marilyn C. Erickson Center for Food Safety, Department of Food Science and Technology, University of Georgia, Griffin, GA, USA

Monique França College of Veterinary Medicine, The University of Georgia, Athens, GA, USA

Alan B. Franklin USDA-APHIS Wildlife Services National Wildlife Research Center, Fort Collins, CO, USA

Henry Giclas Western Growers, Irvine, CA, USA

Amber Ginn Food Science and Human Nutrition Department, University of Florida, Gainesville, FL, USA

Emily S. Gurley Centre for Communicable Diseases, ICDDR, B, Dhaka, Bangladesh

Elizabeth Loh EcoHealth Alliance, New York, NY, USA

Karen Lowell United States Department of Agriculture Natural Resources Conservation Service, Salinas, CA, USA

Stephen P. Luby Center for Innovation in Global Health, Stanford University, Stanford, CA, USA

Zhiyao Luo Food Science and Human Nutrition Department, University of Florida, Gainesville, FL, USA

Catherine Machalaba EcoHealth Alliance, New York, NY, USA

Kris A. Murray EcoHealth Alliance, New York, NY, USA,

Grantham Institute - Climate Change and the Environment, Imperial College London, London, UK

Nazmun Nahar Swiss Tropical Public Health Institute, Basel, Switzerland

Nora Navarro-Gonzalez Servicio de Ecopatología de Fauna Salvaje, Universitat Autònoma de Barcelona, Barcelona, Spain

Nicole M. Nemeth Ontario Veterinary College, University of Guelph, Guelph, ON, Canada

Bennie Osburn Western Institute for Food Safety and Security, School of Veterinary Medicine, University of California, Davis, Davis, CA, USA

Francisco Ruiz-Fons Health & Biotechnology (SaBio) Group, Spanish Wildlife Research Institute (IREC; CSIC-UCLM-JCCM), Ciudad Real, Spain

Susan Sanchez College of Veterinary Medicine, The University of Georgia, Athens, GA, USA

Cheryl Scott Western Institute for Food Safety and Security, School of Veterinary Medicine, University of California, Davis, Davis, CA, USA

María Ugarte-Ruiz VISAVET Health Surveillance Centre, Complutense University of Madrid, Madrid, Spain

Kurt C. VerCauteren USDA-APHIS Wildlife Services National Wildlife Research Center, Fort Collins, CO, USA

Gretchen L. Wall Department of Food Science, Cornell University, Ithaca, NY, USA

Diane Wetherington iDecisionSciences, Seattle, WA, USA

Anita Wright Food Science and Human Nutrition Department, University of Florida, Gainesville, FL, USA

Chapter 1
Overview: Foodborne Pathogens in Wildlife Populations

Marilyn C. Erickson

Abstract Numerous bacterial and parasitic pathogens may be transmitted through food and included in that group are zoonotic pathogens that not only proliferate within domesticated animals but may also be resident within wildlife. As a result of wildlife being a pathogen reservoir and the ability of this animal group to easily intrude on farms, wildlife contributes to the maintenance of infections on domestic farms as well as serves as an environmental source of fresh produce contamination. To discern the degree to which wildlife represents a food safety risk, this overview first summarizes those documented incidents in which contaminated wildlife has been directly or indirectly associated with human illness. It continues with providing a set of tables that document the results of studies directed at assessing the prevalence of bacterial, parasitic, and viral pathogen contamination in mammals, birds, and amphibians and reptiles. To understand the strengths and limitations of those surveillance studies, discussion is included that describes how sample source, cultivation conditions, sample size and number, and specificity of the detection method may impact the data collected. Discussion on factors that contribute to pathogen transmission to wildlife are also presented and include the physiological state of the animal, behavioral features of the animal that contribute to intra- and interspecies interactions, seasonal effects on transmission, and management practices applied to wildlife or domestic animals. The overview concludes with a section directed at discussing other drawbacks to pathogen contamination of wildlife and includes contamination of water sources and wildlife serving as a reservoir for antibiotic resistance and emerging pathogens.

Keywords *Campylobacter* • *Cryptosporidium* • *Escherichia coli* • Foodborne pathogens • Foodborne disease outbreak • *Listeria monocytogenes* • Prevalence • *Salmonella* • *Trichinella* • Wildlife • Zoonosis

M.C. Erickson (✉)
Center for Food Safety, Department of Food Science and Technology, University of Georgia, Griffin, GA 30223, USA
e-mail: mericks@uga.edu

M. Jay-Russell, M.P. Doyle (eds.), *Food Safety Risks from Wildlife*, Food Microbiology and Food Safety, DOI 10.1007/978-3-319-24442-6_1

Introduction

Foodborne pathogen contamination of edible horticultural crops, often consumed raw or minimally processed (e.g., fruits, nuts, and vegetables), has over the past few decades been increasingly linked to foodborne illnesses, outbreaks, and recalls (Batz et al. 2012; Berger et al. 2010). Although there are over 250 pathogens and toxins that can be transmitted by food, 31 are classified as major foodborne pathogens (Scallan et al. 2011), and included among that group are zoonotic pathogens or pathogens that affect multiple animal species. Moreover, those bacterial and parasitic pathogens that contribute to the greatest proportion of illnesses and outbreaks in humans (*Campylobacter jejuni*, nontyphoidal *Salmonella enterica*, Shiga toxin-producing *Escherichia coli*, *Listeria monocytogenes*, *Cryptosporidium*) are largely attributed to their proliferation within domesticated animals. However, wild animals may also serve as a reservoir of zoonotic pathogens affecting humans and domesticated animals. It has been reported that 26 % of human pathogens infect both domestic and wild animals (Cleaveland et al. 2001) and, therefore, there is concern that wildlife contributes to the maintenance of infections on domestic animal farms (Liebana et al. 2003). Given the ease with which wild animals may intrude and defecate in produce fields, this group of animals has also raised concern that they are a likely environmental source for contamination of fresh produce (Jay-Russell 2013; Langholz and Jay-Russell 2013).

Another food safety risk from infection of wild animals by human pathogens is the consumption of their meat when the animal is killed and not properly dressed and cooked. Moreover, another potential consequence of pathogen contamination of wildlife is their potential to serve as a reservoir for emerging diseases. For example, approximately 75 % of all diseases, including zoonoses which have emerged in the last few decades, are of wildlife origin (Jones et al. 2008). Based on the concerns associated with foodborne pathogens in wildlife populations, this chapter will provide an overview of this subject and recount some of the incidents in which contaminated wildlife has been directly or indirectly associated with human illness, summarize some of the data collected on the prevalence of foodborne pathogens in wildlife, briefly address factors that affect prevalence levels in wildlife, and finally touch on other drawbacks to pathogen contamination of wildlife that adversely affect humans. The material presented in this chapter is not intended to be comprehensive but to provide a basic understanding of the subject on which subsequent chapters will expand.

Illnesses/Outbreaks Attributed to Contamination of Wildlife

Direct Association: Consumption of Contaminated Meat

Prior to the domestication of animals, wild animals served as the major source of protein for humans. Today, this proportion has decreased dramatically, but consumption of wild game and reptile meat continues to occur by groups that value

Table 1.1 Examples of reports documenting links between human illnesses/outbreaks and consumption of pathogen-contaminated wildlife

Source	Pathogen	Evidence for linkage between wildlife and human illness	Reference
Deer jerky	*E. coli* O157:H7	PFGE patterns of isolates from the patients, jerky, and source deer were identical	Keene et al. (1997)
Undercooked venison		PFGE pattern of the uncooked venison isolate was indistinguishable from the pattern of the clinical isolate	Rabatsky-Ehr et al. (2002)
Deer sausage		PFGE patterns of isolates from deer sausage and patients were identical	Ahn et al. (2009)
Undercooked venison	*E. coli* O103:H2 and O145:NM	PFGE patterns of isolates from patients and venison were indistinguishable	Rounds et al. (2012)
Uncooked liver from wild boar	Hepatitis E	Two patients eating the liver contracted the illness but none of the liver remained for analysis of pathogen contamination	Matsuda et al. (2003)
Raw deer meat		DNA sequence from leftover frozen deer meat was 99.7–100 % identical to the viruses recovered from the four human patients	Tei et al. (2003)
Wild boar meat		Genotype 3 hepatitis E virus RNA was detected in both patient serum and wild boar meat	Li et al. (2005)
Wild boar meat	*Trichinella spiralis*	47 Thai patients became ill after eating wild boar meat. Encysted *Trichinella* larvae were identified in implicated meat	Marva et al. (2005)

PFGE pulsed-field gel electrophoresis

these animals for subsistence or sport hunting. As a result, there are multiple reports whereby consumption of contaminated meat has been directly linked to human illness (Table 1.1). Additional incidents of infections associated with consumption of reptile meat have also been tabulated in the review of Magnino et al. (2009).

In some cases, zoonotic pathogens (e.g., *Brucella* spp., *Trichinella* spp.) have been controlled in domestic livestock herds in developed countries, but continue to circulate in wild animal populations and cause human infections via consumption of mishandled or undercooked game meat. For example, swine *Trichinella* infection has been virtually eliminated in US swine raised in confinement, but human cases are still reported due to transmission via feral swine, bear, and other wild game meat. Additionally, concerns have been raised regarding the potential for infected wild animals to spread the parasite to domestic swine raised outdoors for "pastured pork," a growing niche market (Burke et al. 2008).

An important point to acknowledge, however, is that with many of these foodborne pathogens, the potential for causing illness and the severity of disease will depend on the strain. For example, genotypes 1 and 2 of hepatitis E virus are restricted to humans and associated with epidemics in developing countries, whereas typically sporadic cases are associated with the zoonotic genotypes 3 and 4 (Meng 2011). As another example, most human illnesses are caused by only three of the serotypes (1/2a, 1/2b, and 4b) of *L. monocytogenes* (Jay-Russell 2013).

Indirect Association: Contamination of Produce Fields

Attention to wild animals serving as a vector for pathogen contamination of produce fields arose following the highly publicized 2006 *Escherichia coli* O157:H7 outbreak associated with ready-to-eat packaged baby spinach that was traced to one field in the central California coast (Jay et al. 2007). In that outbreak, the outbreak strain was isolated from both domestic cattle and feral swine sharing rangeland adjacent to the implicated spinach field. Moreover, evidence of intrusion by the feral swine, including tracks, rooting, or feces in crop fields and adjacent vineyards, was documented.

Other cases that have implicated wild animals as potential sources of contamination of field crops and subsequent infection of humans have been documented. In Finland in 2004, schoolchildren became ill after eating a carrot–white cabbage mixture, with *Yersinia pseudotuberculosis* identified as the likely cause for illness (Kangas et al. 2008). Traceback of the carrots to the processor and farms growing the carrots revealed the presence of this bacterium in one environmental sample from the carrot-peeling processing line and from a pooled sample of common shrew intestines collected from one of the farms. In Alaska, 63 cases of laboratory-confirmed *C. jejuni* infections that occurred in 2008 were associated with the consumption of raw shelled peas (Gardner et al. 2011). Pulsed-field gel electrophoresis (PFGE) patterns of clinical isolates, and pea and Sandhill crane fecal samples taken from the implicated pea farm located near a crane stopover and breeding site, were indistinguishable. Finally, in Oregon in 2011, 14 cases of laboratory-confirmed *E. coli* O157:H7 infections were associated with consuming strawberries purchased at roadside stands or farmers' markets (Laidler et al. 2013). A single farm was identified as the source of the contaminated strawberries, and environmental samples containing visible deer pellets that were collected at the farm were indistinguishable from the outbreak pattern by PFGE.

Prevalence of Foodborne Pathogens in Wildlife

Over the years, numerous studies have been conducted to address the prevalence of bacterial, parasitic, and viral pathogens in wildlife. Initial studies were focused primarily on assessing the degree of contamination resident within an animal's population solely, whereas studies more recently have focused on understanding the factors that contribute to the prevalence in wildlife. For this review, nearly 90 % of the data items included in Tables 1.2, 1.3, and 1.4, covering prevalence of foodborne pathogens in various groups of animals, were obtained from studies conducted in the past

10 years. Results from earlier studies were included to provide an example of a particular animal type or pathogen that may not have been addressed in a recent study. However, there is the possibility that data from older studies are no longer relevant if conditions under which they occurred no longer exist. Additional examples of the prevalence of foodborne pathogens in wildlife have been reviewed in other publications (Simpson 2002; Meerburg and Kijlstra 2007; Benskin et al. 2009; Ferens and Hovde 2011; Langholz and Jay-Russell 2013). In those reviews as well as the data presented in Tables 1.2, 1.3, and 1.4, one observation that is pervasive is the wide range of frequency that foodborne pathogens are detected in wildlife. In the following section, to understand the limitations and strengths of different studies, factors that contribute to pathogen detection in wildlife are discussed.

Factors Affecting Prevalence Levels in Wildlife

Methodology Used for Surveillance

Sample source. One of the common types of samples collected to assess the prevalence of foodborne enteric bacterial pathogens in wildlife is fecal pellets. The assumption in collecting this type sample is that the enteric pathogen either survives in or colonizes the gut of the animal and then is shed with the feces. Studies based on this type of sample, however, may be underestimating the prevalence due to a number of shortcomings. First and foremost is the possibility that contaminated wild animals may only intermittently shed the pathogen as has been reported for pigeons (De Herdt and Devriese 2000). Negative results may also occur when delays in collection of the fecal pellets occur but would vary with the pathogen as they have different degrees of susceptibility to desiccation. Moreover, in collecting fecal waste, there is the assumption that it represents the population at large and that may not be the case, especially if the animals become sick upon infection. Additionally, even a trained biologist may not be able to identify the source of the fecal material on the ground, which may require another method such as wildlife trail cameras or the use of genetic markers to accurately identify feces from different animals. Cloaca or rectum swabs are therefore more accurate in assessing whether carriage of the pathogen by the animal is occurring, but these require capture of the animals. Only on rare occasions are the animal's extremities sampled (Burt et al. 2012) to determine if the animal is serving as a pathogen vector.

Cultivation bias. To detect low levels of bacterial pathogens in a matrix such as feces or food, it is common practice to enrich the sample in a culture broth to increase their numbers and then qualitatively detect their presence using either selective media for colony isolation or a polymerase chain reaction (PCR) assay to screen for the pathogen's DNA. Critical to this approach is the assumption that viable pathogen cells will multiply under the enrichment conditions in the allotted time frame. Unfortunately, it has been observed that *Salmonella* strains vary in their ability to grow in enrichment cultures containing bovine feces (Singer et al. 2009), with strains of serogroups C2 and E more likely to dominate in enrichment culture

Table 1.2 Prevalence of zoonotic pathogens in mammals

Pathogen	Sample description with % prevalence of total number of samples tested in parentheses	Reference
Campylobacter jejuni	Rectal samples from *hares* (4.3 % of 23) in woodlands	Rosef et al. (1983)
	Rectal samples from *rodents* including wood mouse and bank vole (0.0 % of 44) in woodlands	
	Colon, intestinal, or muscle samples from *rodents* including brown rat, house mouse, and yellow-necked mouse (4.5 % of 201) on nine pig farms and five chicken farms	Backhans et al. (2013)
Campylobacter spp.	Fecal samples from *deer* (19.5 % of 113) in Oldman River watershed, Alberta, Canada	Jokinen et al. (2011)
	Fecal samples from *red deer* (0.0 % of 295), *wild boar* (65.5 % of 287), and *other ungulates* including fallow deer and mouflon (0.0 % of 9) in south-central Spain	Díaz-Sánchez et al. (2013)
	Rectal swabs of *wild boars* (43.8 % of 121) and *Sika deer* (0.0 % of 128) in Japan	Sasaki et al. (2013)
Clostridium difficile	Fecal samples from *feral swine* (4.4 % of 161) in North Carolina	Thakur et al. (2011)
	Paws, tail, and snout from *house mouse* (66.0 % of 53) on pig farm	Burt et al. (2012)
	Colons from *rats* (13.1 % of 724) in inner-city neighborhood of Vancouver, Canada	Himsworth et al. (2014)
Enteropathogenic *E. coli* (EPEC), Shiga toxin-producing *E. coli*, enterohemorrhagic *E. coli*	Fecal samples of *roe deer* (17.3 % of 52) and *red deer* (13.6 % of 81) in Belgium	Bardiau et al. (2010)
Shiga toxin-producing *E. coli* (STEC)	Fecal pellets from *rabbits* (20.6 % of 97) during summer on 16 dairy and beef farms	Schaife et al. (2006)
	Tonsil samples from *wild boars* (9.1 % of 153) from Geneva, Switzerland	Wacheck et al. (2010)
	Fecal samples from *red deer* (33.7 % of 264), *wild boar* (3.6 % of 301), and *other ungulates*, including fallow deer and mouflon (33.3 % of 9) in south-central Spain	Díaz-Sánchez et al. (2013)
	Fecal and rectal swabs from *roe deer* (52.5 % of 179), *wild boars* (8.4 % of 262), and *foxes* (1.9 % of 260) from northwest Spain	Mora et al. (2012)
	Fecal samples of *ungulates* (19.4 % of 160) in Idaho	Gilbreath et al. (2009)
	Fecal samples from *roe deer* (73.3 % of 30) and *red deer* (70.0 % of 30) from a national park in Germany	Eggert et al. (2013)

(continued)

Table 1.2 (continued)

Pathogen	Sample description with % prevalence of total number of samples tested in parentheses	Reference
E. coli O157:H7	Colon stool samples from *rodents* including brown rat, wood mice, and house mouse (21.0 % of 19) on beef cattle farm	Čížek et al. (1999)
	Fecal samples from *white-tailed deer* (0.2 % of 1608) in Nebraska	Renter et al. (2001)
	Rectroanal mucosal swabs from *roe deer* (0.0 % of 20), *red deer* (1.5 % of 206), *fallow deer* (0.0 % of 6), and *mouflon* (0.0 % of 11) during hunting season in southwestern Spain	García-Sánchez et al. (2007)
	Fecal samples from *wild boars* (3.3 % of 212) in southwest Spain	Sánchez et al. (2010)
	Buccal swabs, colonic feces, rectal-anal swabs, and tonsils from *feral swine* (40.0 % of 30) on a cattle ranch in California	Jay-Russell et al. (2012)
	Fecal samples from *rodents* (0.2 % of 1043) in 13 agricultural systems (nine produce farms, three cow-calf rangeland operations, and one beef feedlot)	Kilonzo et al. (2013)
E. coli O157:H7, sorbitol-fermenting	Fecal samples from *red deer* (1.1 % of 264) during hunting season in south-central Spain	Díaz et al. (2011)
Non-O157 STEC	Fecal samples from *wild boars* (5.2 % of 212) in southwest Spain	Sánchez et al. (2010)
	Fecal samples in *ruminants*, including red deer, roe deer, fallow deer, and mouflon (23.9 % of 243) in southwest Spain	Sánchez et al. (2009)
	Rectal swabs from several types of *rodents* (4.8 % of 145) in city parks in Buenos Aires, Argentina	Blanco Crivelli et al. (2012)
Atypical EPEC	Fecal samples from *coyotes* (4.9 % of 103) in leafy greens production region at U.S.–Mexico border	Jay-Russell et al. (2014)
Listeria monocytogenes	Tonsil samples from *wild boars* (17.0 % of 153) from Geneva, Switzerland	Wacheck et al. (2010)
Mycobacterium bovis	Tissue from *ferrets* (17.9 % of 548) and *stoats* (1.6 % of 62) in New Zealand	Ragg et al. (1995)
	Tissue rom *hedgehogs* (5.8 % of 69) in tuberculosis-endemic areas of New Zealand	Lugton et al. (1995)

(continued)

Table 1.2 (continued)

Pathogen	Sample description with % prevalence of total number of samples tested in parentheses	Reference
Salmonella spp.	Intestinal samples from *mice* (5.1 % of 175) on six swine farms	Barber et al. (2002)
	Tonsil samples from *wild boars* (12.4 % of 153) from Geneva, Switzerland	Wacheck et al. (2010)
	Fecal samples from *coyotes* (32.0 % of 103) in leafy greens production region at U.S.–Mexico border	Jay-Russell et al. (2014)
	Fecal swabs from *raccoons* in rural (7.8 % of 28), forested (8.7 % of 332), and suburban (5.7 % of 278) areas of western Pennsylvania	Compton et al. (2008)
	Fecal samples from *deer* (0.0 % of 113) in Oldman River watershed, Alberta, Canada	Jokinen et al. (2011)
	Fecal samples from *white-tailed deer* (1.0 % of 500) in southeastern Nebraska	Renter et al. (2006)
	Fecal samples from *rodents* including rats, mice, and voles on 13 production-infected (5.2 % of 135) and non-infected (0.0 % of 68) farms (five pig and eight cattle) and surrounding areas without production animals (0.0 % of 22)	Skov et al. (2008)
	Fecal or cloacal swabs from *Diprotodontia*, including koala, wombats, and possums (1.7 % of 291)	Parsons et al. (2010)
	Fecal samples from *coyotes* (5.0 % of 40), *deer* (1.9 % of 104), *elk* (2.6 % of 39), *wild pigs* (2.4 % of 41), *rabbits* (0.0 % of 57), *raccoons* (0.0 % of 2), and *skunks* (30.7 % of 13) in major produce region of California	Gorski et al. (2011)
	Fecal samples from *feral swine* (5.0 % of 161) in North Carolina	Thakur et al. (2011)
	Fecal and lymph node samples from *wild boars* (41.1 % of 543) at 93 locations in Australia	Cowled et al. (2012)
	Colon, intestinal, or muscle samples from *rodents* including brown rat, house mouse, and yellow-necked mouse (0.0 % of 184) on nine pig farms and five chicken farms	Backhans et al. (2013)
	Fecal samples from *red deer* (0.3 % of 295), *wild boar* (1.2 % of 333), and *other ungulates*, including fallow deer and mouflon (0.0 % of 9) in south-central Spain	Díaz-Sánchez et al. (2013)
	Rectal or cloacal swabs of *deer* (0.0 % of 73) from the Eastern Shore of Virginia	Gruszynski et al. (2013)
	Fecal samples from *rodents* (2.9 % of 1043) in 13 agricultural systems (nine produce farms, three cow-calf rangeland operations, and one beef feedlot)	Kilonzo et al. (2013)
	Intestines, spleens, and livers from *rodents* including rats and mice (10.2 % of 88) on 13 pig farms	Andrés-Barranco et al. (2014)

(continued)

Table 1.2 (continued)

Pathogen	Sample description with % prevalence of total number of samples tested in parentheses	Reference
S. Enteritidis	Liver, spleen, and intestines of four types of *rodents* on *Salmonella*-infected (75.3 % of 483) and non-infected (0.0 % of 232) poultry farms	Henzler and Optiz (1992)
Yersinia enterocolitica	Tonsil samples from *wild boars* (34.6 % of 153) from Geneva, Switzerland	Wacheck et al. (2010)
	Rodents including mice and rats on pig (8.2 % of 110) or chicken (0.0 % of 55) farm	Backhans et al. (2011)
	Colon, intestinal, or muscle samples from *rodents* including brown rat, house mouse, and yellow-necked mouse (4.8 % of 189) on nine pig farms and five chicken farms	Backhans et al. (2013)
Hepatitis E	Sera from *brown rats* (76.8 % of 108) in inner city of Baltimore, Maryland, urban and rural regions of Hawaii, and New Orleans, Louisiana	Kabrane-Lazizi et al. (1999)
	Sera were immunologically assayed for RNA from *wild boar* (12.1 % of 1029), *red deer* (5.3 % of 38), and *roe deer* (0.0 % of 8) in The Netherlands	Rutjes et al. (2010)
	Sera were immunologically assayed from *Yezo deer* (34.8 % of 520) in Hokkaido, Japan	Tomiyama et al. (2009)
	Sera were immunologically assayed from *white-tailed deer* (62.7 % of 142) in Northern Mexico	Medrano et al. (2012)
Nipah virus	Sera from *large flying foxes* (32.8 % of 253) and *small flying foxes* (11.1 % of 117) in Malaysia	Rahman et al. (2013)
	Urine from *Lyle's flying foxes* (1.8 % of 2696) from seven colonies in central Thailand	Wacharapluesadee et al. (2010)
Cryptosporidium spp.	Scats from *wombats* (0.0 % of 55) on stream banks in riparian corridors in Australia	Borchard et al. (2010)
	Fecal samples from *coyotes* (22.2 % of 18), *mountain lions and bobcats* (0.0 % of 11), and *opossums* (25.0 % of 68) in the Monterrey Bay region of California	Oates et al. (2012)
	Colon, intestinal, or muscle samples from *rodents* including brown rat, house mouse, and yellow-necked mouse (11.0 % of 155) on nine pig farms and five chicken farms	Backhans et al. (2013)
	Fecal samples from 11 types of *rodents* (26.0 % of 285) in 13 agricultural systems (nine produce farms, three cow-calf rangeland operations, and one beef feedlot)	Kilonzo et al. (2013)

(continued)

Table 1.2 (continued)

Pathogen	Sample description with % prevalence of total number of samples tested in parentheses	Reference
Giardia spp.	Scats from *wombats* (0.0 % of 55) on stream banks in riparian corridors in Australia	Borchard et al. (2010)
	Fecal samples from *coyotes* (38.9 % of 18), *mountain lions and bobcats* (18.2 % of 11), and *opossums* (14.7 % of 68) in the Monterrey Bay region of California	Oates et al. (2012)
	Colon, intestinal, or muscle samples from *rodents* including brown rat, house mouse, and yellow-necked mouse (13.5 % of 155) on nine pig farms and five chicken farms	Backhans et al. (2013)
	Fecal samples from 11 types of *rodents* (24.2 % of 285) in 13 agricultural systems (nine produce farms, three cow-calf rangeland operations, and one beef feedlot)	Kilonzo et al. (2013)
Toxoplasma gondii	Brain and heart tissue from six types of *rodents* (11.9 % of 101) on three organic pig farms	Kijlstra et al. (2008)
	Sera were immunologically analyzed from *roe deer* (52.0 % of 73) and *red deer* (0.0 % of 7) in Belgium	De Craeye et al. (2011)
	Brain samples from *red foxes* (18.8 % of 304), *roe deer* (5.0 % of 20), and *red deer* (0.0 % of 13) in Belgium	De Craeye et al. (2011)
	Colon, intestinal, or muscle samples from *rodents* including brown rat, house mouse, and yellow-necked mouse (0.0 % of 147) on nine pig farms and five chicken farms	Backhans et al. (2013)
	Sera were immunologically analyzed from *roe deer* (46.4 % of 222) in an agro-system in France	Candela et al. (2014)
Trichinella spp.	Colon, intestinal, or muscle samples from *rodents* including brown rat, house mouse, and yellow-necked mouse (0.0 % of 160) on nine pig farms and five chicken farms	Backhans et al. (2013)

mixtures than strains of serogroups B or C1 (Gorski 2012). To circumvent this limitation due to culture bias, it was recommended that analysis of environmental samples includes multiple enrichment protocols (Gorski 2012); however, there still remains the possibility that the *Salmonella* strain would not be detected if it was incapable of outcompeting indigenous fecal bacteria and growing to sufficient numbers for detection through traditional protocols.

Another complication in the detection of pathogens can occur when using cultural cultivation conditions prior to PCR as was reported for a study of wild mule deer and elk in Idaho (Gilbreath et al. 2009). In this case, loss of the hybridizable *stx* genotype occurred in up to 80 % of subcultured isolates of Shiga toxin-producing *E. coli* (STEC). The question therefore remains as to whether the instability of these genes would have occurred under field conditions and, hence, the risk of human illness associated with these organisms compared to stable STEC isolates.

Table 1.3 Prevalence of zoonotic pathogens in birds

Pathogen	Sample description with % prevalence of total number of samples tested in parentheses	Reference
Campylobacter spp.	Fecal samples from *gulls* (13.7 % of 205) from three coastal locations in Ireland	Moore et al. (2002)
	Cloacal swabs from *yellow-legged gull chicks* (10.4 % of 182) in northeast Spain	Ramos et al. (2010)
	Fecal samples from *gulls* (33.3 % of 3), *geese* (26.2 % of 80), and *ducks* (42.1 % of 38) in Oldman River watershed, Alberta, Canada	Jokinen et al. (2011)
	Fecal samples from *black swan* (45.0 % of 80), *Canada geese* (40.0 % of 80), *duck* (29.0 % of 80), and *gulls* (59.0 % of 80) from New Zealand	Moriarty et al. (2011)
	Oropharyngeal and cloacal swabs from *common tern chicks* (0.6 % of 179) during the breeding season in New Jersey	Rivera et al. (2012)
	Fecal samples from *European starlings* (50.4 % of 113) on dairy cattle farm in NE, Ohio	Sanad et al. (2013)
	Cloacal samples from *griffon vultures* (1.0 % of 97) in eastern Spain	Marin et al. (2014)
Campylobacter jejuni	Fecal samples or cloacal swabs from *graylag geese* (0.0 % of 219), *rock pigeons* (3.0 % of 200), and *mallards* (20.0 % of 5) in Norway	Lillehaug et al. (2005)
	Cloacal samples from feral *pigeons* (69.1 % of 94) in public parks and gardens in Madrid, Spain	Vázquez et al. (2010)
	Fecal samples from *California gulls* (1.2 % of 159) in southern California	Lu et al. (2011)
	Cloacal swab samples from urban *pigeons* (48.3 % of 1800) in coastal area of southern Italy	Gargiulo et al. (2014)
	Fecal and cloacal samples from ten species of *wild birds* (8.1 % of 781) in New Jersey, Delaware, and Pennsylvania	Keller and Shriver (2014)
	Fecal samples from 15 species of *wild birds* (7.4 % of 446) from two ruminant farm sites in Virginia and Maryland	Pao et al. (2014)
	Fecal or cloacal samples from *American crows* (55.1 % of 127) in California	Weis et al. (2014)
Clostridium difficile	Tissue from dead *house sparrows* (65.7 % of 35) on a commercial pig farm in The Netherlands	Burt et al. (2012)
Enteropathogenic *E. coli*	Fecal samples from *ducks* (54.0 % of 50) from a poultry farm in India and cloacal samples from domestic *pigeons* (6.0 % of 100) from seven fanciers in India	Farooq et al. (2009)
	Cloacal or fecal swabs from 15 *avian host sources* (15.3 % of 412) from five locations in British Columbia, Canada	Chandran and Mazumder (2014)
Shiga toxin-producing *E. coli*	Fecal samples from *wild birds*, comprised of 24 species (1.6 % of 244) from cattle and pig farms in Denmark	Nielsen et al. (2004)
	Fecal samples from *ducks* (0.0 % of 50) from a poultry farm in India and cloacal samples from domestic *pigeons* (9.0 % of 100) from seven fanciers in India	Farooq et al. (2009)
	Cloacal or fecal swabs from 15 *avian host sources* (22.6 % of 412) from five locations in British Columbia, Canada	Chandran and Mazumder (2014)

(continued)

Table 1.3 (continued)

Pathogen	Sample description with % prevalence of total number of samples tested in parentheses	Reference
E. coli O157:H7	Fecal samples from *gulls* (0.0 % of 3), *geese* (1.2 % of 80), and *ducks* (2.6 % of 38) in Oldman River watershed, Alberta, Canada	Jokinen et al. (2011)
	Fecal or intestinal contents from *European starlings* (1.2 % of 430) from 150 dairy farms in northern Ohio	Williams et al. (2011)
	Cloacal swab samples from *urban pigeons* (7.8 % of 1800) in coastal area of southern Italy	Gargiulo et al. (2014)
Salmonella spp.	Fecal samples from *birds* (7.9 % of 38) from six swine production facilities in Illinois	Barber et al. (2002)
	Fecal samples or cloacal swabs from *graylag geese* (0.5 % of 219), *rock pigeons* (0.0 % of 200), and *mallards* (0.0 % of 5) in Norway	Lillehaug et al. (2005)
	Fecal samples from *barn swallows* (0.0 % of 500+) in northern, central, and southern Sweden	Haemig et al. (2008)
	Cloacal swabs from *birds* (55 species) at or near *Salmonella*-infected (1.5 % of 185) and non-infected (0.0 % of 1004) cattle and pig farms in Denmark and surrounding areas without production animals (0.0 % of 278)	Skov et al. (2008)
	Fecal or cloacal swabs of *birds* (0.0 % of 689) in Australia	Parsons et al. (2010)
	Cloacal swabs from *yellow-legged gull chicks* (17.0 % of 182) in northeast Spain	Ramos et al. (2010)
	Gastrointestinal tract samples from *European starlings* (2.5 % of 81) in three cattle-concentrated animal feeding operations	Carlson et al. (2011)
	Fecal samples from *birds* (6.6 % of 105) in major produce region of California	Gorski et al. (2011)
	Fecal samples from *gulls* (66.7 % of 3), *geese* (10.0 % of 80), and *ducks* (7.9 % of 38) in Oldman River watershed, Alberta, Canada	Jokinen et al. (2011)
	Fecal samples from *black swan* (0.0 % of 80), *Canada geese* (0.0 % of 80), *duck* (0.0 % of 80), and *gulls* (0.0 % of 80) from New Zealand	Moriarty et al. (2011)
	Oropharyngeal and cloacal swabs from *common tern chicks* (0.6 % of 179) during the breeding season in New Jersey	Rivera et al. (2012)
	Rectal or cloacal, or carapace swabs of *geese* (0.0 % of 7) and *gulls* (29.8 % of 47) from the Eastern shore of Virginia	Gruszynski et al. (2013)
	Cloacal samples from *griffon vultures* (52.6 % of 97) in eastern Spain	Marin et al. (2014)
	Fecal samples from 47 different species of *birds* (4.0 % of 672) on 41 pig farms in Northeast Spain	Andrés-Barranco et al. (2014)
	Cecal samples from *migratory birds*, including brown-headed cowbirds, common grackles, and cattle egrets (14.9 % of 376) during fall migration in Texas	Callaway et al. (2014)
	Fecal swabs from *waterfowl*, including Franklin's gull, kelp gull, grey gull, and Andean goose (6.1 % of 758) from eight sites in five Chilean regions	Fresno et al. (2013)
	Fecal samples from *gulls* (17.2 % of 360) from three landfill sites and on the Eastern shore of Virginia	Gruszynski et al. (2014)
	Fecal samples from 15 species of *wild birds* (0.2 % of 446) from two ruminant farm sites in Virginia and Maryland	Pao et al. (2014)

Table 1.3 (continued)

Pathogen	Sample description with % prevalence of total number of samples tested in parentheses	Reference
Salmonella Oranienburg	Fecal samples from *wild turkeys* (22.5 % of 71) from a home garden fertilized with raw horse manure	Jay-Russell et al. (2013)
S. Typhimurium	Cloacal swab samples from urban *pigeons* (0.9 % of 1800) in coastal area of southern Italy	Gargiulo et al. (2014)
Cryptosporidium spp.	Fecal samples from *gulls* (0.0 % of 205) from three coastal locations in Ireland	Moore et al. (2002)
	Fecal samples from *Canada geese* (23.4 % of 209) from ten sites in Ohio and Illinois	Zhou et al. (2004)
	Fecal samples from *black swan* (2.5 % of 80), *Canada geese* (5.0 % of 80), *duck* (1.3 % of 80), and *gulls* (0.0 % of 80) from New Zealand	Moriarty et al. (2011)
	Fecal samples from *gulls* (0.0 % of 145) in the Monterrey Bay region of California	Oates et al. (2012)
	Fecal samples from *common tern chicks* (72.2 % of 54) during breeding season in New Jersey	Rivera et al. (2012)
Giardia	Fecal samples from *gulls* (2.1 % of 145) in the Monterrey Bay region of California	Oates et al. (2012)

Sample size and number. Depending on the wild animal and its typical fecal mass, the prevalence of zoonotic enteric pathogens may be underestimated. This situation may occur when fecal amounts per assay are less than 0.10 g and pathogen shedding intensity is low. Under these conditions, there occurred an artificial downward bias for the prevalence by well over 50 % (Atwill et al. 2012). Such a situation would explain why double sampling improved the detection of methicillin-resistant *Staphylococcus aureus* carriage in 4 different types of wild animals in Spain (Concepción Porrero et al. 2013).

Surveillance of wildlife in many studies has been conducted with samples obtained by trapping the animals or collecting samples from hunters. Although such sampling is assumed to be representative of a population, Hoye et al. (2010) suggested that it likely involved selection bias, making it difficult to develop statistically valid estimates of pathogen prevalence. Hence, to enhance the design and interpretation of wildlife surveys, these investigators also provided estimates of the number of animals that should be sampled to achieve the study's objective (establishing absence of infection or an estimate of pathogen prevalence).

Specificity of detection method. A number of methods for detecting enteric foodborne pathogens in wildlife have been used and vary in their specificity relative to the organism present. Culture-based assays, for example, are often only capable of specifying the bacteria growing on a specific agar by its genus (i.e., *Campylobacter* spp., *Salmonella* spp., etc.), hence it is not possible to know if the pathogen is pathogenic. More recent studies that are conducted often employ advanced assays to characterize the phenotypic and genotypic properties of organisms isolated from wildlife so that these isolates can be compared to isolates associated with human illness. Another purpose for molecular characterization of wildlife isolates is for comparison to isolates obtained from domestic animals or to isolates obtained over

Table 1.4 Prevalence of zoonotic pathogens in amphibians and reptiles

Pathogen	Sample description with % prevalence of total number of samples tested in parentheses	Reference
Campylobacter spp.	Cloacal samples from *European pond turtle* (0.0 % of 83) and *red-eared terrapin* (0.0 % of 117) in 11 natural ponds in eastern Spain	Marin et al. (2013)
	Fecal samples from *bullfrogs* (0.0 % of 164) from Belgium and The Netherlands	Martel et al. (2013)
Mycoplasma spp.	Oral and cloacal swabs from *tortoises* (36.7 % of 30) in Italy	Lecis et al. (2011)
Salmonella spp.	Cloacal swabs of free-ranging *alligators* (2.8 % of 71) in southeast Texas and south Louisiana	Scott and Foster (1997)
	Fecal samples from *reptiles* (41.5 % of 94) and *amphibians* (0.0 % of 72) in Spain	Briones et al. (2004)
	Fecal or cloacal swabs of *frogs* (0.0 % of 106), *lizards* (10.7 % of 298), *crocodiles* (3.0 % of 33), *snakes* (0.0 % of 48), and *turtles* (0.0 % of 64) in Australia	Parsons et al. (2010)
	Oral and cloacal swabs from *tortoises* (10.0 % of 30) in Italy	Lecis et al. (2011)
	Cloacal swabs and cecal contents from *cane toads* (41.4 % of 58) in Grenada	Drake et al. (2013)
	Cloacal and ventral swabs as well as washes from *frogs* (1.2 % of 331), *lizards* (9.0 % of 59), *newts* (0.0 % of 5), *salamanders* (0.0 % of 6), *snakes* (59.0 % of 39), and *toads* (5.0 % of 20) in a produce-growing region of the California Central Coast	Gorski et al. (2013)
	Cloacal swab samples from *caimans* (13.9 % of 21), *turtles* (21.2 % of 32), *green iguanas* (15.2 % of 23), *other lizards* (25.2 % of 38), and *snakes* (24.5 % of 37) from French Guiana	Gay et al. (2014)
	Fecal samples from *bullfrogs* (0.0 % of 164) from Belgium and The Netherlands	Martel et al. (2013)
	Cloacal swabs from *wild green iguanas* (57.4 % of 47) in Grenada	Sylvester et al. (2014)

extended periods of time. Hence, isolates obtained from different animals, but having similar or identical molecular profiles, can be evidence that transmission between the two groups occurred (Williams et al. 2011), whereas isolates with similar profiles obtained at different time points can be evidence that they are persistent in the environment (Gorski et al. 2011). Detection of viruses and parasites in animals, however, often rely on immunological assays to detect antibodies in the serum that have been expressed when the pathogen invades the animal's system. The drawback to immunological assays, however, is that immunity may extend for periods long after the pathogen is eliminated from the animal. Examples of studies that have employed serotyping and molecular characterization of isolates recovered from wildlife as a means to measure their potential to serve as a reservoir of infection for humans or animals are listed in Table 1.5.

Table 1.5 Selected studies employing serotyping or molecular characterization of wildlife isolates to assess their contribution as reservoirs of foodborne pathogens to humans or domesticated animals

Pathogen	Observations and overall conclusions of study	Reference
Campylobacter spp.	Although culture-based and genus-specific PCR-detection assays attributed 45 % positive samples from gulls as *Campylobacter* spp., sequence analysis revealed that most were novel *Campylobacter* spp. and would not likely present a major public health risk. A smaller proportion of the sequences (2–5 %) were closely related to pathogenic species (*C. jejuni* and *C. lari*) at a 99 % or greater sequence identity	Lu et al. (2011)
Campylobacter jejuni	Serotype distribution of isolates collected from multiple species of wildlife was significantly different from known distributions in broilers and humans. Macrorestriction profiles of two serotypes (O:12 and O:38) isolated from wildlife and other sources indicated propagation in a wide group of animal species but were not detected in humans or broilers. The typing methods used revealed that wildlife is of limited importance as a reservoir of infection	Petersen et al. (2001)
	Despite this pathogen being genotypically diverse and host restricted, PFGE analysis revealed that there were several shared genotypes between dairy cattle and starling isolates recovered from the same farm. Moreover, multilocus sequence typing (MLST) analysis revealed that cattle and starlings also shared many clonal complexes. Overall, these results highlighted that starlings serve as potential reservoirs for *C. jejuni*	Sanad et al. (2013)
Escherichia coli O157:H7	PFGE patterns of *E. coli* O157:H7 isolates recovered from a wild boar and from a human patient with diarrhea living in the same geographic area were indistinguishable	Sánchez et al. (2010)
	Isolates obtained from fecal samples of dairy cattle and European starlings collected on different farms were subjected to multiple-locus variable-nucleotide tandem repeat analysis (MLVA). Indistinguishable subtypes between starlings and cattle on different farms were found, supporting the hypothesis that these birds contributed to the transmission of *E. coli* O157:H7 between dairy farms	Williams et al. (2011)
Salmonella spp.	22 different serotypes of *Salmonella* were isolated from wild amphibians and reptiles. Many of the isolates were of subtypes IIIa and IIIb, which are less frequently associated with human infections than subtype I	Gorski et al. (2013)
	Twenty-two serotypes of *Salmonella* were isolated from gull fecal samples, with the most common serovars (Infantis [21 %], Typhimurium [12.9 %] and Newport [11.3 %]) also recognized by CDC's FoodNet as being in the top ten laboratory-confirmed *Salmonella* serotypes (CDC 2013). PFGE patterns from gull isolates that were clustered in time and space are evidence that gulls are colonized by *Salmonella* in the environment for short periods of time, but during those times, would be capable of contaminating nearby tomato fields	Gruszynski et al. (2014)
Toxoplasma gondii	Although seroprevalence of deer for *T. gondii* was 45.2 %, this pathogen was only found in 3.0 % of deer brain tissue when analyzed by PCR. In contrast, PCR detected this pathogen in 18.8 % of red fox brain tissues. Genotyping of the fox samples revealed that all but one was type II, which is the predominant genotype among patients who acquired toxoplasmic infection in Europe (Ajzenberg et al. 2009)	De Craeye et al. (2011)

Host Attributes That Impact Contamination by Pathogens

Physiological state of host. Prevalence of foodborne pathogens within a wildlife population is often not uniform but is influenced by the physiological state of the individuals. One phenotypic variable that differentiates a population into distinct groups is age, with younger animals being more susceptible to infection than adults. As examples, carriage of *Clostridium difficile* was more common in younger urban Norway rats than in their adult counterparts (Himsworth et al. 2014), and *Campylobacter*'s prevalence in feral swine (Jay-Russell et al. 2012) and in Canada geese (Keller and Shriver 2014) was greater in younger versus older animals. Similarly, when prevalence in both domestic and wild animals was investigated using multiple logistic regression models, it was determined that young animals were approximately twice as likely to shed *Cryptosporidium* and *Giardia* in their feces than adults (Oates et al. 2012). In these cases, establishment of the pathogen in the young animal's gut maybe due to the presence of an immature gut microbiota which, when mature, would in older animals outcompete the pathogen and prevent colonization. This relationship, however, has not been observed in all cases. For example, the percentage of carriers of pathogenic *E.coli* (EPEC, EHEC, and STEC) by wild cervids (red and roe deer) did not differ between adult and juvenile/sub-adult animals (Bardiau et al. 2010), whereas in pteropid bats, a greater number of adults were seropositive for Nipah virus than juveniles or pups (Rahman et al. 2013).

Another characteristic of individuals within an animal population that is associated with different degrees of pathogen prevalence is the sex of the animal. In the case of feral swine sampled in Geneva, Switzerland, 71 % of females carried one or more foodborne pathogens compared to 53 % of males (Wacheck et al. 2010). Similarly, more feral swine females were positive for *Campylobacter* than were males in a study conducted in California (Jay-Russell et al. 2012). In contrast, in another study conducted in California that included both domestic and wild animals, but no feral swine, males were 1.2 times more likely to be *Giardia* spp.-positive than were females (Oates et al. 2012). Sex, however, was not a notable variable for prevalence of pathogenic *E. coli* in roe and red deer (Bardiau et al. 2010), nor was it associated with the seroprevalence of Nipah virus in bats (Rahman et al. 2013). A higher rate of seropositivity to Nipah virus was observed in nursing bats which was attributed to the increased stress that they have experienced in reproductive and nursing activities, which in turn likely increased their risk for infection (Rahman et al. 2013). Pathogen prevalence differences between sexes may also be attributed to behavioral differences that occur between the sexes. For example, in a wild pig population in Australia, transmission of *Salmonella* was more common between males than females and was attributed to the previous observations that adult male pigs have larger home ranges than females, and were more often found associating in small male groups in the study area (Cowled et al. 2012).

In general, susceptibility of animals to infection by foodborne pathogens increases with diminished health or increased stress, both of which compromise the immune system. For example, when wild animals are sampled during the hunting season and would be under increased stress, there is a greater likelihood that the pathogen would be present if the animal was recently exposed to the pathogen. Hence, studies that rely on this method of collection may be measuring prevalence that would not be typical throughout the year. Good health and decreased stress, however, do not always translate into reduced pathogen prevalence in wildlife. In the wild pig population in Australia, better conditioned (fatter) pigs were associated with an increased probability of infection (Cowled et al. 2012). To explain this statistic, the authors suggested that the better body conditions of these pigs actually enabled them to travel farther and forage more effectively and widely for food, and in turn be exposed to more pathogens.

Another factor that affects the efficacy of an animal's immune system to combat colonization by foodborne pathogens is the exposure dose. For example, European starlings transiently excreted *E. coli* O157:H7 following a low-dose inoculation, but when exposed to a population greater than 5.5 log, shedding occurred in 50 % of the birds for more than 3 days (Kauffman and LeJeune 2011). Similarly, exposures to high pathogen dosages have resulted in both rats and pigeons fecally shedding *E. coli* O157:H7 for longer periods of time than if exposed to smaller dosages (Čížek et al. 2000).

Although immunity to pathogens is usually considered beneficial from an individual standpoint, when the animal population is only partially immune and exposed to a new source of the pathogen, the pathogen may actually survive within the population for longer periods of time and increase the risk of spread to non-infected animals. Such a situation has been proposed as the scenario leading to the outbreak of Nipah virus in Malaysia during 1998–1999 (Epstein et al. 2006). More specifically, it was hypothesized that Nipah virus-infected fruit bats were attracted to fruit trees surrounding a large intensive pig farm and led to an initial infection that died out quickly. In the subsequent year, reintroduction of the virus into a partially immune population resulted in prolonged circulation on the farm, and when these infected pigs were sold from the affected farm and transported to other areas where there was a high density of smaller intensive pig farms and a high human density, a large outbreak occurred in humans, stimulating an investigation.

Hosts' behavioral features that contribute to intra- and interspecies interactions and pathogen transmission. Contamination of wildlife by foodborne pathogens requires that the wild animal first be exposed to a pathogen source which is often related to the animal's behavior patterns and food choices. Once pathogen transmission to the wild animal has occurred, that animal may then serve as a vehicle for intra- or inter-species transmission to other non-infected animals, but the extent to which that occurs will be dependent on the animal's behavioral patterns and whether the infection is self-limiting or not. Multiple examples illustrating the relationship between behavioral attributes and the observed or perceived potential for pathogen transmission between wildlife are presented in Table 1.6.

In general, pathogenic *E.coli* is found in many wild animals at a low prevalence due to limited intra-species interactions (Nielsen et al. 2004). Moreover, when wild animals are contaminated, the animal has likely been living close to domestic animals whose infection is perpetuated by their high-density living conditions (Díaz-

Table 1.6 Selected examples of behavioral attributes of wildlife that affected their intra- and interspecies interactions and the degree of foodborne pathogen transmission

Examples	Reference
Salmonella was detected in wild birds on pig and cattle farms carrying *Salmonella*-positive production animals and only during the periods when *Salmonella* was detected in the production animals. Presence of *Salmonella* in wild birds significantly correlated to their migration pattern and food preference. More specifically, partially migratory or short-to-medium distance migratory birds were at a higher risk of contracting *Salmonella* infections than non-migrating (resident) birds. Birds feeding on insects and invertebrates in the summer were at a higher risk of infection compared to birds feeding on seeds and grains. A slightly higher risk of *Salmonella* detection occurred for birds foraging on the ground in comparison to aerial foraging or foraging in the vegetation	Skov et al. (2008)
Campylobacter occurrence in gull chicks was directly related to their degree of refuse consumption, whereas *Salmonella* prevalence did not reflect any dietary relationship. Gulls are attracted to garbage dumps, untreated sewage, and manure where pathogens may be consumed or physically attached to the birds, allowing the animals to spread the pathogen to other resting areas, including agricultural fields	Ramos et al. (2010)
In ten concentrated animal feeding operations (CAFOs), *Salmonella* contamination of cattle feed and water troughs occurred where starlings also frequented. Probability of contamination of cattle feed by *Salmonella* increased as the number of starlings in feed troughs increased. *Salmonella* contamination in water troughs increased asymptotically as the numbers of starlings on CAFOs increased. Although carriage of *Salmonella* was documented in starlings, the serotype did not match the serotypes in the water and feed samples, suggesting that fecal material adhering to their feet and feathers was the likely mode of dissemination of the pathogen	Carlson et al. (2011)
Toxoplasma gondii was detected in red foxes at a greater frequency than in deer. Foxes are carnivores and are considered opportunistic feeders. Due to these feeding habits, foxes are more likely to have greater exposure to both parasites' tissue cysts and oocysts compared to deer, which largely feed on grass, leaves, young shoots, and berries	De Craeye et al. (2011)
Feral swine were observed to be infected with *Campylobacter* strains that were more often associated with cattle. Feral swine were observed sharing pastures, food, and water with cattle. Rooting in cow pats and wallowing in riparian areas of the pastures where cattle waste was deposited likely contributed to the *Campylobacter* contamination of feral swine	Jay-Russell et al. (2012)
Pathogen clustering occurred in deer mice. Frequent licking of extremities and close contact between deer mice may facilitate fecal–oral transmission and maintenance of infection within a close-knit population	Kilonzo et al. (2013)
Seroprevalence for *Toxoplasma gondii* was less in juvenile females of roe deer than in juvenile males. It was speculated that young males differed from young females by being more exploratory that increased their likelihood of consumption of vegetation on contaminated soil	Candela et al. (2014)

The highest prevalence of *C. jejuni* was in ruddy turnstones (a shorebird that undergoes long-distance complete migrations). The second highest prevalence occurred in Red Knots and Semipalmated Sandpipers, both of which undergo long-distance migrations. These birds often intermingle on stopover sites, providing opportunities for bacterial exchange. In comparison, lower prevalences were in the Snow Geese, another migratory bird, and in resident Canada Geese	Keller and Shriver (2014)
It was determined statistically that starling roosts were spatially associated with an increased prevalence of *E. coli* O157:H7 infection in dairy cattle, a greater diversity of distinguishable MLVA types, and a higher number of isolates with MLVA types from starling-bovine clades versus bovine-only clades. These data were supported by the behavioral patterns of starlings in which starlings tend to fly directly from their night roosts to preferred dairies in the morning; however, on the return flights, the birds often stop at farms along the flight routes	Swirski et al. (2014)
Campylobacter jejuni were detected in high frequency in American crows. This bird forages in a variety of settings, including dumps, animal feedlots, pastures, and urban areas, and therefore has the potential to transfer pathogens from waste sites to other uncontaminated areas	Weis et al. (2014)

Sánchez et al. 2013). Hence, when conditions of low stocking density were encountered in rangeland beef cattle, it was speculated that those conditions were responsible for the lack of interspecies transmission of *Cryptosporidium* from infected cattle to susceptible rodents (Kilonzo et al. 2013).

In cases when pathogen prevalence is low in a wildlife population, the contribution of these animals to persistence on source farms and to transmission between farms is minimal compared to animals, such as red deer and feral swine, whose gut is colonized by *E. coli* O157:H7 (Díaz-Sánchez et al. 2013) and *Campylobacter* (Jay-Russell et al. 2012), respectively. In either case, however, infected wild animals that are extremely mobile amplify the likelihood of transmission by disseminating the pathogen through uncontrolled routes. An example is migratory birds that often associate with cattle (brown-headed cowbirds, common grackles, and cattle egrets); such birds become infected with *Salmonella* and *E. coli* O157:H7 while migrating (Callaway et al. 2014). In other situations, wildlife behavior has become less of an issue as expansion of agriculture into wildlife habitats has created new opportunities for spillover of pathogens from domestic animals to wildlife or vice versa (Jones et al. 2013). A good example would be the emergence of Nipah virus in Malaysia where intensification of the pig industry combined with fruit production occurred in an area already populated by Nipah virus-infected fruit bats (Epstein et al. 2006).

Climatic (seasonal) impacton pathogen prevalence in wildlife. Many surveys on pathogen prevalence in wildlife have been conducted by collecting samples at different times of the year or over multiple years to determine if fluxes in prevalence occur in response to climatic or environmental changes. Through knowledge of seasonal preferences, pathogen transmission dynamics may be better understood and could assist in defining effective interventions for disease management.

One study addressing a seasonal preference in disease transmission of Nipah virus was conducted in Central Thailand, with differences observed between the Malaysian and Bangladesh strains (Wacharapluesadee et al. 2010). The Bangladesh strain was almost exclusively detected during April to June, whereas the Malaysian strain was found dispersed during December to June; however, the cause for these differences could not be determined. In another study, seasonal shedding patterns were observed in wild rodents, with fewer rodents trapped during the spring and summer months shedding *Cryptosporidium* oocysts than rodents trapped during autumn (Kilonzo et al. 2013). In this case, higher prevalences in autumn may have been linked to the breeding cycles of the animals, since most of the animals give birth during the warmer months of the year and begin to disperse in autumn (Ziegler et al. 2007). Behavior may have also contributed to the seasonal prevalence patterns observed in Canada geese, with prevalence of pathogenic *E.coli* being positively correlated with prevailing warmer seasonal temperatures, being higher in the spring and summer and lower during the fall and winter (Kullas et al. 2002). In this case, it was hypothesized that during the fall and winter, the daily movement patterns of the birds largely occurred on dry upland harvested grain fields located outside of town and away from habitats contaminated with mammalian sources of *E. coli*. In contrast, during the spring and summer, the birds did not move far from their nests during breeding and these areas consisted of small water impoundments and littoral zones that easily become fouled. A similar scenario was offered as an explanation for the dominance of *Salmonella*-positive birds or rodents detected in the winter compared to the summer in that wildlife moved closer to farms in winter in search of food and shelter (Skov et al. 2008).

Long-term climatic fluxes may also impact on the degree of pathogen prevalence in wildlife over successive years. Moreover, the impact of this variable may be more evident in wildlife where population turnover is greater. For example, the sharp declines in *T. gondii* seropositive roe deer from one year to another were partly explained by the replacement of seropositive individuals with new seronegative ones (Candela et al. 2014).

Management practices applied to wildlife or domestic animals that influence pathogen prevalence in wildlife. To minimize the transmission of enteric pathogens from wildlife to production sites (both animal and produce), efforts to control access of wildlife into those sites have been studied but with variable results. Furthermore, the type of management will vary depending upon whether the wild animal species is an invasive species causing environmental and economic damage or if it is considered an endangered or threatened species and is protected. When animals such as starlings that are not protected are targeted, as they are in the United States and Australia, management focuses on lethal control that includes chemical toxicants and shooting. Even in these situations, however, these methods are not always foolproof as evident in an Australian study that found a chemical toxicant to be ineffective because of poor bait acceptance (Bentz et al. 2007). When the target wild animal for control is a species of conservation concern, managing that animal's activities becomes more complex. Examples of non-lethal management techniques include non-lethal chemical repellants (Glahn et al. 1989), exclusionary devices (Khan et al. 2012), and frightening devices (Berge et al. 2007). As an example of the effectiveness of exclusionary devices, bats contacted date palm sap at a 2 % frequency when the food source was protected by a bamboo, *dhoincha*, jute stick, or polythene skirt compared to a frequency of 83 % when the date palm sap was not protected by a skirt (Khan et al. 2012). Enhanced efficacy of any of these management tools, however, generally requires that they be used in tandem or switched on a regular basis (Berge et al. 2007). In addition, direct management tools are not always the most effective means by which to reduce prevalence or sharing of pathogens between domestic and wild animals. For example, Mentaberre et al. (2013) determined in Northeastern Spain that cattle removal was more efficient than the culling of wild boar by hunting or trapping as a means of reducing the prevalence of shared serotypes of *Salmonella*.

In many cases, management of the pathogen prevalence of wildlife requires a systems approach. For example, to understand potential inter-species transmission pathways among wild and domestic ungulates in Kenya, *E.coli* collected from feces were genetically compared (VanderWaal et al. 2014). Under the assumption that when two individuals shared the same genetic subtype of this organism, they were part of the same transmission chain, the zebra was identified as an animal bridging distinct transmission networks. Therefore, these investigators hypothesized that interventions targeted at the zebra would diminish transmission among discrete networks.

Other Drawbacks to Pathogen Contamination of Wildlife That Impact Humans

Contamination of Water Sources

Although the focus of this chapter to this point has been primarily aimed at the contribution of wildlife to the direct spread of enteric foodborne pathogens to either animal production facilities or produce fields, wildlife may also indirectly impact these sites through contamination of water sources that would subsequently be used in agricultural production. In fact, the number of studies addressing this latter route of contamination is much greater than those addressing the direct routes for pathogen transmission. Support for this route of contamination stems first from the results of studies revealing that enteric foodborne pathogens are in surface and irrigation waters. For example, a study of ten irrigation ponds in Georgia revealed that nine of the ponds were contaminated with *C. jejuni* at some point during the year, with an overall prevalence of 19.3 % (Gu et al. 2013b), *E. coli O157* was found in all ponds occasionally, but mainly in summer and fall (Gu et al. 2013a), and *Salmonella* was found in 39 % of pond samples (Aminabadi et al. 2013). The presence of pathogens in surface water is not unique to Georgia as *Salmonella* was also detected in 7.1 % of surface waters in a major produce region of California (Gorski et al. 2011) and in 94 % of surface irrigation water sources in New York (Jones et al. 2014). In a later study of water and sediment from leafy green produce farms and streams on the Central California coast, *Salmonella* was detected in 6.2 % of water and 4.3 % of sediment samples, and *E. coli* O157 was detected in 13.8 % of water and 1.7 % of sediment samples (Benjamin et al. 2013). In all these cases, it is presumed that contamination of these water sources by wildlife could occur either directly or through storm runoff of adjacent contaminated lands.

An additional line of evidence for pathogen contamination of waterways and irrigation ponds by wildlife is through two avenues of exploration. First, studies have documented either directly or indirectly (through global positioning collars) that wildlife accesses water sources and engages in behavior that would lead to contamination of the water (Hampton et al. 2006; Cooper et al. 2010). In the second approach, pathogen isolates obtained from water and wildlife fecal samples have been compared by serotyping and molecular subtyping to determine their similarity. Similar serotypes of *Salmonella* have been isolated from both water and wildlife samples (Gorski et al. 2011; Jokinen et al. 2011; Aminabadi et al. 2013). Molecular typing by restriction fragment length polymorphism of *C. jejuni* isolates from water has revealed clustering with duck and geese isolates (Jokinen et al. 2011). Not all studies, however, have established a relationship between pathogen isolates in water and wildlife. For example, Gorski et al. (2013) detected *Salmonella* in both wildlife and associated water samples; however, the PFGE of the isolates did not match. In another study, *Campylobacter* was detected in shorebird excreta but was not found in the water samples collected from locations presumed to be impacted by these birds (Ryu et al 2014). These latter results suggest that large numbers of animals may be needed to impact the water quality especially if the animal has low resident populations of a given pathogen. Large resident pathogen populations in a waterway, however, are not always indicative of a large public health risk. Many pathogen parasite lineages, such as *Cryptosporidium* and *Giardia*, are host specific and not zoonotic to the human population at large. However, such parasites could still cause opportunistic infections in animals and humans. Hence, the significance of wildlife contamination of water sources with foodborne pathogens must be determined on a case-by-case basis.

Reservoir for Antibiotic Resistance

Over the past two decades, a growing concern has arisen regarding antimicrobial resistance in pathogenic and commensal bacteria. These concerns extend to wildlife as antibiotic-resistant pathogenic bacteria have been associated with wildlife (Bardiau et al. 2010; Drake et al. 2013; Fresno et al. 2013; Gorski et al. 2013; Sasaki et al. 2013; Sylvester et al. 2014) and is evidence of transmission from environmental sources or animals that have been exposed to antibiotic therapy. Unfortunately, the presence of antibiotic-resistant pathogens in wildlife only serves to perpetuate this human health problem, especially as limited options are available to effectively control these animal populations.

Reservoir for Emerging Pathogens

Pathogens, including those of foodborne origin, have an extremely high evolutionary potential given their large populations, high genetic variation, and short generation times. Given that approximately 75 % of emerged diseases, including zoonoses, were of wildlife origin (Jones et al. 2013), efforts are being intensified to focus on these animals as the driving forces (i.e., climate change, agricultural expansion, urbanization, and habitat destruction) continue to have an impact. The emergence of Nipah virus, for example, revealed the interplay between several of those driving forces (Daszak et al. 2013; Hayman et al. 2013). Another example for which wildlife has appeared to play a role in the evolution of foodborne pathogens is with enterohemorrhagic *E. coli* (EHEC) O157 (Jenke et al. 2012). By sequencing O157:H7/H- isolates, it was determined that deer occupied an intermediate position between O55:H7 and both sorbitol-fermenting (SF) and non-SF O157 branches. Based upon a study of Díaz et al. (2011), it also appeared that free-ranging red deer has been a possible reservoir of Stx-negative derivatives of SF O157:H7.

Summary

Additional research and monitoring of foodborne pathogen carriage by wildlife is needed to better elucidate transmission cycles, temporal–spatial fluctuations, and emerging strains with the goal of reducing potential risks to public health. Given the large number of wildlife species as well as the large number of foodborne pathogens, this task may be daunting. To date, numerous studies have already been conducted to investigate the prevalence of foodborne pathogens in many of the wildlife species that could interface with humans or their agricultural activities. To ensure that available resources applied to future surveillance have the greatest impact, prioritizing the pathogens and wildlife as to their need for surveillance using risk assessment systems, such as the Wildtool described by Tavernier et al. (2011), should be considered.

References

Ahn CK, Russo AJ, Howell KR et al (2009) Deer sausage: a newly identified vehicle of transmission of *Escherichia coli* O157:H7. J Pediatr 155:587–589

Ajzenberg D, Year H, Marty P et al (2009) Genotype of 88 *Toxoplasma gondii* isolates associated with toxoplasmosis in immunocompromised patients and correlation with clinical findings. J Infect Dis 199:1155–1167

Aminabadi P, Smith L, Adams MP et al (2013) Evaluation of foodborne pathogens in aquatic wildlife and irrigation ponds in Southeastern Georgia. Abstr Annu Mtg Int Assoc Food Prot, Charlotte, NC, P3–125

Andrés-Barranco S, Vico JP, Garrido V et al (2014) Role of wild bird and rodents in the epidemiology of subclinical Salmonellosis in finishing pigs. Foodborne Pathog Dis 11:689–697

Atwill ER, Jay-Russell M, Li X et al (2012) Methodological and epidemiological concerns when comparing microbial food safety food risks from wildlife, livestock, and companion animals. In: Timm RM (ed) Proceedings of the 25th vertebrate pest conference. Western Center for Food Safety, University of California, Davis, CA, pp 101–103

Backhans A, Fellström C, Lambertz ST (2011) Occurrence of pathogenic *Yersinia enterocolitica* and *Yersinia pseutotuberculosis* in small wild rodents. Epidemiol Infect 139:1230–1238

Backhans A, Jacobson M, Hansson I et al (2013) Occurrence of pathogens in wild rodents caught on Swedish pig and chicken farms. Epidemiol Infect 141:1885–1891

Barber DA, Bahnson PB, Isaacson R et al (2002) Distribution of *Salmonella* in swine production ecosystems. J Food Prot 65:1861–1868

Bardiau M, Grégoire F, Muylaert A et al (2010) Enteropathogenic (EPEC), enterohaemorrhagic (EHEC) and verotoxigenic (VTEC) *Escherichia coli* in wild cervids. J Appl Microbiol 109:2214–2222

Batz MB, Hoffman S, Morris JG Jr (2012) Ranking in the disease burden of 14 pathogens in food sources in the United States using attribution data from outbreak investigations and expert elicitation. J Food Prot 75:1278–1291

Benjamin L, Atwill ER, Jay-Russell M et al (2013) Occurrence of generic *Escherichia coli*, *E. coli* O157 and *Salmonella* spp. in water and sediment from leafy green produce farms and streams on the Central California coast. Int J Food Microbiol 165:65–76

Benskin CMH, Wilson K, Jones K et al (2009) Bacterial pathogens in wild birds: a review of the frequency and effects of infection. Biol Rev 84:349–373

Bentz T, Lapidge S, Dall D et al (2007) Managing starlings in Australia – can DRC-1339 be the answer? In: Witmer GW, Pitt WC, Fagerstone KA (eds) Managing vertebrate invasive species: proceedings of an international symposium. USDA/APHIS/WS, National Wildlife Research Center, Fort Collins, CO, pp 361–364

Berge AJ, Delwiche MJ, Gorenzel WP et al (2007) Sonic broadcast unit for bird control in vineyards. Appl Eng Agric 23:819–825

Berger CN, Sodha SV, Shaw RK et al (2010) Fresh fruit and vegetables as vehicles for the transmission of human pathogens. Environ Microbiol 12:2385–2397

Blanco Crivelli X, Rumi MV, Carfagnini JC et al (2012) Synanthropic rodents as possible reservoirs of shigatoxigenic *Escherichia coli* strains. Front Cell Infect Microbiol 2:Article 134

Borchard P, Wright IA, Eldridge DJ (2010) Wombats and domestic livestock as potential vectors of *Cryptosporidium* and *Giardia* in an agricultural riparian area. Aust J Zool 58:150–153

Briones V, Téllez S, Goyache J et al (2004) *Salmonella* diversity associated with wild reptiles and amphibians in Spain. Environ Microbiol 6:868–871

Burke R, Masuoka P, Murrell KD (2008) Swine *Trichinella* infection and geographic information system tools. Emerg Infect Dis 14:1109–1111

Burt SA, Siemeling L, Kuijper EJ et al (2012) Vermin on pig farms are vectors for *Clostridium difficile* PCR ribotypes O78 and O45. Vet Microbiol 160:256–258

Callaway TR, Edrington TS, Nisbet DJ (2014) Isolation of *Escherichia coli* O157:H7 and *Salmonella* from migratory brown-headed cowbirds (*Molothrus ater*), common grackles (*Quiscalus quiscula*), and cattle egrets (*Bubulcus ibis*). Foodborne Pathog Dis 11:791–794

Candela MG, Serrano E, Sevila J et al (2014) Pathogens of zoonotic and biological importance in roe deer (*Capreolus capreolus*): Seroprevalence in an agro-system population in France. Res Vet Sci 96:254–259

Carlson JC, Franklin AB, Hyatt DR et al (2011) The role of starlings in the spread of *Salmonella* within concentrated animal feeding operations. J Appl Ecol 48:479–486

CDC [Centers for Disease Control and Prevention] (2013) Table 5. FoodNet number and incidence of *Salmonella* infections by serotype 2013. Foodborne Diseases Active Surveillance Network (FoodNet). http://www.cdc.gov/foodnet/data/trends/tables/2013/table5.html. Accessed 5 Nov 2014

Chandran A, Mazumder A (2014) Occurrence of diarrheagenic virulence genes and genetic diversity in *Escherichia coli* isolates from fecal material of various avian hosts in British Columbia, Canada. Appl Environ Microbiol 80:1933–1940

Čížek A, Alexa P, Literák I et al (1999) Shiga toxin-producing *Escherichia coli* O157 in feedlot cattle and Norwegian rats from a large-scale farm. Lett Appl Microbiol 28:435–439

Čížek A, Literák I, Scheer P (2000) Survival of *Escherichia coli* O157 in faeces of experimentally infected rats and domestic pigeons. Lett Appl Microbiol 31:349–352

Cleaveland S, Laurenson MK, Taylor LH (2001) Diseases of humans and their domestic mammals: pathogen characteristics, host range and the risk of emergence. Philos Trans R Soc Lond B Biol Sci 356:991–999

Compton JA, Baney JA, Donaldson SC et al (2008) *Salmonella* infections in the common raccoon (*Procyon lotor*) in Western Pennsylvania. J Clin Microbiol 46:3084–3086

Concepción Porrero M, Mentaberre G, Sánchez S et al (2013) Methicillin resistant *Staphylococcus aureus* (MRSA) carriage in different free-living wild animal species in Spain. Vet J 198:127–130

Cooper SM, Scott HM, de la Garza GR et al (2010) Distribution and interspecies contact of feral swine and cattle on rangeland in South Texas: Implications for disease transmission. J Wildl Dis 46:152–164

Cowled BD, Ward MP, Laffan SW et al (2012) Integrating survey and molecular approaches to better understand wildlife disease ecology. PLoS One 7, e46310

Daszak P, Zambrana-Torrelio C, Bogich TL et al (2013) Interdisciplinary approaches to understanding disease emergence: the past, present, and future drivers of Nipah virus emergence. Proc Natl Acad Sci 110:3681–3688

De Craeye S, Speybroeck N, Ajzenberg D et al (2011) *Toxoplasma gondii* and *Neospora caninum* in wildlife: common parasites in Belgian foxes and Cervidae? Vet Parasitol 178:64–69

De Herdt P, Devriese L (2000) Pigeons. In: Tully TN, Lawton MPC, Dorrestein GM (eds) Avian medicine. Butterworth-Heinemann Ltd., Oxford, pp 320–322

Díaz S, Vidal D, Herrera-León S et al (2011) Sorbitol-fermenting, β-glucuronidase-positive, Shiga toxin-negative *Escherichia coli* O157:H7 in free-ranging red deer in south-central Spain. Foodborne Pathog Dis 8:1313–1315

Díaz-Sánchez S, Sánchez S, Herrera-León S et al (2013) Prevalence of Shiga toxin-producing *Escherichia coli*, *Salmonella* spp. and *Campylobacter* spp. in large game animals intended for consumption: Relationship with management practices and livestock influence. Vet Microbiol 163:274–281

Drake M, Amadi V, Zieger U et al (2013) Prevalence of *Salmonella* spp. in cane toads (*Bufo marinus*) from Grenada, West Indies, and their antimicrobial susceptibility. Zoonoses Public Health 60:437–441

Eggert M, Stüber E, Heurich M et al (2013) Detection and characterization of Shiga toxin-producing *Escherichia coli* in faeces and lymphatic tissue of free-ranging deer. Epidemiol Infect 141:251–259

Epstein JH, Field HE, Luby S et al (2006) Nipah virus: impact, origins, and causes of emergence. Curr Infect Dis Rep 8:59–65

Farooq S, Hussain I, Mir MA et al (2009) Isolation of atypical enteropathogenic *Escherichia coli* and Shiga toxin 1 and 2f-producing *Escherichia coli* from avian species in India. Lett Appl Microbiol 48:692–697

Ferens WA, Hovde CJ (2011) *Escherichia coli* O157:H7: animal reservoir and sources of human infection. Foodborne Pathog Dis 8:465–487

Fresno M, Barrera V, Gornall V et al (2013) Identification of diverse *Salmonella* serotypes, virulotypes, and antimicrobial resistance phenotypes in waterfowl from Chile. Vector Borne Zoonotic Dis 12:884–887

García-Sánchez A, Sánchez S, Rubio R et al (2007) Presence of Shiga toxin-producing *E. coli* O157:H7 in a survey of wild artiodactyls. Vet Microbiol 121:373–377

Gardner TJ, Fitzgerald C, Xavier C et al (2011) Outbreak of campylobacteriosis associated with consumption of raw peas. Clin Infect Dis 53:26–32

Gargiulo A, Russo TP, Schettini R et al (2014) Occurrence of enteropathogenic bacteria in urban pigeons (*Columba livia*) in Italy. Vector Borne Zoonotic Dis 14:251–255

Gay N, Le Hello S, Weill F-X et al (2014) *Salmonella* serotypes in reptiles and humans, French Guiana. Vet Microbiol 170:167–171

Gilbreath JJ, Shields MS, Smith RL et al (2009) Shiga toxins, and the genes encoding them, in fecal samples from native Idaho ungulates. Appl Environ Microbiol 75:862–865

Glahn JF, Mason JR, Woods DR (1989) Dimethyl anthranilate as a bird repellant in livestock feed. Wildl Soc Bull 17:313–320

Gorski L (2012) Selective enrichment media bias the types of *Salmonella enterica* strains isolated from mixed strain cultures and complex enrichment broths. PLoS One 7, e34722

Gorski L, Parker CT, Liang A et al (2011) Prevalence, distribution, and diversity of *Salmonella enterica* in a major produce region of California. Appl Environ Microbiol 77:2734–2748

Gorski L, Jay-Russell MT, Liang AS et al (2013) Diversity of pulsed-field gel electrophoresis pulsotypes, serovars, and antibiotic resistance among *Salmonella* isolates from wild amphibians and reptiles in the California Central Coast. Foodborne Pathog Dis 10:540–548

Gruszynski K, Pao S, Kim C et al (2013) Evaluating wildlife as a potential source of *Salmonella* serotype Newport (JJPX01.0061) contamination for tomatoes on the Eastern shore of Virginia. Zoonoses Public Health 61:202–207

Gruszynski K, Pao S, Kim C et al (2014) Evaluating gulls as potential vehicles of *Salmonella enterica* serotype Newport (JJPX01.0061) contamination of tomatoes grown on the Eastern shore of Virginia. Appl Environ Microbiol 80:235–238

Gu G, Luo Z, Cevallos-Cevallos JM et al (2013a) Factors affecting the occurrence of *Escherichia coli* O157 contamination in irrigation ponds on produce farms in the Suwanee River Watershed. Can J Microbiol 59:175–182

Gu G, Luo Z, Cevallos-Cevallos JM et al (2013b) Occurrence and population density of *Campylobacter jejuni* in irrigation ponds on produce farms in the Suwanee River Watershed. Can J Microbiol 59:339–346

Haemig PD, Hernandez J, Waldenström J et al (2008) Barn swallows (*Hirundo rustica*) test negative for *Salmonella*. Vector Borne Zoonotic Dis 8:451–453

Hampton J, Spencer PBS, Elliot AD et al (2006) Prevalence of zoonotic pathogens from feral pigs in major public drinking water catchments in Western Australia. Ecohealth 3:103–108

Hayman DTS, Bowen RA, Cryan PM et al (2013) Ecology of zoonotic infectious diseases in bats: current knowledge and future directions. Zoonoses Public Health 60:2–21

Henzler DJ, Optiz HM (1992) The role of mice in the epizootiology of *Salmonella* Enteritidis infection on chicken layer farms. Avian Dis 36:625–631

Himsworth CG, Patrick DM, Mak S et al (2014) Carriage of *Clostridium difficile* by wild urban Norway rats (*Rattus norvegicus*) and black rats (*Rattus rattus*). Appl Environ Microbiol 80:1299–1305

Hoye BJ, Munster VJ, Nishiura H et al (2010) Surveillance of wild birds for avian influenza virus. Emerg Infect Dis 16:1827–1834

Jay MT, Cooley M, Carychao D et al (2007) *Escherichia coli* O157:H7 in feral swine near spinach fields and cattle, Central California coast. Emerg Infect Dis 13:1908–1911

Jay-Russell MT (2013) What is the risk from wild animals in food-borne pathogen contamination of plants? CAB Rev 8:No. 040

Jay-Russell MT, Bates A, Harden L et al (2012) Isolation of *Campylobacter* from feral swine (*Sus scrofa*) on the ranch associated with the 2006 *Escherichia coli* O157:H7 spinach outbreak investigation in California. Zoonoses Public Health 59:314–319

Jay-Russell MT, Madigan JE, Bengson Y et al (2013) *Salmonella* Oranienburg isolated from horses, wild turkeys and an edible home garden fertilized with raw horse manure. Zoonoses Public Health 61:64–71

Jay-Russell M, Hake AF, Bengson Y et al (2014) Prevalence and characterization of *Escherichia coli* and *Salmonella* strains isolated from stray dog and coyote feces in a major leafy greens production region at the United States-Mexico border. PLoS One 9, e113433

Jenke C, Leopold SR, Weniger T et al (2012) Identification of intermediate in evolutionary model of enterohemorrhagic *Escherichia coli* O157. Emerg Infect Dis 18:582–588

Jokinen C, Edge TA, Ho S et al (2011) Molecular subtypes of *Campylobacter* spp., *Salmonella enterica*, and *Escherichia coli* O157:H7 isolated from faecal and surface water samples in the Oldman River watershed, Alberta, Canada. Water Res 45:1247–1257

Jones KE, Patel NG, Levy MA et al (2008) Global trends in emerging infectious diseases. Nature 451:990–993

Jones BA, Grace D, Kock R et al (2013) Zoonosis emergence linked to agricultural intensification and environmental change. Proc Natl Acad Sci 110:8399–8404

Jones LA, Worobo RW, Smart CD (2014) Plant-pathogenic oomycetes, *Escherichia coli* strains, and *Salmonella* spp. frequently found in surface water used for irrigation of fruit and vegetable crops in New York State. Appl Environ Microbiol 80:4814–4820

Kabrane-Lazizi Y, Fine JB, Elm J et al (1999) Evidence for widespread infection of wild rats with hepatitis E virus in the United States. Am J Trop Med Hyg 61:331–335

Kangas S, Takkinen J, Hakkinen M et al (2008) *Yersinia pseudotuberculosis* O:1 traced to raw carrots, Finland. Emerg Infect Dis 14:1959–1961

Kauffman MD, LeJeune J (2011) European starlings (*Sturnus vulgaris*) challenged with *Escherichia coli* O157 can carry and transmit the human pathogen to cattle. Lett Appl Microbiol 53:596–601

Keene WE, Sazie E, Kok J et al (1997) An outbreak of *Escherichia coli* O157:H7 infections traced to jerky made from deer meat. J Am Med Assoc 277:1229–1231

Keller JI, Shriver WG (2014) Prevalence of three *Campylobacter* species, *C. jejuni*, *C. coli*, and *C. lari*, using multilocus sequence typing in wild birds of the Mid-Atlantic region, USA. J Wildl Dis 50:31–41

Khan SU, Gurley ES, Jahangir Hossain M et al (2012) A randomized controlled trial of interventions to impede date palm sap contamination by bats to prevent Nipah virus transmission in Bangladesh. PLoS One 7, e42689

Kijlstra A, Meerburg B, Cornelissen J et al (2008) The role of rodents and shrews in the transmission of *Toxoplasma gondii* to pigs. Vet Parasitol 156:183–190

Kilonzo C, Li X, Vivas EJ et al (2013) Fecal shedding of zoonotic food-borne pathogens by wild rodents in a major agricultural region of the Central California coast. Appl Environ Microbiol 79:6337–6344

Kullas H, Coles M, Rhyan J et al (2002) Prevalence of *Escherichia coli* serogroups and human virulence factors in faeces of urban Canada geese (*Branta canadensis*). Int J Environ Health Res 12:153–162

Laidler MR, Tourdjman M, Buser GL et al (2013) *Escherichia coli* O157:H7 infections associated with consumption of locally grown strawberries contaminated by deer. Clin Infect Dis 57:1129–1134

Langholz JA, Jay-Russell MT (2013) Potential role of wildlife in pathogenic contamination of fresh produce. Hum Wildl Interact 7:140–157

Lecis R, Paglietti B, Rubino S et al (2011) Detection and characterization of *Mycoplasma* spp. and *Salmonella* spp. in free-living European tortoises (*Testudo hermanni*, *Testudo graeca*, and *Testudo marginata*). J Wildl Dis 47:717–724

Li T-C, Chijiwa K, Sera N et al (2005) Hepatitis E virus transmission from wild boar meat. Emerg Infect Dis 11:1958–1960

Liebana E, Garcia-Migura L, Clouting C et al (2003) Molecular fingerprinting evidence of the contribution of wildlife vectors in the maintenance of *Salmonella* Enteritidis infection in layer farms. J Appl Microbiol 94:1024–1029

Lillehaug A, Monceyron Jonassen C, Bergsjø B et al (2005) Screening of feral pigeon (*Colomba livia*), mallard (*Anas platyrhynchos*) and graylag goose (*Anser anser*) populations for *Campylobacter* spp., *Salmonella* spp., avian influenza virus and avian paramyxovirus. Acta Vet Scand 46:193–202

Lu J, Ryu H, Santo Domingo JW et al (2011) Molecular detection of *Campylobacter* spp. in California gull (*Larus californicus*) excreta. Appl Environ Microbiol 77:5034–5039

Lugton IW, Johnstone AC, Morris RS (1995) *Mycobacterium bovis* infection in New Zealand hedgehogs (*Erinaceus europaeus*). New Zeal Vet J 43:342–345

Magnino S, Colin P, Dei-Cas E et al (2009) Biological risks associated with consumption of reptile products. Int J Food Microbiol 134:163–175

Marin C, Ingresa-Capaccioni S, González-Bodi S et al (2013) Free-living turtles are a reservoir for *Salmonella* but not for *Campylobacter*. PLoS One 8, e72350

Marin C, Palomeque M-D, Marco-Jiménez F et al (2014) Wild griffon vultures (*Gyps fulvus*) as a source of *Salmonella* and *Campylobacter* in eastern Spain. PLoS One 9, e94191

Martel A, Adriaensen C, Sharifian-Fard M et al (2013) The absence of zoonotic agents in invasive bullfrogs (*Lithobates catesbeianus*) in Belgium and The Netherlands. Ecohealth 10:344–347

Marva E, Markovics A, Gdalevich M et al (2005) Trichinellosis outbreak. Emerg Infect Dis 11:1979–1981

Matsuda H, Okada K, Takahashi K et al (2003) Severe hepatitis E virus infection after ingestion of uncooked liver from a wild boar. J Infect Dis 188:944

Medrano C, Boadella M, Barrios H et al (2012) Zoonotic pathogens among white-tailed deer, Northern Mexico, 2004–2009. Emerg Infect Dis 18:1372–1374

Meerburg BG, Kijlstra A (2007) Role of rodents in transmission of *Salmonella* and *Campylobacter*. J Sci Food Agric 87:2774–2781

Meng X-J (2011) From barnyard to food table: the omnipresence of hepatitis E virus and risk for zoonotic infection and food safety. Virus Res 161:23–30

Mentaberre G, Porrero MC, Navarro-Gonzalez N et al (2013) Cattle drive *Salmonella* infection in the wildlife-livestock interface. Zoonoses Public Health 60:510–518

Moore JE, Gilpin D, Crothers E et al (2002) Occurrence of *Campylobacter* spp. and *Cryptosporidium* spp. in seagulls (*Larus* spp.). Vector Borne Zoonotic Dis 2:111–114

Mora A, López C, Dhabi G et al (2012) Seropathotypes, phylogroups, Stx subtypes, and intimin types of wildlife-carried, Shiga toxin-producing *Escherichia coli* strains with the same characteristics as human-pathogenic isolates. Appl Environ Microbiol 78:2578–2585

Moriarty EM, Karki N, Mackenzi M et al (2011) Faecal indicators and pathogens in selected New Zealand waterfowl. New Zeal J Mar Freshwat Res 45:679–688

Nielsen EM, Skov MN, Madsen JJ et al (2004) Verocytotoxin-producing *Escherichia coli* in wild birds and rodents in close proximity to farms. Appl Environ Microbiol 70:6944–6947

Oates SC, Miller MA, Hardin D et al (2012) Prevalence, environmental loading, and molecular characterization of *Cryptosporidium* and *Giardia* isolates from domestic and wild animals along the Central California coast. Appl Environ Microbiol 78:8762–8772

Pao S, Hagens BE, Kim C et al (2014) Prevalence and molecular analyses of *Campylobacter jejuni* and *Salmonella* spp. in co-grazing small ruminants and wild-living birds. Livestock Sci 160:163–171

Parsons SK, Michael Bull C, Gordon DM (2010) Low prevalence of *Salmonella enterica* in Australian wildlife. Environ Microbiol Rep 2:657–659

Petersen L, Nielsen EM, Engberg J et al (2001) Comparison of genotypes and serotypes of *Campylobacter jejuni* isolated from Danish wild mammals and birds and from broiler flocks and humans. Appl Environ Microbiol 67:3115–3121

Rabatsky-Ehr T, Dingman D, Marcus R et al (2002) Deer meat as the source for a sporadic case of *Escherichia coli* O157:H7 infection, Connecticut. Emerg Infect Dis 8:525–527

Ragg JR, Moller H, Waldrup KA (1995) The prevalence of bovine tuberculosis (*Mycobacterium bovis*) infections in feral populations of cats (*Felis catus*), ferrets (*Mustela furo*) and stoats (*Mustela erminea*) in Otago and Southland, New Zealand. New Zeal Vet J 43:333–337

Rahman SA, Hassan L, Epstein JH et al (2013) Risk factors for Nipah virus infection among pteropid bats, peninsular Malaysia. Emerg Infect Dis 19:51–60

Ramos R, Cerdà-Cuéllar M, Ramírez F et al (2010) Influence of refuse sites on the prevalence of *Campylobacter* spp. and *Salmonella* serovars in seagulls. Appl Environ Microbiol 76:3052–3056

Renter DG, Sargeant JM, Hygnstorm SE et al (2001) *Escherichia coli* O157:H7 in free-ranging deer in Nebraska. J Wildl Dis 37:755–760

Renter DG, Gnad DP, Sargeant JM et al (2006) Prevalence and serovars of *Salmonella* in the feces of free-ranging white-tailed deer (*Odocoileus virginianus*) in Nebraska. J Wildl Dis 42:699–703

Rivera WA, Husic JS, Gaylets CE et al (2012) Carriage of bacteria and protozoa in the intestinal tract of common tern chicks. Waterbirds 35:490–494

Rosef O, Gondrosen B, Kapperud G et al (1983) Isolation and characterization of *Campylobacter jejuni* and *Campylobacter coli* from domestic and wild mammals in Norway. Appl Environ Microbiol 46:855–859

Rounds JM, Rigdon CE, Muhl LJ et al (2012) Non-O157 Shiga toxin-producing *Escherichia coli* associated with venison. Emerg Infect Dis 18:279–282

Rutjes SA, Lodder-Verschoor F, Lodder WJ et al (2010) Seroprevalence and molecular detection of hepatitis E virus in wild boar and red deer in The Netherlands. J Virol Meth 168:197–206

Ryu H, Grond K, Verheijen B et al (2014) Intestinal microbiota and species diversity of *Campylobacter* and *Helicobacter* spp. in migrating shorebirds in Delaware Bay. Appl Environ Microbiol 80:1838–1847

Sanad YM, Closs G Jr, Kumar A et al (2013) Molecular epidemiology and public health relevance of *Campylobacter* isolated from dairy cattle and European starlings in Ohio, USA. Foodborne Pathog Dis 10:229–236

Sánchez S, García-Sanchez A, Martínez R et al (2009) Detection and characterisation of Shiga toxin-producing *Escherichia coli* other than *Escherichia coli* O157:H7 in wild ruminants. Vet J 180:384–388

Sánchez S, Martínez R, García A et al (2010) Detection and characterisation of O157:H7 and non-O157 Shiga toxin-producing *Escherichia coli* in wild boars. Vet Microbiol 143:420–423

Sasaki Y, Goshima T, Mori T et al (2013) Prevalence and antimicrobial susceptibility of foodborne bacteria in wild boars (*Sus scrofa*) and wild deer (*Cervus nippon*) in Japan. Foodborne Pathog Dis 10:985–991

Scallan E, Hoekstra RM, Angulo FJ et al (2011) Foodborne illness acquired in the United States – major pathogens. Emerg Infect Dis 17:7–15

Schaife HR, Cowan D, Finney J et al (2006) Wild rabbits (*Oryctolagus cuniculus*) as potential carriers of verocytotoxin-producing *Escherichia coli*. Vet Rec 159:175–178

Scott T, Foster BG (1997) *Salmonella* spp. in free-ranging and farmed alligators (*Alligator mississippiensis*) from Texas and Louisiana, U.S.A. Aquaculture 156:179–181

Simpson VR (2002) Wild animals as reservoirs of infectious diseases in the UK. Vet J 163:128–146

Singer RS, Mayer AE, Hanson TE et al (2009) Do microbial interactions and cultivation media decrease the accuracy of *Salmonella* surveillance systems and outbreak investigations? J Food Prot 72:707–713

Skov MN, Madsen JJ, Rahbek C et al (2008) Transmission of *Salmonella* between wildlife and meat-production animals in Denmark. J Appl Microbiol 105:1558–1568

Swirski AL, Pearl DL, Williams ML et al (2014) Spatial epidemiology of *Escherichia coli* O157:H7 in dairy cattle in relation to night roosts of *Sturnus vulgaris* (European starling) in Ohio, USA (2007–2009). Zoonoses Public Health 61:427–435

Sylvester WRB, Amadi V, Pinckney R et al (2014) Prevalence, serovars and antimicrobial susceptibility of *Salmonella* spp. from wild and domestic green iguanas (*Iguana iguana*) in Grenada, West Indies. Zoonoses Public Health 61:436–441

Tavernier P, Dewulf J, Roelandt S et al (2011) Wildtool, a flexible, first-line risk assessment system for wildlife-borne pathogens. Eur J Wildl Res 57:1065–1075

Tei S, Kitajima N, Takahashi K et al (2003) Zoonotic transmission of hepatitis E virus from deer to human beings. Lancet 362:371–373

Thakur S, Sandfoss M, Kennedy-Stoskopf S et al (2011) Detection of *Clostridium difficile* and *Salmonella* in feral swine population in North Carolina. J Wildl Dis 47:774–776

Tomiyama D, Inoue E, Osawa Y et al (2009) Serological evidence of infection with hepatitis E virus among wild Yezo-deer, *Cervus nippon yesoensis*, in Hokkaido, Japan. J Viral Hepat 16:524–528

VanderWaal KL, Atwill ER, Isbell LA et al (2014) Quantifying microbe transmission networks for wild and domestic ungulates in Kenya. Biol Conserv 169:136–146

Vázquez B, Esperón F, Neves E et al (2010) Screening for several potential pathogens in feral pigeons (*Columba livia*) in Madrid. Acta Vet Scand 52:Article 45

Wacharapluesadee S, Boongird K, Wanghongsa S et al (2010) A longitudinal study of the prevalence of Nipah virus in *Pteropus lylei* bats in Thailand: evidence for seasonal preference in disease transmission. Vector Borne Zoonotic Dis 10:183–190

Wacheck S, Fredriksson-Ahomaa M, König M et al (2010) Wild boars as an important reservoir for foodborne pathogens. Foodborne Pathog Dis 7:307–312

Weis AM, Miller WA, Byrne BA et al (2014) Prevalence and pathogenic potential of *Campylobacter* isolates from free-living, human-commensal American crows. Appl Environ Microbiol 80:1639–1644

Williams ML, Pearl DL, LeJeune JT (2011) Multiple-locus variable nucleotide tandem repeat subtype analysis implicates European starlings as biological vectors for *Escherichia coli* O157:H7 in Ohio, USA. J Appl Microbiol 111:982–988

Zhou L, Kassa H, Tischler ML et al (2004) Host-adapted *Cryptosporidium* spp. in Canada geese (*Branta canadensis*). Appl Environ Microbiol 70:4211–4215

Ziegler PE, Wade SE, Schaaf SL et al (2007) *Cryptosporidium* spp. from small mammals in the New York City watershed. J Wildl Dis 43:586–596

Chapter 2
Emerging Viral Zoonoses from Wildlife Associated with Animal-Based Food Systems: Risks and Opportunities

Kris A. Murray, Toph Allen, Elizabeth Loh, Catherine Machalaba, and Peter Daszak

Abstract Zoonotic viruses of wildlife origin have caused the majority of recent emerging infectious diseases (EIDs) that have had significant impacts on human health or economies. Animal consumption-based food systems, ranging from the harvest of free-ranging wild species (hereafter, wild harvest systems) to the in situ stocking of domestic or farmed wild animals (hereafter, animal production systems), have been implicated in the emergence of many of these viruses, including HIV, Ebola, SARS, and highly pathogenic avian influenza (HPAI).

Keywords Animal production systems • Biodiversity • Bushmeat • Climate change • Ebola virus • Ecosystem • Emerging infectious diseases • Food systems • Highly pathogenic avian influenza • Viral zoonosis

Introduction

Zoonotic viruses of wildlife origin have caused the majority of recent emerging infectious diseases (EIDs) that have had significant impacts on human health or economies (Morse et al. 2012; Jones et al. 2008). Animal consumption-based food systems, ranging from the harvest of free-ranging wild species (hereafter, wild harvest systems) to the in situ stocking of domestic or farmed wild animals (hereafter, animal production systems), have been implicated in the emergence of many of these viruses, including

K.A. Murray
EcoHealth Alliance, 460 West 34th St., New York, NY, USA

Grantham Institute - Climate Change and the Environment, Imperial College London, Exhibition Road, London SW72AZ, UK

T. Allen • E. Loh • C. Machalaba • P. Daszak (✉)
EcoHealth Alliance, 460 West 34th St., New York, NY, USA
e-mail: daszak@ecohealthalliance.org

M. Jay-Russell, M.P. Doyle (eds.), *Food Safety Risks from Wildlife*, Food Microbiology and Food Safety, DOI 10.1007/978-3-319-24442-6_2

HIV, Ebola, SARS, and highly pathogenic avian influenza (HPAI) (Karesh et al. 2012b; Zambrana-Torrelio et al. 2012).

At the same time, wild harvest and animal production systems form a fundamental component of food systems more broadly (Milner-Gulland and Bennett 2003; FAOSTAT 2014). Food forms the foundation of human societies, promoting health and wellbeing, and sustaining growing populations (Tilman et al. 2011). The role of wild harvest and animal production systems in the emergence of human and domestic animal diseases thus presents something of a paradox, where ecosystem services meet ecosystem disservices, sometimes with catastrophic consequences.

Here we review the current status of EIDs, and in particular viral zoonoses originating in wildlife, as they relate to wild harvest and animal production systems. We conclude that both systems present considerable proximal and distal risks for disease emergence through a number of mechanisms. The reasons are that they frequently entail or promote human contact with a diversity of wildlife species, unusual assemblages of high numbers and densities of animals, rapid and widespread transportation networks and large-scale environmental perturbation, which are all key risk factors of disease emergence (Daszak et al. 2000; Patz et al. 2004).

More broadly, the costs of wild harvest and animal production systems to global environments are enormous and mounting. The resultant biodiversity loss due to overhunting, and ecosystem loss or degradation due to the expansion of areas suitable for livestock, are major environmental and societal challenges in themselves (Steinfeld et al. 2006; Milner-Gulland and Bennett 2003). In addition to the threat of disease emergence, these impacts contribute to novel and damaging negative feedback costs and a direct toll on other aspects of human health and wellbeing (Raudsepp-Hearne et al. 2010; Schröter et al. 2005; McMichael et al. 2007). To understand the risks, we must look to the combination between direct and indirect risk factors that together shape the disease risks of food systems; for example, the act of consuming a wild animal in addition to the upstream factors, such as deforestation, that can more broadly increase the availability of wildlife for food. To mitigate the risks, we concur with previous authors that opportunities exist to manipulate food systems to provide win-win or more equitable solutions for conservation and health (Tilman et al. 2011; Nelson et al. 2009; McMichael et al. 2007) and to develop or inform preventative policy for better public health and ecosystem health outcomes.

This review will focus on two distinct animal-based food systems that encompass potentially very different risk pathways for disease emergence. First, "animal production systems" are typified by in situ stocking and raising of animals (both domestic and in some cases wild species) at small to very large scales and from low to very high densities. Secondly, "wild harvest systems" typically involve direct harvest of wild, free-roaming animal species (including, for example, "bushmeat"). In some cases these systems, and the disease risks associated with them, are nested or overlap. For example, wild-harvested species can be marketed through outlets that are associated with sophisticated sale and distribution systems (Milner-Gulland and Bennett 2003). Conversely, domesticated or farmed wild species may also be released or allowed to roam shepherded or freely in landscapes, providing opportunities for contact with wild species (Kilpatrick et al. 2009). Nevertheless, we feel that distinguishing between wild harvest and in situ produced is useful when considering the role of animal-based food systems in the emergence of viral zoonoses.

Scale of Animal Production Systems and Their Importance to Human Health

Since 1950, there have been enormous increases in the production of, and trade in, domestic livestock species used for food (particularly chickens and pigs) (Godfray et al. 2010; FAOSTAT 2014) (Fig. 2.1). Although the density and composition of domestic species varies dramatically globally (Fig. 2.2), livestock systems alone

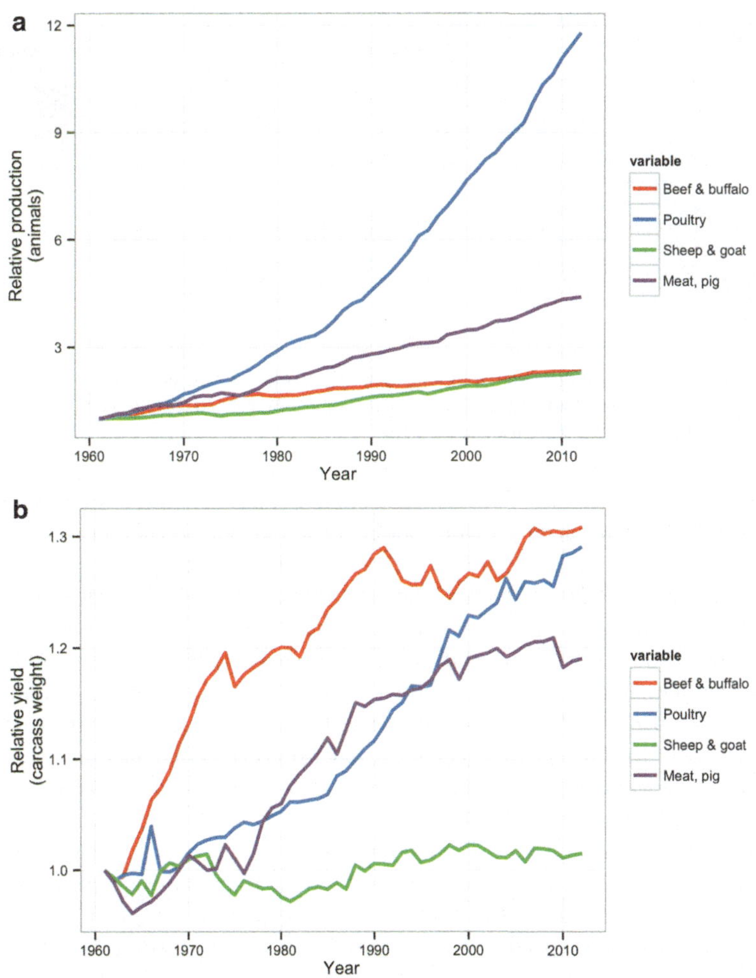

Fig. 2.1 Global trends in livestock production (**a**) and yield (**b**), in relative number of animals and relative carcass weight (hectograms/carcass), respectively, from 1962 to 2014. Livestock production data from Food and Agriculture Organization of the United Nations (FAO), accessible at http://faostat3.fao.org)

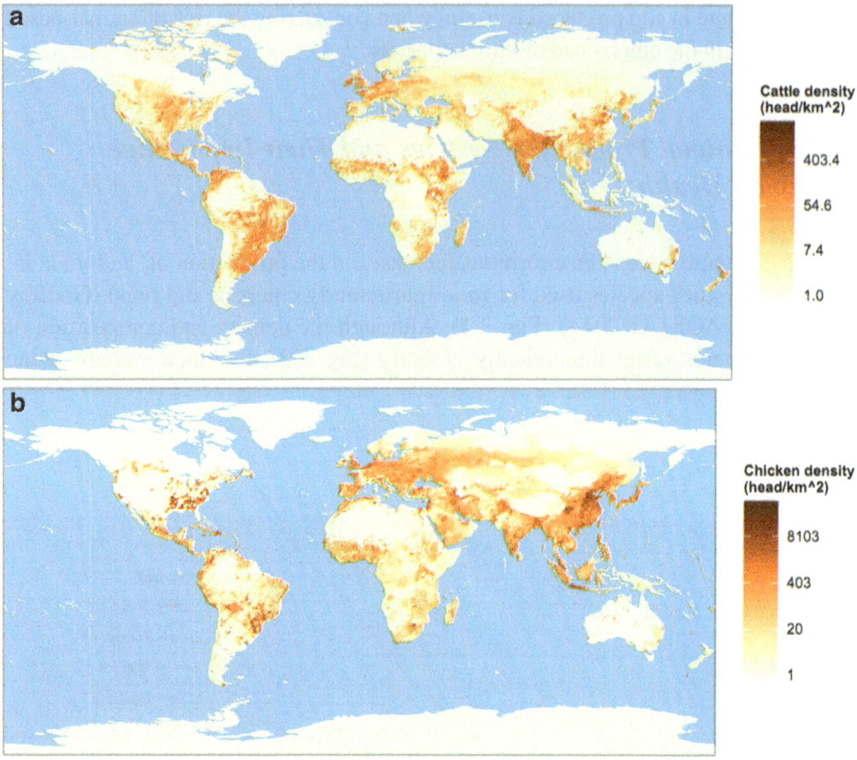

Fig. 2.2 Estimated global distribution of livestock (population density, head/km^2). (**a**) Cattle. (**b**) Chickens. Data from Food and Agriculture Organization of the United Nations (FAO), Gridded Livestock of the World 2.0 (Robinson et al. 2014)

now account for more than 30 % of the Earth's ice-free terrestrial area (Steinfeld et al. 2006). In addition to growing populations, demand for higher volume and higher quality diets has driven these increases, and per capita production has also increased (FAOSTAT 2014). In 2013 there were approximately 3.5 individual poultry and 0.5 common production mammals (cattle, sheep, and pigs) raised, on average, for every one of nearly seven billion people globally (calculated from FAOSTAT 2014). Facilitating this growth, the global capacity to raise both more animals and more animals per unit of land area has increased, marking an increase in efficiency and intensity of food production. The highest densities and efficiencies are achieved with the aid of technological advances that were developed and are primarily used in the developed world (Tilman et al. 2011).

These increases in animal production match or exceed human population growth, which has itself almost tripled over the same period. To put this in context, population growth has been so dramatic over the last century that 7–14 % of all humans ever born remain alive today (Bradshaw and Brook 2014; Westing 2010; PRB 2014). The same statistics for many domestic animals likely exceed this. The sheer scale and global reach of food production systems means that they are also major drivers of ecosystem change. In addition to the direct risks posed by high stocking densities and sophisticated transportation and trade networks, it is the associated environmental and demographic factors of animal production systems that present some of the biggest challenges from a disease emergence perspective (McMichael et al. 2007).

Scale of Wild Harvest Systems and Their Importance to Human Health

Despite the growth and scale of animal production systems globally, the direct acquisition and consumption of wild meat still forms an important component of local economies and diets, and in many instances holds cultural significance and other preference determinants that enhance its value.

Relative to other sources of meat, the contribution of wild-harvested meat to household diets, nutrition and local livelihoods is highest for the rural poor, who are often underserved by local animal production systems and/or have limited ability to raise animals for a variety of reasons (e.g., environmental constraints, lack of technology) (Brashares et al. 2011). Some populations are essentially dependent on wild-harvested food to meet basic nutritional requirements; for example, in some parts of the Congo Basin, protein from bushmeat comprises as much as 94 % of total protein of the household diet (Fa et al. 2003). In Madagascar, restricting access to bushmeat would reportedly result in a significant increase of anemia cases among children, with the poorest households worst-affected (Golden et al. 2011).

Although people have been hunting wild species for food for millennia, there has been a marked increase in the harvest of wild species over the last several decades (Milner-Gulland and Bennett 2003; Ziegler 2010). For example, the development of industrial logging in Republic of Congo has led to a 69 % increase in the population of logging towns and a 64 % increase in bushmeat supply (Poulsen et al. 2009). The emergence of market-based economies and the commercialization of wild-harvest animals in urban centers have further increased demand. The scale of the trade in wild-harvest meat has also changed considerably due to advances in hunting practices, population growth, and increasing accessibility to remote areas (Nyaki et al. 2014). The trade of wild-harvested meat for food can now be viewed as a continuum ranging from subsistence-based rural consumption to commercial hunting for the international trade in wild animal meat and products (e.g., exotic food and traditional medicine), with this leading to dramatic price point differences (Brashares et al. 2011; Chaber et al. 2010). For example, bushmeat traders have reported pricing per kg of wild meat in Paris markets at up to double the price of domestic meat for sale in French supermarkets (Chaber et al. 2010). Similarly, in New York, USA, smoked duiker (an antelope) from Ghana can be readily attained, although at up to 25 times the cost of the same species sold near its source (Brashares et al. 2011). Such "urban" demand, which also occurs locally, places a premium on wild species and often permits local hunters to earn incomes comparable to or higher than local wages for other occupations, sustaining commercial hunting on a large scale (Schulte-Herbrüggen et al. 2013). Market factors thus provide a significant additional incentive for wild harvest beyond the protein needs of an individual hunter or family.

The monetary incentives for importation of wild-harvested meat have resulted in extensive trade networks, and an expansion of the public health risk of zoonotic disease spillover. An estimated 5 tons of bushmeat is smuggled through Paris Roissy-Charles de Gaulle airport from Africa per week in passenger baggage (Chaber et al. 2010). Even if the prevalence of potentially zoonotic pathogens in the animals traded is low, and viability of microbes much reduced after time in the trade, the sheer volume of bushmeat traded internationally, and lack of traceability through illegal or clandestine trade, suggests a significant public health risk. Measuring or controlling this risk is made more difficult because the global distribution of wild-harvested food is highly variable, poorly reported, and difficult to map (Fig. 2.3).

Emerging Infectious Diseases Associated with Food Systems

Infectious diseases that appear in a new host (e.g., humans) for the first time or markedly increase in incidence or geographic range, or cause disease with apparently novel clinical patterns are often referred to as emerging infectious diseases (EIDs) (Taylor et al. 2001). Historically, many human diseases are thought to have arisen as a result of the environmental and demographic changes attributable to the advent of food systems (agriculture and/or animal domestication) (Pearce-Duvet 2006; Wolfe et al. 2007). Such "civilization diseases" include some likely acquired directly from domesticated species (e.g., measles, pertussis) or indirectly, either

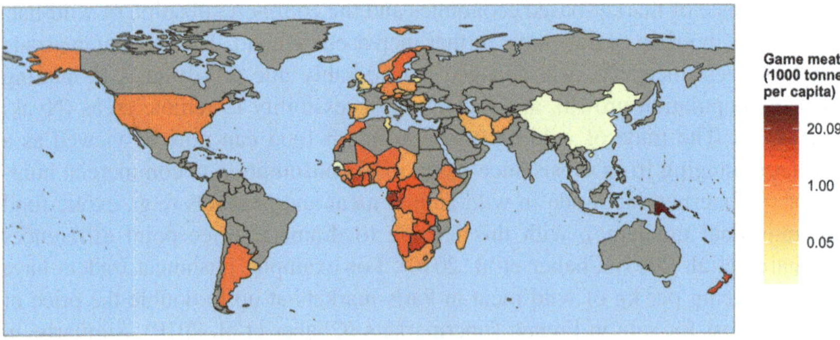

Fig. 2.3 Estimated global production of game meat in tons per capita, yearly average 2000–2009. Livestock production data from Food and Agriculture Organization of the United Nations (FAO), accessible at http://faostat3.fao.org). These data likely reflect a widespread lack of reporting of wild-harvested meat

because domestic animals provided a more stable route of infection for pathogens to enter human populations from wildlife (e.g., smallpox) or due to the influence of environmental perturbation in elevating the risk of pathogen transmission to humans from wildlife hosts and vectors (e.g., *falciparum* malaria) (Pearce-Duvet 2006). All of the diseases mentioned above were at one time EIDs, highlighting how some of the emerging diseases of the recent past and present will almost certainly become the diseases of humanity in the future. Understanding the origins and drivers of EIDs is thus of considerable and growing public health interest (Morse et al. 2012).

Demographic, behavioral, ecological, and climatic changes have all been variably cited as drivers of historical and contemporary disease emergence (Patz et al. 2004, 2008; Smith et al. 2007; Wolfe et al. 2005a, b; Daszak et al. 2000; Morse 1995; Taylor et al. 2001; Foley et al. 2005; Jones et al. 2008). The increasing impact of an exponentially rising human population has led to an increase in these drivers over time which likely explains why the frequency of disease emergence appears to have increased in recent decades, even after correcting for increased capacity and effort to detect them (Pike et al. 2014; Jones et al. 2008). Systems in equilibrium are probably the least likely systems to give rise to EIDs.

The current scale and continued expansion of wild-harvest and animal production systems thus present ongoing opportunities for diseases to emerge into the human population.

Viral Zoonoses of Wildlife Associated with Animal-Based Food Systems

Although rarely observed (approximately 1 per year globally) (Jones et al. 2008), zoonotic viruses that originate in wildlife and are associated with food systems punch above their weight in terms of their potential human, animal, and economic impacts. Some of the best recognized examples include HIV, SARS, Ebola, and Avian Influenza A viruses (Karesh et al. 2012b; Zambrana-Torrelio et al. 2012; Hahn 2000a, b; Heymann 2004a, b), but they also include diseases that have caused significant regional or more local impacts, such as Japanese Encephalitis virus (Mackenzie et al. 2004), a number of rodent-borne hantaviruses (e.g., Junin, Laguna, Machupo viruses) (Epstein 1995; Young et al. 1998a, b; Johnson et al. 1997a, b; Webb et al. 1967a, b), Lassa virus (Ter Meulen et al. 1996), a number of bat-borne viruses (e.g., Nipah, Menanagle viruses) (Calisher et al. 2006; Pulliam et al. 2011; Luby et al. 2006), and monkeypox virus (Parker et al. 2007) (Table 2.1).

Across the spectrum of animal-based food systems described above, there are a range of common features or activities (e.g., capture and handling, butchering, trade, transport, and consumption) that provide opportunities for pathogens to move from wildlife into humans, whether directly or indirectly via a domestic animal link or via vectors. The processes involved, however, can be complex. Below we use the diseases listed in Table 2.1 as examples to decompose these risks into three fundamental com-

Table 2.1 Table detailing food-system-associated disease emergence events, including context, transmission modes, and risk factors identified in literature review

		Circumstances of emergence				Confirmed or probable transmission routes (Jones et al. 2008 categories)					Distal risk factors (documented, probable, unknown)						
Food system	Pathogen or disease	Reservoir species	Location of transmission	Outcome of transmission	Contact event type/ food system risk behavior	Direct contact	Air-borne	Vector-borne	Oral (ingest)	Envir./ fomite	Road building	Deforestation/ agricultural conversion	Climate/ climatic change	Trade	Increasing human connectivity	Agricultural intensification (high population densities/ food storage)	References
Wild harvest	Ebola	Unknown (fruit bats suspected)	DRC, Sudan, Guinea, Sierra Leone	Localized and regional outbreaks	Human–wildlife; hunting and consumption	Yes			Yes	Yes	Greater wildlife access	Increasing niche overlap, stressed food security, greater wildlife access	Shifting species ranges, stressed food security	Local trade in bushmeat	Human to human or international spread		Leroy et al. (2009); Pigott et al. (2014)
	SARS	Horseshoe bats (Rhinolophus spp. infection also found in masked palm civets, pigs)	Guangzhou, Guangdong Province, China	Far-reaching epidemic with person-to-person transmission	Human–wildlife; handling	Yes	Possible, but unlikely				Greater wildlife access	Greater wildlife access		Extensive trade of wild caught species, interspecies mixing in market setting	Human to human or international spread		Li et al. (2005a, b); Wang et al. (2006); Shi and Hu (2008)
	HIV	Chimpanzee, Sooty Mangabey	Africa	Repeated spillover, global pandemic	Human–wildlife; hunting and consumption	Yes					Greater wildlife access	Greater wildlife access			Human to human or international spread		Hahn (2000a, b); Wolfe et al. (2005a, b)
	Lassa virus	Wild rodents (Mastomys natalensis, Natal multi-mammate mouse)	Endemic in Guinea, Liberia, Sierra Leone, and regions of Nigeria	Regional epidemics, with little human-to-human transmission	Human–wildlife; hunting, handling, consumption	Yes	Yes (aero-solized rodent excreta)		Yes (rodent excreta)			Elevated reservoir numbers (greater availability as food)		Local trade in bushmeat		Elevated reservoir numbers around grain storage	Heymann (2004a, b)

Animal production	Influenza A virus	Aquatic birds, poultry, swine. Infection found in other mammals.	Widespread. Azerbaijan, Cambodia, China, Djibouti, Egypt, Indonesia, Iraq, Lao People's Democratic Republic, Myanmar, Nigeria, Pakistan, Thailand, Turkey, Vietnam.	Localized outbreaks	Wildlife–domestic animal–human; farming activities	Yes	Yes		Yes			Land clearance for domestic production	Shifting species ranges (poultry–wildlife contact), stressed food security (increased poultry numbers and densities)	Increased global food demand	Increased pandemic risk	Agricultural intensi-fication	Heymann (2004a, b)
	Japanese encephalitis virus	Pigs (mostly asymptomatic amplifying host), wading birds (reservoir host), mosquito vector	East, South-East, South Asia	Epidemics, but without human-to-human transmission (humans are dead-end hosts)	Vector-domestic animal, vector-human; farming activities			Yes (mos-quito)				Land clearance and flooding for rice and pig production (increased vector population, domestic animal amplifier)	Shifting species ranges (mosquito), stressed food security (increased pig numbers and densities)	Increased food demand		Agricultural intensi-fication	Campbell et al. (2011); Centers for Disease Control (2013)

(continued)

Table 2.1 (continued)

Food system	Pathogen or disease	Circumstances of emergence				Confirmed or probable transmission routes (Jones et al. 2008 categories)					Distal risk factors (documented, probable, unknown)						References
		Reservoir species	Location of transmission	Outcome of transmission	Contact event type/ food system risk behavior	Direct contact	Air-borne	Vector-borne	Oral (ingest)	Envir./ fomite	Road building	Deforestation/ agricultural conversion	Climate/ climatic change	Trade	Increasing human connectivity	Agricultural intensification (high population densities/ food storage)	
	Nipah virus	Fruit bats (Pteropus spp.)	Malaysia, Bangladesh	Localized outbreaks	Wildlife–domestic animal–human; farming activities (also consumption of contaminated date palm sap in Bangladesh)	Yes			Yes			Increasing niche overlap (bat hosts losing habitat)	Shifting species ranges (bat hosts), stressed food security (increased pig numbers and densities, mixed cropping)	Increased food demand		Agricultural intensi-fication	Luby et al. (2006); Rahman et al. (2010); Pulliam et al. (2011)
	Menangle virus	Fruit bats (suspected reservoir host), pigs (amplifier host)	Australia (Menangle)	Localized spill-over	Wildlife–domestic animal–human; farming activities	Yes						Increasing niche overlap (bat hosts losing habitat)	Shifting species ranges (bat contact with pigs)	Increased food demand		Agricultural intensi-fication	Kirkland et al. (2001); Center for Food Security and Public Health (2007)
Agricultural	Junin virus (Argentine hemorrhagic fever)	Rodents (Calomys musculinus, dry lands vesper mouse; Calomys laucha, small vesper mouse)	Buenos Aires	Localized outbreaks (endemic in region)	Wildlife excreta–human; agricultural activities	Yes	Yes (aerosolized rodent excreta)		Yes (rodent excreta)			Elevated reservoir numbers				Elevated reservoir numbers around food storage	Maiztegui (1975); Public Health Agency of Canada (2010)

	Laguna Negra virus (Hantavirus pulmonary syndrome)	Calomys laucha (vesper mouse)	Bolivia, Paraguay, Chile	Localized outbreaks	Wildlife excreta–human; agricultural activities		Yes (aerosolized rodent excreta)					Elevated reservoir numbers				Elevated reservoir numbers around food storage	Johnson et al. (1997a, b); Young et al. (1998a, b)
	Machupo virus (Bolivian hemorrhagic fever)	Calomys callosus (vesper mouse)	Bolivia	Localized outbreaks	Wildlife excreta–human; agricultural activities		Yes (aerosolized rodent excreta)					Elevated reservoir numbers				Elevated reservoir numbers around food storage	Webb et al. (1967a, b)
	Monkeypox virus	Unknown, multiple (rodents, including Gambian rats, striped mice, dormice, rope and tree squirrels; primates)	Democratic Republic of Congo, Sudan, elsewhere in Congo Basin and western Africa. Cases also seen in Midwestern United States	Localized outbreaks	Wildlife–human; handling, contact with infected animals and people or bodily fluids thereof	Yes						Elevated reservoir numbers		Extensive trade of wild caught species	Human to human or international spread		World Health Organization (2011)
	Lassa virus	Wild rodents (Mastomys natalensis, Natal multimammate mouse)	Endemic in Guinea, Liberia, Sierra Leone, and regions of Nigeria	Regional epidemics, with little human-to-human transmission	Wildlife excreta–human; agricultural activities	Yes	Yes (aerosolized rodent excreta)		Yes (rodent excreta)			Elevated reservoir numbers				Elevated reservoir numbers around food storage	Heymann (2004a, b)

Color shadings relate to importance to the event, based on our judgment from our literature review. Darker factors are more proximate or directly related to the event ponents—the types of contact events associated with them, the various transmission pathways that are involved, and the upstream distal risk factors that promote the former to facilitate emergence. These together help highlight the activities and conditions common to food systems that may promote disease emergence.

Contact

Contact events provide the "proximal" risk interfaces that allow disease transmission. Contact events can occur in many different contexts but their common feature is that they provide the opportunity for the transmission of a pathogen. Transmission interfaces could include: human–wildlife, human–vector, human–domestic animal, human–human, wildlife–vector, wildlife–domestic animal, and vector–domestic animal contact. The diversity of types of contact that have been relevant historically for the emergence of viral zoonoses from wildlife associated with food systems is summarized in Table 2.1. In wild harvest systems, contact events have typically occurred directly between a person and a range of wildlife species ultimately used for food (HIV, Ebola, SARS, Lassa, Monkeypox), via contact activities such as hunting, handling, butchering, consumption, and trade. In animal production systems, people have become infected most commonly from contact with domestic animals that had first been exposed to wildlife pathogens (e.g., HPAI influenza, Nipah, Menangle), where tending and treating domestic animals for illness resulted in human infection. Japanese Encephalitis represents an example where humans are infected when bitten by mosquito vectors, which acquire and maintain infection after feeding on wild hosts or infected domestic species (e.g., pigs). For diseases more diffusely associated with agricultural activities, infection often occurs via contact with virus present in wildlife excreta (e.g., hantaviruses) or fomites (see Table 2.1 and references therein).

Transmission Routes

While contact events serve as the fundamental infection interface, different types of contact may carry very different levels of "riskiness" depending largely on the mode of transmission of a given pathogen. An important challenge in understanding the risks of viral zoonoses due to food systems is identifying the relevant transmission routes that allow for pathogen transmission between wild animal reservoirs, vectors, domestic animals, and humans. Transmission routes can be classified into five broad but distinct categories and used to analyze patterns of disease emergence (Loh et al. 2015). These include direct contact (i.e., skin-to-skin contact; scratches; animal bites; contact with body fluids, organs, and tissues; direct large droplet >5 μm exposure), airborne (i.e., via dust particles and airborne small droplets <5 μm), vector-borne (i.e., by biting or mechanical transfer by arthropods), oral (i.e., consumption of contaminated food or water; ingestion of arthropods), and contamination (i.e., indirect contact with soil or vegetation, contact with water, indirect transmission by contaminated inanimate objects). Direct contact is the most common transmission pathway cited for diseases associated with food systems, although airborne transmission of virus associated with aerosolized wildlife excreta is also relatively common (Table 2.1). Nevertheless, the range of transmission pathways implicated in the emergence of viral zoonoses from food systems is relatively diverse, with each transmission pathway represented at least once.

Distal Factors

While the type of contact, mediated by the various transmission pathways, represents the proximal risk factor for spillover (i.e., where and how transmission takes place), other factors may promote or reduce the likelihood that contact events occur in the first place or result in pathogen transmission, thereby altering the risk of emergence. These distal or upstream risk factors also include any condition or activity along any transmission pathway that intensifies the contact rate, increases the prevalence or diversity of available pathogens to be transmitted, or elevates the likelihood of successful disease transmission given contact (Murray and Daszak 2013; Lloyd-Smith et al. 2009). Distal risk factors could also include other "enabling" factors, such as climate or other environmental factors. The key distal factors that have been associated with the emergence of viral zoonoses from food systems are summarized in Table 2.1. Broadly speaking, large-scale ecosystem and environmental change, including deforestation, land-use change and conversion for agriculture, have been commonly implicated in disease emergence within food systems. Examples include the rodent-borne arenaviruses (Lassa, Junin, Machupo, Laguna Negra) that are often facilitated by agricultural land conversion, HIV which is thought to have emerged as a result of the changes in forest access and human connectivity attributable to industrial development, and Nipah and Menangle viruses which are thought to have emerged due to increasing niche overlap and contact between reservoirs and domestic animal species (see Table 2.1 and references therein). In addition to human-induced ecosystem changes, there are a range of social and demographic factors that have also played roles as distal risk factors, including the trade of wildlife species within markets with sophisticated transport networks and in which inter-species mixing has occurred (SARS, monkeypox), or increased domestic animal stocking densities (agricultural intensification) to meet growing human food demands while at the same time enhancing conditions for viral amplification (HPAI, Japanese Encephalitis) (see Table 2.1 and references therein).

If the diversity of previous disease emergence mechanisms is anything to go by, forecasting disease risks within food systems should not rely solely on historical precedence. While decomposing the risks of disease emergence into subcategories of proximal and distal risk factors can seem trivial, particularly for the well-known examples examined here (Table 2.1), the real utility and application of this approach is for forecasting future risks (see Sect. 4.3 below). Such horizon scanning exercises are critical for anticipating the risks associated with the growth in scale and magnitude of food systems into the future. For example, in industrial food systems, airborne transmission may potentially be an under-recognized pathway as a recent study found a million-fold elevated concentration of aerosolized invisible dust in a poultry barn fan compared to the outside air (Leibler et al. 2009). This could have implications for both human and animal health in addition to the spread of true airborne diseases such as Foot and Mouth disease (FMD), influenza, or Q fever.

Future Trends in Food Acquisition and Production Systems: Implications for Viral Zoonoses

Both the acquisition of food from wild sources as well as the scale and intensity of animal production systems are projected to continue increasing over the coming decades (McMichael et al. 2007; The World Bank June 2012; Zambrana-Torrelio et al. 2012). This presents challenges for disease emergence and for environmental stability as increasing global populations demand higher dietary quantity and quality leading to continued land-use change and deforestation, expansion of global trade and travel networks and potential secondary impacts through climate change, biodiversity loss, and other outcomes.

Wild Harvest Trends

Harvesting wild animals for food and other uses has been increasing in the recent past, and is likely to continue its growth as one of the greatest threats to biodiversity (Fa et al. 2002). This follows increasing reliance on wild animals to meet dietary needs for protein under conditions of food insecurity in many regions, especially developing countries in the tropics (Fa et al. 2003). Exploitation of wildlife for food will likely be facilitated by increasing land-use change and deforestation activities, whatever their purpose, particularly in more remote regions where these activities make forests more accessible to hunters and create new markets for bushmeat (Poulsen et al. 2009). Climate change is also expected to threaten food security in many regions, again promoting greater reliance on wild harvest species in some regions (Nkem et al. 2010). This is set against a background of exponentially increasing global air travel which already poses a significant risk to global health via the transportation of pathogens (Hufnagel et al. 2004), and is likely to promote increased global trade in wild-harvested meat.

The development of roads may represent one of the most significant ways of increasing opportunities for wild harvest. Roads are considered critical infrastructure developments that can improve access to technologies, healthcare and education, forming a key component of many countries' development plans. Approximately 60% more roads are projected by 2050 compared to 2010, mostly in developing countries (Dulac 2013), potentially making road building one of the most significant drivers of future environmental change (Laurance et al. 2014). Road building has already increased the risk of some diseases associated with human development (e.g., agricultural intensification), with an increase in number of cases of human hantavirus reported following the completion of a highway through the Brazilian Amazon (Medeiros et al. 2010). Road building, particularly on such a large scale, will almost certainly further facilitate bushmeat hunting in the most biodiverse regions of the planet and change the scale at which people are able to move wild animals out of newly exploited areas and into commodity chains, thereby increasing public health risks.

Animal Production Trends

Global food production is forecast to approximately double by 2050 to meet the food demands of a global population that is expected to plateau at around nine billion people (Godfray et al. 2010; Tilman et al. 2011). The biggest growth will be seen in domestic animal products, with predictions suggesting an increase in annual demand for meat of 6–23 kg per person per year worldwide by 2050. The largest increases will be in Latin America, the Caribbean, South East Asia and the Pacific, and demand per person will more than double in sub-Saharan Africa (Thornton and Herrero 2010). Food production is expected to more intensely compete with the acquisition of other products from the environment such as land, water and energy, contributing to loss of ecosystem services and biodiversity, including some related to health (Tilman et al. 2011; McMichael et al. 2007). Deforestation and associated human activities related to domestic animal production, for example, will continue to alter the structure and species composition of ecosystems and increase contact rates between humans, wildlife, vectors and domestic animals, resulting in disease emergence (Murray and Daszak 2013).

Food production will also continue to contribute to, and be strongly affected by, climate change (Godfray et al. 2010; McMichael et al. 2007), particularly in developing and less developed countries, and this will coincide with changes in disease risk. For example, climate change may influence some key elements of the avian influenza A transmission cycle. Climate change is expected to influence migration patterns of migratory bird species that are the natural reservoirs for many AI viruses, alter transmission dynamics and affect the survival of virus outside of hosts, all of which have the potential to shift disease risks for this important group of viruses (Gilbert et al. 2008). In addition, the link between domestic duck production, which is expected to grow in scale and extent to build food security in Asia, and the persistence of HPAI H5N1 is often synchronously linked to the production of rice. The strong seasonal component of this system means that climate change has the potential to impact the distribution and persistence of HPAI in other more indirect ways as well (Gilbert et al. 2008).

The increasing intensification of food production, marked by high animal densities and stressful conditions, may facilitate rapid spread of diseases among immunocompromised and genetically similar animals, potentially compromising food security and posing zoonotic disease risks. In addition to the risk of wildlife origin zoonoses making their way into humans via a domestic animal intermediary, the widespread use of antimicrobials in food production, primarily for non-therapeutic growth promotion in livestock and aquaculture production, may introduce rapid selection pressure for resistant bacterial and viral strains and further contribute to disease risks. While drug-resistant EIDs are more common in non-zoonotic EIDs than zoonotic EIDs (Zambrana-Torrelio et al. 2012), greater use of growth-promoting antimicrobials in animal production and human exposure via food as well as antimicrobials disseminated into the environment from animal production waste may potentially increase human susceptibility to infections (Marshall and Levy 2011).

Additionally, as intensification occurs, biosecurity measures become all the more necessary. For example, a lag in biosecurity practices during increases in poultry production has been attributed to the evolution of HPAI H5N1 in poultry flocks, which caused extensive impacts to the poultry and public health sectors, leading to mortality or culling of over 200 million birds, as well as several hundred human deaths (Karesh et al. 2012a). The lack of adequate infrastructure for biosecurity measures in low-income nations where bushmeat currently serves as a major form of subsistence nutrition thus presents vulnerability around potential intensified livestock production to shift protein sources. Agricultural practices may also pose risks to wildlife, including flow of pathogens between livestock and wild species, in addition to the more usual culprits of habitat destruction or degradation.

Looking Forward: Intervention and Risk Mitigation Options

The range of both proximal and distal risk factors associated with disease emergence from food systems makes effective disease management a complex and daunting proposition. However, this also provides opportunities for mitigation and adaptation with a view to better managing food systems to reduce environmental and biodiversity impacts in addition to disease risks in the future. For proximal risk factors associated with specific contact events, better safety and biosecurity standards will be a core part of any strategy to reduce disease risks from wild harvest and animal production systems. However, the more distal drivers of disease emergence (e.g., land-use change) or global changes that occur in step with, or that directly facilitate, the expansion of food systems present a much more nebulous and diffuse range of risks. Managing these underlying drivers may ultimately provide solutions for sustainability and public health threats. We propose that direct mitigation of disease transmission is thus only ever going to be a part of what urgently needs to be a much more encompassing, proactive strategy targeting the distal risks of disease emergence (Murray et al. 2012). This requires a novel response that could be rooted in holistic cost–benefit analyses of total ecosystem services (Costanza et al. 2014).

Win-Win Solutions for Conservation and Health?

The number of hungry people globally has declined by more than 200 million since 1990, despite the addition of almost two billion people over the same period (FAO 2014). This largely can be attributed to ongoing improvements and increases in global food production and supply systems and global efforts to improve food security (FAO 2014). These improvements have improved human health more broadly by decreasing malnourishment, increasing life expectancy and reducing child mortality (Raudsepp-Hearne et al. 2010; Godfray et al. 2010). Health gains will of course continue to be an ongoing human objective, with food security being a central part of the development agenda (FAO 2014). The health benefits of food production, however, need to be weighed against the health and environmental costs, including those associated with ecosystem degradation (McMichael et al. 2007). There have been calls for concerted redistribution of excess food and deployment of food production technologies to areas of the world most in need (Tilman et al. 2011). These strategies might have secondary benefits to global health by reducing food demands in some regions, leading to reduced environmental and total area designated for food production.

Health and conservation goals and actions have not always aligned, with history of some rash disease control efforts unnecessarily resulting in harm to wildlife and domestic animal populations, and when conservation frameworks (e.g., the Convention on International Trade in Endangered Species of Wild Fauna and Flora) do not directly consider disease risks in their decision making. To more effectively address both public health and conservation concerns, it is necessary to improve synergy between the two communities with integrated, science-based approaches. This need is especially urgent in the food safety realm, where nutritional dependencies demand sustainable access to food sources. The UN's post-2015 Sustainable Development Goals set the stage for poverty reduction, food access/security, health, and environmental balance, potentially providing opportunities for integrated solutions that could be applied to food safety challenges related to wildlife and food systems.

The underlying drivers of disease emergence from wildlife are also the same main pressures that drive biodiversity loss as identified by the UN Convention on Biological Diversity Global Biodiversity Outlook 4, namely habitat loss, degradation and fragmentation, overexploitation of wildlife, unsustainable production in agriculture and other industries, and impact of invasive species (Secretariat of the Convention on Biological Diversity 2014). In addition, emerging viruses are not only threats to humans, but may also be pathogenic to susceptible wild host species. There is thus a compelling opportunity for co-benefits for conservation and public health through collaborative efforts.

The Policy Landscape

Despite the globalization of food supply systems, there is no central global governance structure for foodborne or food-associated disease risks, and there is no precise estimate of foodborne or food-associated disease incidence or burden. To address this, the World Health Organization (WHO) is undergoing an assessment of the global impacts of foodborne illnesses through its Department of Food Safety and Zoonoses. While the FAO-WHO Codex Alimentarius provides benchmark international trade standards to promote food safety, the guidance is voluntary; the U.S., for example, does not require its producers and suppliers to adhere to its rigorous standards. The lack of a central authority for wildlife health has translated into limited infrastructure for disease surveillance and control around the safety of bushmeat in both source and demand settings. As a result, efforts have largely focused on reactive responses to disease emergence events, rather than prevention of disease risks. The World Organization for Animal Health (OIE) regulates trade of livestock for priority diseases, which include some potential zoonoses (e.g., HPAI), but does not address wildlife trade/pathogens specifically in its World Trade Organization-enforced sanitary standards. There is no comparable regulation for wildlife diseases, although in the USA, the U.S. Centers for Disease Control and Prevention specifically restricts imports of certain turtles in response to salmonellosis, bats in response to Nipah virus, African rodents in response to monkeypox, civets in response to SARS, and non-human primates (Smith et al. 2012).

Risk Analysis

Greater knowledge of disease emergence risks from wildlife can inform identification of key areas for intervention. Risk assessment is commonly conducted in food safety to identify vulnerabilities in the food supply, but more fully protecting health requires determining and addressing upstream or distal risks of viral emergence from harvested wild meat. Employing risk analysis tools can assist in science-based policies by anticipating and identifying ways to mitigate risk, as well as identifying priority knowledge gaps for research investments to refine future analyses. The structure of a formal risk analysis can help provide continuity and objectivity in the process, involving problem description, hazard identification, risk assessment, risk management, implementation and review, and risk communication throughout. More proactive risk analysis efforts can systematically identify critical control points for conservation and health benefits, and congruence among both where synergies can be maximized.

For example, the OIE-IUCN Guidelines for Wildlife Disease Risk Analysis promote analysis of disease risk in an ecosystem, rather than single-species, context (World Organisation for Animal Health (OIE) & International Union for Conservation of Nature (IUCN) 2014). This perspective can help determine conservation risks as well as zoonotic disease risks. While uncertainty and complexity inherently exist in wildlife disease risk analysis (Jakob-Hoff et al. 2014), useful information can be gained, especially for viral disease threats where initial knowledge on transmission pathways and pathogen dynamics can enable best practices to reduce risks while more information is gathered.

Realistic Interventions

Harvesting of wild meat holds a critical position in the diets, economies, and cultures of millions of people globally. Current governance and enforcement structures are therefore unlikely to be fully effective and in many cases unwarranted for reducing local demand (e.g., for local populations living in or on the periphery of forests with few suitable alternatives). In this context, some interventions may be low-resource and high-yield, such as working with hunters and foresters to convey risks of collecting deceased wildlife carcasses and encourage reporting of animal morbidity or mortality that can inform disease surveillance efforts (Rouquet et al. 2005; Olson et al. 2012). These interventions to prevent initial spillover are especially important given the challenges of influencing human behaviors when controlling human outbreaks. For example, the UN recently reported the dismissal of a local chief in Sierra Leone for failing to report secret burials that may have violated regulations intended to contain the spread of Ebola (UN Mission for Ebola Emergency Response (UNMEER) 19 November 2014). However, it seems inevitable that reducing demand for bushmeat will be fundamentally necessary to safeguard species from overhunting and extinction and to mitigate the disease risks. Reducing demand will be easiest for populations with access to alternative food sources. High demand and pricing for wild-harvest species may influence hunting practices, including expanding volumes and time of year spent hunting, whereas previously hunting pressure has been naturally limited by hunting for subsistence, traditional techniques, seasonality, and cultural taboos on harvesting certain species (Lindsey et al. 2012).

Strong regulations can be established to prohibit and provide disincentives for legal and illegal sale of bushmeat to overcome growing demand as a luxury product. High taxation levies may sufficiently raise the price to reduce demand and provide revenue for enforcement and surveillance efforts (see Courchamp et al. 2006). The clandestine nature of the illegal wildlife trade remains a challenge for tracking and enforcement, but high penalties have not yet been enacted in many settings; steeper penalties may provide stronger disincentives to participation in the illegal wildlife trade, such that even if zero volume cannot be realistically achieved, a large reduction in volume will still have large benefits from a risk reduction viewpoint. Additionally, development projects that encroach into wildlife habitat can be managed to ensure they do not fuel demand for bushmeat. Governments can demand responsibility on the part of corporations to provide alternative food sources for employees and set policies to provide deterrents for bushmeat consumption. Governments could require wildlife disease risk analysis processes to be undertaken for proposed development projects to more proactively weigh risks and ensure risk prevention or mitigation measures are conducted. This type of analysis could be included within existing Health Impact Assessment (HIA) structures, because, while some HIAs include risk of zoonotic diseases from domestic animals and other vectors, few adequately address the range of potential zoonotic pathogens in their intended scope.

Can the Farming of Wildlife Become a Safe Alternative to Wild-Harvest Meat?

The farming of wildlife for food may reduce pressures on wild populations, and is increasingly becoming a way to sustain demand in the face of increasing prices of wild-caught individuals. For example, porcupines, snakes, frogs, tigers, and a range of other wildlife species are farmed in Southern China for food and medicine (Abbott and van Kooten 2011). While this has been debated widely as a tool for conservation (e.g., the farming of tigers to reduce poaching), it has not been proposed as a strategy to reduce the public health risks of the wildlife trade. We propose that the farming of wildlife species could reduce the risk of zoonotic disease spillover if similar health and biosecurity measures are applied to farmed wildlife as to livestock. In this scenario, specific known zoonoses are tested for, treated or infected animals removed from a farm's founder wild-caught stock, resulting in reduced risk of zoonotic pathogen "spillover" to ranchers, traders, or butchers. Biosecurity measures will be critical to reducing risk because the intensive production of species that potentially carry novel zoonotic agents could result in increased pandemic risk. For example, civets have long been farmed in some parts of Africa (Eniang and Daniel 2007; Tolosa and Regassa 2007), and prior to the SARS outbreak in China in 2002, civets were farmed increasingly in China. While the role of civets in the emergence of SARS is not fully understood, it is thought that they may have acted as amplifier hosts, expanding transmission and evolution of a bat-origin SARS-like coronavirus (Wang and Eaton 2007).

Wildlife ranching (typically lower density, semi-free ranging stock) may provide a more suitable production option in areas where more conventional and higher intensity animal production is not supported. For example, regions with tsetse fly infestations affect cattle production through high morbidity and mortality burden from trypanosomiasis; while wildlife appear to carry infection, they are not highly susceptible to it (Steverding 2008). In theory, wildlife ranching may provide a contained environment where disease may be controlled through adoption of effective biosecurity measures. For example, Zambia's wildlife ranching is subject to inspection of animals or meat by veterinarians prior to sale (Lindsey et al. 2013). However, the sensitivity of this approach in detecting disease risks is not known; visual inspections by veterinarians may not recognize all illness in animals, especially asymptomatic infections that wildlife may be carrying, and viral pathogens are often not evident in meat without laboratory screening. Challenges around traceability in the market chain also introduce risk if free-ranging and ranched animals cannot be distinguished.

Ranch-raised wild animals may also potentially come into greater contact with wildlife (e.g., if ranches are at the periphery of protected areas), potentially shifting the dynamics of population genetics and pathogen flow. Since the main risk pathways from viral zoonoses originating in wildlife associated with animal production systems come from the spillover of wildlife pathogens to domestically farmed species, more research is needed on disease risks in wildlife farms versus in free-ranging wildlife, as well as the development of formal guidelines on biosecurity and other practices to reduce risks to and from native wildlife, such as guidelines on proximity to conserved areas.

Conclusions, Gaps, and Future Research Needs

Several key research gaps remain that limit our ability to recognize and prioritize needs for viral threat reduction related to wildlife. Firstly, we lack knowledge of most of the viral pathogens that are circulating in wildlife (most of which have not yet been discovered and characterized (Anthony et al. 2013)), and how those pathogens are evolving in relation to our changing pressures on the environment. Secondly, we lack criteria to fully determine zoonotic potential of viral agents that are detected. Progress in these research areas is important for identifying practices that drive disease transmission risks and for prioritizing critical control points in risk analyses and risk reduction efforts (Morse et al. 2012).

Current surveillance systems for viral zoonoses are highly reactive, largely capturing threats once they have emerged in humans or have caused extensive livestock or wildlife morbidity or mortality. While current systems are inadequate for prevention and early detection, existing programs may be leveraged as a starting point (Murray et al. 2012). For example, many countries conduct wild bird surveillance for avian influenza, but screening is typically limited to only a subset of HA and NA subtypes, limiting knowledge of viral diversity circulating in populations (Hoye et al. 2010). Targeted surveillance for broader indicators of viral diversity (e.g., whole genome sequencing or at least typing for all 8 AI gene segments) can provide baseline monitoring to capture changes, including risk potential, over time. Coordinated global research priorities, such as set forward by the OIE-FAO OFFLU global network of expertise on influenza (http://www.offlu.net/), can provide an international platform for systematic surveillance approaches and data aggregation and identify high-priority investment areas to maximize surveillance resources.

To sufficiently respond to viral disease threats that are identified by surveillance, a coordinated, multi-disciplinary system is needed. The currently siloed mandates of intergovernmental organizations and government departments limit the actionable utility of data. To move forward at a global level, investments made toward achievement of OIE Performance of Veterinary Service (PVS) and the WHO's International Health Regulations might expand capabilities related to pathogen surveillance in wildlife.

Partners from the biodiversity community also have a strong role for participation, through conservation efforts that are increasingly recognizing the risks of infectious disease agents to wildlife populations (e.g., great ape die-offs from infection with Ebola). The UN's Convention on Biological Diversity recently "recognized the value of a One Health approach" toward shared health and biodiversity benefits at its 12th Conference of the Parties in October 2014, and also has addressed sustainable use of biodiversity in terms of bushmeat and sustainable wildlife management, providing a possible entry for work on both topics by CBD member countries.

Additionally, critical areas of need for collaboration can be identified under the CBD-WHO Joint Work Programme on Biodiversity and Human Health. On a national level, including through integration into CBD members' National Biodiversity Strategies and Action Plans, laboratories can be modified or constructed to serve human and animal health screening needs, avoiding potential duplication of resources, and enabling closer collaboration among human and animal health authorities and researchers (Murray et al. 2012). A phase change in the broadening of health toward an ecosystem perspective is needed to truly maximize cross-disciplinary synergies.

The USAID Emerging Pandemic Threats PREDICT program has developed viral pathogen discovery programs in wildlife at high-spillover risk interfaces in 20 developing countries that are "hotspots" for disease emergence (http://www.vetmed.ucdavis.edu/ohi/predict/). The protocols could be implemented more widely, including in national surveillance systems. Surveillance can be targeted to assess risks at food-associated interfaces, such as wildlife hunting, markets where bushmeat is present, and restaurants serving wildlife.

In addition to the benefits for strengthening public health capacity and infrastructure, there is a strong overall cost argument to detecting and preventing viral disease emergence from bushmeat and other wildlife sources. A recent study using ground-truthed data for viral discovery in bat species estimated that around 300,000 viruses exist in mammalian wildlife, 85 % of which could be detected through investments of approximately US$1.4 billion. Aiming for 100 % detection would be more expensive ($6.8 billion) due to diminishing returns on viral discovery, but even this figure is less than the cost of some major single outbreaks (e.g., SARS) (Anthony et al. 2013), and far less than the total costs of emerging zoonotic diseases over the past two decades, estimated to be in the order of hundreds of billions of dollars (The World Bank June 2012; Karesh et al. 2012a). Globally coordinated, mitigative responses that reduce the risks and frequency of diseases emerging in the first place and are implemented now are forecast to save approximately US$3.5 billion per year over a 100-year time horizon in comparison to a business-as-usual approach to EID response (Pike et al. 2014).

References

Abbott B, van Kooten GC (2011) Can domestication of wildlife lead to conservation? The economics of tiger farming in China. Ecol Econ 70(4):721–728. doi:10.1016/j.ecolecon.2010.11.006

Anthony SJ, Epstein JH, Murray KA et al (2013) A strategy to estimate unknown viral diversity in mammals. mBio 4(5):e00513–e00598

Bradshaw CJA, Brook BW (2014) Human population reduction is not a quick fix for environmental problems. Proc Natl Acad Sci 111(46):16610–16615. doi:10.1073/pnas.1410465111

Brashares JS, Golden CD, Weinbaum KZ et al (2011) Economic and geographic drivers of wildlife consumption in rural Africa. Proc Natl Acad Sci 108(34):13931–13936

Calisher CH, Childs JE, Field HE et al (2006) Bats: important reservoir hosts of emerging viruses. Clin Microbiol Rev 19(3):531–545

Campbell GL, Hills SL, Fischer M, Jacobson JA, Hoke CH, Hombach JM et al (2011) Estimated global incidence of Japanese encephalitis: a systematic review. Bull World Health Organ 89(766–774):774A–774E

Center for Food Security & Public Health (2007) Menangle virus infection. University, IS

Centers for Disease Control (2013). Japanese Encephalitis. http://www.cdc.gov/japaneseencephalitis/. Accessed 25 Nov 2014

Chaber AL, Allebone Webb S et al (2010) The scale of illegal meat importation from Africa to Europe via Paris. Conserv Lett 3(5):317–321

Costanza R, de Groot R, Sutton P et al (2014) Changes in the global value of ecosystem services. Glob Environ Chang 26:152–158. doi:10.1016/j.gloenvcha.2014.04.002

Courchamp F, Angulo E, Rivalan P et al (2006) Rarity value and species extinction: the anthropogenic allee effect. PLoS Biol 4(12), e415. doi:10.1371/journal.pbio.0040415

Daszak P, Cunningham A, Hyatt A (2000) Emerging infectious diseases of wildlife – threats to biodiversity and human health. Science 287:443–449

Dulac J (2013) Global land transport infrastructure requirements: Estimating road and railway infrastructure capacity and costs to 2050. International Energy Agency, Paris

Eniang EA, Daniel W (2007) Revisiting the age-long African civet farming in Ethiopia against the backdrop of increasing animal welfare concerns: what strategies for sustainable utilization and conservation? Walia 25:34–40

Epstein PR (1995) Emerging diseases and ecosystem instability: new threats to public health. Am J Public Health 85(2):168–172

Fa JE, Peres CA, Meeuwig J (2002) Bushmeat exploitation in tropical forests: an intercontinental comparison. Conserv Biol 16(1):232–237

Fa JE, Currie D, Meeuwig J (2003) Bushmeat and food security in the Congo Basin: linkages between wildlife and people's future. Environ Conserv 30(01):71–78

FAO (2014) State of food insecurity in the world: in brief. Food and Agriculture Organization of the United Nations, Rome

FAOSTAT (2014) Food and agriculture organization of the United Nations Statistics Division. http://faostat3.fao.org/browse/FB/*/E

Foley JA, DeFries R, Asner GP et al (2005) Global consequences of land use. Science 309(5734):570–574

Gilbert M, Slingenbergh J, Xiao X (2008) Climate change and avian influenza. Rev Sci Tech 27(2):459

Godfray HCJ, Beddington JR, Crute IR et al (2010) Food security: the challenge of feeding 9 billion people. Science 327(5967):812–818

Golden CD, Fernald LC, Brashares JS et al (2011) Benefits of wildlife consumption to child nutrition in a biodiversity hotspot. Proc Natl Acad Sci 108(49):19653–19656

Hahn BH (2000a) AIDS as a zoonosis: scientific and public health implications. Science 287(5453):607–614. doi:10.1126/science.287.5453.607

Hahn BH (2000b) AIDS as a zoonosis: scientific and public health implications. Science 287:607–614

Heymann DL (2004) Control of communicable diseases manual. American Public Health Association Publications, Washington, DC

Hoye BJ, Munster VJ, Nishiura H et al (2010) Surveillance of wild birds for avian influenza virus. Emerg Infect Dis 16(12):1827–1834. doi:10.3201/eid1612.100589

Hufnagel L, Brockmann D, Geisel T (2004) Forecast and control of epidemics in a globalized world. Proc Natl Acad Sci U S A 101(42):15124–15129

Jakob-Hoff RM, MacDiarmid SC, Lees C et al (2014) Manual of procedures for wildlife disease risk analysis. World Organisation for Animal Health in association with the International Union for Conservation of Nature and the Species Survival Commission, Paris

Johnson M, Bowen MD, Ksiazek TG et al (1997a) Laguna Negra virus associated with HPS in western Paraguay and Bolivia. Virology 238(1):115–127. doi:10.1006/viro.1997.8840

Johnson M, Bowen MD, Ksiazek TG, Williams RJ, Bryan RT, Mills JN et al (1997b) Laguna Negra virus associated with HPS in western Paraguay and Bolivia. Virology 238:115–127

Jones KE, Patel NG, Levy MA et al (2008) Global trends in emerging infectious diseases. Nature 451:990–994

Karesh WB, Dobson A, Lloyd-Smith JO et al (2012a) Ecology of zoonoses: natural and unnatural histories. Lancet 380(9857):1936–1945. doi:10.1016/S0140-6736(12)61678-X

Karesh WB, Loh E, Machalaba C (2012b) Food safety: a view from the wild side. In: Choffnes ER, Relman DA, Olsen L, Hutton R, Mack A (eds) Improving food safety through a one health approach: workshop summary. National Academies Press, Washington, DC, http://www.ncbi.nlm.nih.gov/books/NBK114500/

Kilpatrick AM, Gillin CM, Daszak P (2009) Wildlife–livestock conflict: the risk of pathogen transmission from bison to cattle outside Yellowstone National Park. J Appl Ecol 46(2):476–485. doi:10.1111/j.1365-2664.2008.01602.x

Kirkland PD, Love RJ, Philbey AW, Ross AD, Davis RJ, Hart KG (2001) Epidemiology and control of Menangle virus in pigs. Aust Vet J 79:199–206

Laurance WF, Clements GR, Sloan S et al (2014) A global strategy for road building. Nature 513:229–232

Leibler JH, Otte J, Roland-Holst D et al (2009) Industrial food animal production and global health risks: exploring the ecosystems and economics of avian influenza. Ecohealth 6(1):58–70

Leroy EM, Epelboin A, Mondonge V, Pourrut X, Gonzalez J-P, Muyembe-Tamfum J-J et al (2009) Human Ebola outbreak resulting from direct exposure to fruit bats in Luebo, Democratic Republic of Congo, 2007. Vector Borne Zoonotic Dis 9:723–728

Li W, Shi Z, Yu M, Ren W, Smith C, Epstein JH et al (2005a) Bats are natural reservoirs of SARS-like coronaviruses. Science 310:676–679

Li Y, Huang X, Yu ITS, Wong TW, Qian H (2005b) Role of air distribution in SARS transmission during the largest nosocomial outbreak in Hong Kong. Indoor Air 15:83–95

Lindsey P, Balme G, Becker M et al (2012) Illegal hunting and the bush-meat trade in savanna Africa: drivers, impacts and solutions to address the problem. Panthera/Zoological Society of London/Wildlife Conservation Society report, New York, p 74

Lindsey PA, Barnes J, Nyirenda V et al (2013) The Zambian wildlife ranching industry: scale, associated benefits, and limitations affecting its development. PLoS One 8(12), e81761. doi:10.1371/journal.pone.0081761

Lloyd-Smith JO, George D, Pepin KM et al (2009) Epidemic dynamics at the human–animal interface. Science 326(5958):1362–1367. doi:10.1126/science.1177345

Loh E, Olival KJ, Zambrana-Torellio C et al (2015) Targeting transmission pathways for emerging zoonotic disease surveillance and control. Vector Borne Zoonotic Dis 15(7):432–437

Luby SP, Rahman M, Hossain MJ et al (2006) Foodborne transmission of Nipah virus, Bangladesh. Emerg Infect Dis 12:1888–1894. doi:10.3201/eid1212.060732

Mackenzie JS, Gubler DJ, Petersen LR (2004) Emerging flaviviruses: the spread and resurgence of Japanese encephalitis, West Nile and dengue viruses. Nat Med 10:98–109

Maiztegui JI (1975) Clinical and epidemiological patterns of Argentine haemorrhagic fever. Bull World Health Organ 52:567–575

Marshall BM, Levy SB (2011) Food animals and antimicrobials: impacts on human health. Clin Microbiol Rev 24(4):718–733. doi:10.1128/CMR.00002-11

McMichael AJ, Powles JW, Butler CD et al (2007) Food, livestock production, energy, climate change, and health. Lancet 370(9594):1253–1263

Medeiros D, Rosa ES, Marques AA et al (2010) Circulation of hantaviruses in the influence area of the Cuiabá-Santarém Highway. Mem Inst Oswaldo Cruz 105(5):665–671

Milner-Gulland E, Bennett EL (2003) Wild meat: the bigger picture. Trends Ecol Evol 18(7):351–357

Morse SS (1995) Factors in the emergence of infectious-diseases. Emerg Infect Dis 1(1):7–15

Morse SS, Mazet JA, Woolhouse M et al (2012) Prediction and prevention of the next pandemic zoonosis. Lancet 380(9857):1956–1965

Murray KA, Daszak P (2013) Human ecology in pathogenic landscapes: two hypotheses on how land use change drives viral emergence. Curr Opin Virol 3(1):79–83. doi:10.1016/j.coviro.2013.01.006

Murray KA, Skerratt LF, Speare R et al (2012) Cooling off health security hot spots: getting on top of it down under. Environ Int 48:56–64. doi:10.1016/j.envint.2012.06.015

Nelson E, Mendoza G, Regetz J et al (2009) Modeling multiple ecosystem services, biodiversity conservation, commodity production, and tradeoffs at landscape scales. Front Ecol Environ 7(1):4–11

Nkem J, Kalame FB, Idinoba M et al (2010) Shaping forest safety nets with markets: adaptation to climate change under changing roles of tropical forests in Congo Basin. Environ Sci Policy 13(6):498–508

Nyaki A, Gray SA, Lepczyk CA et al (2014) Local scale dynamics and local drivers of bushmeat trade. Conserv Biol 28(5):1403–1414

Olson SH, Reed P, Cameron KN et al (2012) Dead or alive: animal sampling during Ebola hemorrhagic fever outbreaks in humans. Emerg Health Threats J 5. doi:10.3402/ehtj.v5i0.9134

Parker S, Nuara A, Buller RML et al (2007) Human monkeypox: an emerging zoonotic disease. Future Microbiol 2:17–34

Patz JA, Daszak P, Tabor GM et al (2004) Unhealthy landscapes: policy recommendations on land use change and infectious disease emergence. Environ Health Perspect 112(10):1092–1098

Patz JA, Olson SH, Uejio CK et al (2008) Disease emergence from global climate and land use change. Med Clin North Am 92(6):1473–1491. doi:10.1016/j.mcna.2008.07.007

Pearce-Duvet JM (2006) The origin of human pathogens: evaluating the role of agriculture and domestic animals in the evolution of human disease. Biol Rev 81(3):369–382

Pigott DM, Golding N, Mylne A, Huang Z, Henry AJ, Weiss DJ et al (2014) Mapping the zoonotic niche of Ebola virus disease in Africa. eLife 3

Pike J, Bogich T, Elwood S et al (2014) Economic optimization of a global strategy to address the pandemic threat. Proc Natl Acad Sci U S A 111(52):18519–18523. doi:10.1073/pnas.1412661112

Poulsen JR, Clark CJ, Mavah G et al (2009) Bushmeat supply and consumption in a tropical logging concession in Northern Congo. Conserv Biol 23(6):1597–1608. doi:10.1111/j.1523-1739.2009.01251.x

PRB (2014) How many people have ever lived on earth? Population Reference Bureau. http://www.worldometers.info/world-population/#total

Public Health Agency of Canada (2010) Junin Virus. http://www.phac-aspc.gc.ca/lab-bio/res/psds-ftss/junin-eng.php. Accessed 25 Nov 2014

Pulliam JRC, Epstein JH, Dushoff J et al (2011) Agricultural intensification, priming for persistence and the emergence of Nipah virus: a lethal bat-borne zoonosis. J R Soc Interface 9(66):89–101

Rahman SA, Hassan SS, Olival KJ, Mohamed M, Chang LY, Hassan L et al (2010) Characterization of Nipah virus from naturally infected Pteropus vampyrus bats, Malaysia. Emerg Infect Dis 16:1990–1993

Raudsepp-Hearne C, Peterson GD, Tengö M et al (2010) Untangling the environmentalist's paradox: why is human well-being increasing as ecosystem services degrade? BioScience 60(8):576–589. doi:10.1525/bio.2010.60.8.4

Robinson TP, Wint GRW, Conchedda G et al (2014) Mapping the global distribution of livestock. PLoS One 9:e96084–e96084

Rouquet P, Froment JM, Bermejo M et al (2005) Wild animal mortality monitoring and human Ebola outbreaks, Gabon and Republic of Congo, 2001–2003. Emerg Infect Dis 11(2):283–290. doi:10.3201/eid1102.040533

Schröter D, Cramer W, Leemans R et al (2005) Ecosystem service supply and vulnerability to global change in Europe. Science 310(5752):1333–1337. doi:10.1126/science.1115233

Schulte-Herbrüggen B, Cowlishaw G, Homewood K et al (2013) The importance of bushmeat in the livelihoods of West African cash-crop farmers living in a faunally-depleted landscape. PLoS One 8(8), e72807

Secretariat of the Convention on Biological Diversity (2014) Global biodiversity outlook 4. Secretariat of the Convention on Biological Diversity, Montreal, QC

Shi Z, Hu Z (2008) A review of studies on animal reservoirs of the SARS coronavirus. Virus Res 133:74–87

Smith KF, Sax DF, Gaines SD et al (2007) Globalization of human infectious disease. Ecology 88(8):1903–1910

Smith KM, Anthony SJ, Switzer WM et al (2012) Zoonotic viruses associated with illegally imported wildlife products. PLoS One 7(1), e29505. doi:10.1371/journal.pone.0029505

Steinfeld H, Gerber P, Wassenaar T et al (2006) Livestock's long shadow: environmental issues and options. Food and Agriculture Organization of the United Nations, Rome

Steverding D (2008) The history of African trypanosomiasis. Parasite Vectors 1(1):3. doi:10.1186/1756-3305-1-3

Taylor LH, Latham SM, Woolhouse MEJ (2001) Risk factors for human disease emergence. Philos Trans R Soc Lond Ser B Biol Sci 356(1411):983–989

Ter Meulen J, Lukashevich I, Sidibe K et al (1996) Hunting of peridomestic rodents and consumption of their meat as possible risk factors for rodent-to-human transmission of Lassa virus in the Republic of Guinea. Am J Trop Med Hyg 55(6):661–666

The World Bank (2012) People, pathogens, and our planet: economics of one health. Report No. 69145-GLB. The World Bank, Washington, DC

Thornton PK, Herrero M (2010) The Inter-linkages between rapid growth in livestock production, climate change, and the impacts on water resources, land use, and deforestation. Policy Research Working Paper 5178. The World Bank

Tilman D, Balzer C, Hill J et al (2011) Global food demand and the sustainable intensification of agriculture. Proc Natl Acad Sci 108(50):20260–20264

Tolosa T, Regassa F (2007) The husbandry, welfare and health of captive African civets (Vivera civetica) in western Ethiopia. Anim Welfare 16(1):15–19

UN Mission for Ebola Emergency Response (UNMEER) (2014) External situation report, 19 Nov 2014

Wang LF, Eaton BT (2007) Bats, civets and the emergence of SARS. Curr Top Microbiol Immunol 315:325–344

Wang LF, Shi Z, Zhang S, Field H, Daszak P, Eaton BT (2006) Review of bats and SARS, pp 1834–1840

Webb PA, Johnson KM, Mackenzie RB, Kuns ML (1967) Some characteristics of machupo virus, causative agent of bolivian hemorrhagic fever. Am J Trop Med Hyg 16(4):531

Westing AH (2010) All the many humans ever: an update. BioScience 60(10):777

Wolfe ND, Daszak P, Kilpatrick AM (2005) Bushmeat hunting, deforestation, and prediction of zoonotic disease emergence. Emerg Infect Dis 11(12):1822–1827

Wolfe ND, Dunavan CP, Diamond J (2007) Origins of major human infectious diseases. Nature 447(7142):279–283

World Health Organization (2011) Monkeypox. http://www.who.int/mediacentre/factsheets/fs161/en/. Accessed 25 Nov 2014

World Organisation for Animal Health (OIE) & International Union for Conservation of Nature (IUCN) (2014) Guidelines for Wildlife Disease Risk Analysis. OIE, in association with the IUCN and the Species Survival Commission, Paris

Young JC, Mills JN, Enria DA et al (1998a) New world hantaviruses. Br Med Bull 54(3):659–673. doi:10.1093/oxfordjournals.bmb.a011718

Young JC, Mills JN, Enria D, Dolan NE, Khan S, Ksiazek TG (1998b) New world hantaviruses. Br Med Bull 54:659–673

Zambrana-Torrelio C, Murray KA et al (2012) One Health and hotspots of food-borne EIDs. In: Choffnes ER, Relman DA, Olsen L, Hutton R, Mack A (eds) Improving food safety through a one health approach: workshop summary. National Academies Press, Washington, DC, http://www.ncbi.nlm.nih.gov/books/NBK114500/

Ziegler S (2010) Application of food balance sheets to assess the scale of the bushmeat trade in Central Africa. Traffic bulletin, vol 22. TRAFFIC International, Cambridge

Chapter 3
A European Perspective on the Transmission of Foodborne Pathogens at the Wildlife–Livestock–Human Interface

Nora Navarro-Gonzalez, María Ugarte-Ruiz, Lucas Domínguez, and Francisco Ruiz-Fons

Abstract There are many unique aspects and peculiarities regarding the transmission of foodborne pathogens at the wildlife–livestock–human interface in Europe, which include the diversity of farming systems, wildlife and habitats, as well as the consumption habits of the European human population. However, it can be generalized that zoonotic diseases acquired from wildlife (or directly related to wildlife) are mainly linked to the consumption of undercooked venison, hunting or handling infected game carcasses. Hunting has always been an integral part of the cultures and traditions of European rural societies, with an estimated greater than seven million hunters practicing this activity for recreational, social, and/or consumptive purposes (Brainerd, http://fp7hunt.net/Portals/HUNT/Hunting_Charter.pdf, 2007). Recently, there has been a growing consumer demand for hunted meat and cured, fermented, and dried game products, which have become more popular and accessible in the European market (Cenci-Goga et al., Meat Sci 90:599–606, 2012; Obwegeser et al., Vet Micro 159:149–154, 2012). Schulp et al. (Ecol Econ 105:292–305, 2014) report that 38 of the 97 European game species, including birds and mammals, are consumed, with the red deer (*Cervus elaphus*), the roe deer (*Capreolus capreolus*), the European hare (*Lepus europaeus*), the common pheasant (*Phasianus*

N. Navarro-Gonzalez (✉)
Servicio de Ecopatología de Fauna Salvaje, Universitat Autònoma de Barcelona, Barcelona, Spain
e-mail: nnavarrog@ucdavis.edu

M. Ugarte-Ruiz • L. Domínguez
VISAVET Health Surveillance Centre, Complutense University of Madrid, Madrid, Spain

F. Ruiz-Fons
Health & Biotechnology (SaBio) Group, Spanish Wildlife Research Institute (IREC; CSIC-UCLM-JCCM), Ciudad Real, Spain

M. Jay-Russell, M.P. Doyle (eds.), *Food Safety Risks from Wildlife*, Food Microbiology and Food Safety, DOI 10.1007/978-3-319-24442-6_3

colchicus), and the wild boar (*Sus scrofa*) being the main game food species, since they are hunted in all countries and have the largest harvest numbers. Considerable research has been conducted on foodborne pathogens in the most common wild ungulate species in several European countries; however, relatively little information is available on hares and other lagomorphs, as well as wild game birds.

Keywords Antimicrobial resistance • Cattle • European Union • Foodborne pathogens • Game meat • Human–wildlife interactions • Iberian ibex • Livestock • Wild boar • Wildlife • Zoonosis

Introduction

There are many unique aspects and peculiarities regarding the transmission of foodborne pathogens at the wildlife–livestock–human interface in Europe, which include the diversity of farming systems, wildlife and habitats, as well as the consumption habits of the European human population. However, it can be generalized that zoonotic diseases acquired from wildlife (or directly related to wildlife) are mainly linked to the consumption of undercooked venison, hunting or handling infected game carcasses. Hunting has always been an integral part of the cultures and traditions of European rural societies, with an estimated greater than seven million hunters practicing this activity for recreational, social, and/or consumptive purposes (Brainerd 2007). Recently, there has been a growing consumer demand for hunted meat and cured, fermented, and dried game products, which have become more popular and accessible in the European market (Cenci-Goga et al. 2012; Obwegeser et al. 2012). Schulp et al. (2014) report that 38 of the 97 European game species, including birds and mammals, are consumed, with the red deer (*Cervus elaphus*), the roe deer (*Capreolus capreolus*), the European hare (*Lepus europaeus*), the common pheasant (*Phasianus colchicus*), and the wild boar (*Sus scrofa*) being the main game food species, since they are hunted in all countries and have the largest harvest numbers. Considerable research has been conducted on foodborne pathogens in the most common wild ungulate species in several European countries; however, relatively little information is available on hares and other lagomorphs, as well as wild game birds.

There are two common European farming systems in which there is likely wildlife–livestock interaction. These two systems are either intensive or extensive/free-range, and although there are many intermediate categories, this classification complements the aims of this chapter. In the case of intensive farming animals are always indoors or in confinement, and the farms employ consistent biosecurity measures. Direct contact between intensively farmed livestock and large wildlife species is limited, so most wildlife contact is with small mammals and birds. Large fauna can also be exposed to livestock pathogens from intensive farming through manure or uncontrolled waste, as well as to aerosol-borne pathogens, e.g., *Coxiella burnetii*. In contrast, extensive farming consists of a more integrated system in which animals have access to outdoor pastures and other natural resources, at least during some periods of the year, and this farming system is increasing in Europe due to consumer preferences. In this case, livestock and wildlife share natural resources such as pastures, water, and salt licks. Figure 3.1 shows some examples of proximity and interaction between wildlife and free-range livestock in Europe. Importantly, although extensively produced livestock may be in contact with several wildlife species, there is a greater likelihood of pathogen transmission if contact occurs with wild species of similar taxons, as pathogens are more likely to be efficiently transmitted between related host species.

Interactions between wild and domestic animals in Europe are believed to be increasing due to: (1) the concentration of large populations of wild animals in small, delimited natural areas due to the high distribution and density of humans (Gummow 2010); (2) European wildlife politics and consumer preferences moving the animal breeding industry from more intensive to more extensive farming systems (Gortázar et al. 2007); (3) wildlife populations becoming more abundant because of the implementation of game management through feeding, translocations, and fencing (Gortázar et al. 2007); and (4) the increase in recent decades of forested

Fig. 3.1 Wildlife and livestock living in proximity share habitat and resources, which may lead to interactions and direct contact. Panel (**a**) A domestic goat approaches an ibex resting on rocks in the French Alps (Photo by D. Gauthier). Panel (**b**) An ibex intermingles with a herd of sheep in the French Alps (Photo by D. Gauthier). Panel (**c**) Wild boars trapped in proximity to a cattle paddock in a Mediterranean mountain range in Spain (Photo by G. Mentaberre). Panel (**d**) A mouflon grazes in proximity to cattle in the eastern Pyrenees in Spain (Photo by J. R. Lopez-Olvera)

areas at the expense of agricultural areas (Martin et al. 2011), thereby providing additional refuge and food resources to wildlife. In recent decades, there has been large population increases in most European ungulate species (Apollonio et al. 2010), thus enhancing the likelihood of interaction with livestock and the risk of transmission of shared pathogens.

Beyond game management, wildlife is being increasingly farmed, thereby creating a new risk situation. A large variety of wild species are being farm-reared for direct meat consumption and for releases into the wild, mainly for game purposes. This management practice is common in most European countries, and restocking numbers are not negligible. In the United Kingdom, for example, several tens of millions of game birds are reared and released each year (Mustin et al. 2011). Such practices can increase the prevalence of certain pathogens as well as enhance their transmission to other species in the wild (Díaz-Sánchez et al. 2012a; Horigan et al. 2014).

Behavioral changes within European societies also play a significant role in the potential for exposure to wildlife and shared pathogens. Livestock zoonotic pathogens co-evolved through the human history of domestication, and now we are most likely confronting a new era of co-evolution of wildlife–livestock–human shared pathogens within a One Health context. Due to the increasing importance of the flourishing wild game industry in Europe, direct and indirect transmission of pathogens from wildlife to humans is occurring through both occupational and recreational exposure (Gortázar et al. 2006; Massei et al. 2011; Hälli et al. 2012; Saito et al. 2012). The public perception that game meat is a sustainable, healthy, and ecologically friendly product raises the potential for an increase in foodborne pathogen transmission associated with wildlife (Hoffman and Wiklund 2006; Ramanzin et al. 2010). The increasing trend of outdoor tourism and sports activities, and the colonization of peri-urban semi-natural areas by human settings, are increasing worldwide (Bradley and Altizer 2007; Cahill et al. 2012), further elevating the likelihood of exposure to pathogens carried by wildlife. Additionally, socioeconomic changes in developed societies may indirectly increase exposure to zoonotic pathogens in wildlife populations (Godfrey and Randolph 2011).

Paulsen et al. (2012) recently described how the marketing pathways of game meat differ between European countries and by species, as do the ante-mortem and perimortal phases of the game meat chain. In particular, the location of the shot wound, the time from killing to evisceration, hygiene practiced during skinning, cutting and evisceration, and the time to refrigeration all affect the spread and multiplication of bacteria. Membré et al. (2011) suggest that improving hunting practices across European countries and encouraging good hygienic practices would enhance the microbiological quality of large wild game meat. The proper training of hunters may be an important contributor to improving the hygienic quality and safety of game meat.

In this chapter, we address the main risk factors associated with wildlife–livestock–human interactions that can enhance the transmission of zoonotic pathogens.

Main Foodborne Pathogens in the European Union

Bacteria

Non-typhoidal *Salmonella* spp.

As a multi-host pathogen, *Salmonella* plays an important role in any wildlife–livestock interface situation. Several factors likely contribute to the high frequency and prevalence in which this pathogen can be found in wildlife and livestock: (1) the broad range of hosts that can be colonized by non-typhoidal *Salmonella* including mammals, birds, amphibians, and reptiles; (2) the presence of healthy carriers able to shed the pathogen in feces over prolonged periods of time; (3) the ability of insects and other invertebrates to act as vectors; (4) the high environmental persistence of this microorganism; and (5) the existence of two mechanisms of transmission, namely the feco-oral and respiratory/tonsilar pathways, the latter best described in cattle and swine (De Jong and Ekdahl 1965; Fedorka-Cray et al. 1995).

An additional factor is the human being, as human salmonellosis has for years been one of the most important foodborne zoonosis in the European Union (EU), with Enteritidis and Typhimurium being the most frequently reported serovars. However, the trend in salmonellosis incidence in recent years is decreasing in the EU (EFSA and ECDC 2014). Although control programs for *Salmonella* in fowl have been successful in decreasing the occurrence of this pathogen in eggs, it can still be found in a variety of foodstuffs, including meat and vegetables.

As reviewed by Paulsen et al. (2012), *Salmonella* is likely to be present in the game meat chain of most, or even all, EU member countries and can be isolated at virtually every step in the food chain, although usually at a low prevalence. In 2011, an outbreak of human salmonellosis in France was traced back to wild boar meat following a celebration by hunters (http://ecdc.europa.eu/en/escaide/materials/presentations/escaide2011_session_14_2_nogareda.pdf). When consulting the online Rapid Alerts System for Food and Feed of the European Commission (https://webgate.ec.europa.eu/rasff-window/portal/), it is reported that *Salmonella* is sporadically present in wild boar products, although sometimes contaminated products originate from outside the EU.

Hilbert et al. (2012) concluded that the *Salmonella* wildlife–human cycle is complex. The transmission dynamics of *Salmonella* at the wildlife–livestock interface depends on the hosts (wild and domestic) and serovars present in each agroecosystem. There are regional differences in the prevalence of *Salmonella*, both in livestock and wildlife (e.g., higher prevalence in southern countries vs. northern countries) (Paulsen et al. 2012). Also, serovars differ in their host range (they can be host-restricted, host-adapted or ubiquitous), and in their ability to cause systemic disease. Thus, the epidemiology of *Salmonella* in a given location is strongly dependent on the number, frequency, and pathogenicity of circulating strains.

Wild hosts differ in their respective roles in the spread of *Salmonella*, for example, wild boars shed *Salmonella* more frequently than wild ruminants (Paulsen et al. 2012). Chiari et al. (2013) and Navarro-Gonzalez et al. (2012) determined that young wild boars are more frequently infected with *Salmonella* than older boars. Navarro-Gonzalez et al. (2012, 2014a) studied the crossover of *Salmonella enterica* between wild ungulates and sympatric free-ranging livestock in Spain. In an area with a considerable prevalence of *Salmonella* in cattle, the Iberian ibex (*Capra pyrenaica*) and wild boars differ in their role in the epidemiology of *Salmonella* despite sharing their habitat with cattle. Wild boars have a high *Salmonella* prevalence and serovar richness (diversity), especially when co-habiting with cattle. Additionally, the probability of interspecies transmission between livestock and wild boar increases as the herd size of sympatric cattle increases. Mentaberre et al. (2013) validated the transmission of *Salmonella* serovars Meleagridis and Anatum between cattle and wild boars, and vice versa, occurring in the same location. In contrast, the Iberian ibex has a very low prevalence of *Salmonella* (Navarro-Gonzalez et al. 2014a) and no serovars were shared with cattle in their study. It was postulated that the difference in *Salmonella* prevalence in wild boar versus Iberian ibex populations may be due to different feeding habits and space use. In Thuringia (Germany), the pig-adapted serovar *S.*Cholerasuis causes clinical salmonellosis in wild boars and the identification of epidemiologic groups strongly suggests an exchange of this serovar between wild boars and domestic pigs (Methner et al. 2010). Interestingly, the results of a study by Vieira-Pinto et al. (2011) in Northern Portugal revealed common sources of infection and circulation of *Salmonella* strains between wild boars, wild rabbits (*Oryctolagus cuniculus*), and domestic pigs. In contrast, Díaz-Sánchez et al. (2013) did not observe an association between the presence of livestock and the prevalence of *Salmonella* in wild boars and red deer from South-Central Spain, with the overall prevalence of *Salmonella* being low. In Italy, Chiari et al. (2013) observed a high diversity and prevalence of *Salmonella* in wild boars but the same serovars were not isolated from livestock in official surveys carried out in the same area and period of time. These are examples of how each wildlife–livestock interface situation can be unique epidemiologically, but the research also suggests that spillover can be sporadic in space and time, and hence difficult to observe.

Besides being a concern for public health and animal husbandry, *Salmonella* can cause disease in many wildlife species and its consequences should be considered when introducing livestock into a natural environment or when keeping livestock outdoors. Glawischnig et al. (2000) described an outbreak of salmonellosis due to *S.* Dublin in Alpine chamois (*Rupicapra rupicapra*) that was transmitted by cattle. *S.* Dublin was also present in a freely accessible water trough. The difficulty of finding sick animals in the wild and arriving at a final diagnosis hinders the determination of the actual importance of *Salmonella*, its implication in disease outbreaks and mortality, and in the conservation of endangered species. It is known that this pathogen has been involved in fatal outbreaks in wild birds, especially passerines, and has caused a decline in local wild bird populations (Gaffuri and Holmes 2012). Furthermore, the presence of *Salmonella* antibodies in Alpine chamois has been linked to reduced fertility (Pioz et al. 2008), revealing an additional deleterious effect of the presence of this pathogen at the livestock–wildlife interface.

Shiga Toxin-Producing *Escherichia coli* (STEC)

Shiga toxin-producing *E. coli* are a subset of *E. coli* that can cause severe disease in humans, but are carried asymptomatically in a variety of animals. In Europe, most reported human STEC infections are sporadic cases and frequently the source of infection is food of bovine and ovine origin, but vegetables, drinking water, and direct animal contact (e.g., petting zoos) have also been implicated (EFSA and ECDC 2014). The trend for STEC infections in the EU during 2008–2012 significantly increased, with STEC O157 being the most frequently reported serogroup. However, the highly publicized STEC O104:H4 outbreak in 2011 in Europe, mostly Germany, was not linked to a known animal or animal product (EFSA and ECDC 2014). In fact, the fenugreek sprouts implicated as the food vehicle originated from outside the EU, and the strain had characteristics of both STEC and enteroaggregative *E. coli* pathotypes.

Young cattle (especially between 3 and 24 months of age) are important reservoirs of STEC (EFSA 2009), but sheep, goats, and wild ruminants are also reservoirs (EFSA and ECDC 2013a). Nonruminant species that carry STEC are believed to be transient hosts that only excrete the organism for a short period after infection. Roe deer and red deer are considered in Belgium to be potential sources of STEC (Bardiau et al. 2010). In Spain, red deer are considered to be a natural reservoir of *E. coli* O157:H7 (García-Sánchez et al. 2007; Díaz-Sánchez et al. 2013), although the prevalence is generally low (1.5–3.3 %). Other wild ruminants can carry STEC, such as fallow deer (*Dama dama*), mouflon (*Ovis musimon*), Alpine chamois, Alpine ibex (*Capra ibex*), and Iberian ibex (Sánchez et al. 2009; Obwegeser et al. 2012; Navarro-Gonzalez et al. 2015). Also lagomorphs, e.g., wild rabbits and Iberian hares (Martínez et al. 2011), or carnivores, such as the red fox (*Vulpes vulpes*), have been reported as carriers of STEC (Mora et al. 2012). Wild boars have been repeatedly described as carriers of STEC strains that are potential human pathogens (Miko et al. 2009; Wacheck et al. 2010), including *E. coli* O157:H7 (Mora et al. 2012; Sánchez et al. 2010). However, results of STEC prevalence surveys in wildlife are variable throughout Europe. For example, in Swedish wildlife, STEC is practically absent; Wahlström et al. (2003) only found one positive wild boar out of 791 samples from wild mammals and birds, and this bacterium was not isolated from either Norwegian cervids (Lillehaug et al. 2005) or wild ruminants and marmots from the Italian Alps (Caprioli et al. 1991).

These ecological studies indicate that environmental factors together with contact with primary reservoirs, especially domestic animal reservoirs, must play a role in the epidemiology of this pathogen. In some instances, there is consistent evidence of transmission between livestock and wildlife, but many other risk factors remain unknown for different species and habitats. A good example is reported by Scaife et al. (2006), who determined the prevalence of STEC ranged from 9.5 to 40% in rabbit populations living in proximity to farms with infected cattle. This investigation was initiated following an outbreak of STEC O157 infection in humans after visiting a wildlife park where wild rabbits co-habiting with infected cattle had apparently introduced the pathogen (Pritchard et al. 2001). Also, wild birds and rodents at farms can carry STEC strains indistinguishable by pulsed-field gel electrophoresis (PFGE) from those obtained from cattle, although at a low prevalence (Nielsen et al. 2004). However, a STEC outbreak in the UK was linked to rooks scavenging troughs at a farm at which cattle and wild rabbits were STEC-negative (Ejidokun et al. 2006). In the Austrian Alps, STEC strains of the same molecular subtypes have been simultaneously isolated from Alpine chamois and cattle that shared a common pasture during the summer (Freidl et al. 2011). Díaz-Sánchez et al. (2013) determined there was a positive relationship between the occurrence of *stx* genes in red deer feces and the presence of livestock and deer density, whereas the occurrence of *stx* genes in wild boars was related to sex. However, the link between livestock and wildlife is not always so obvious and is not the sole factor in explaining STEC infection in wildlife (Navarro-Gonzalez et al. 2015).

To elucidate risk factors associated with the carriage of STEC, future research should consider concomitant sampling of domestic and sympatric wildlife populations, preferably during the summer–autumn months since there may be a seasonal shedding pattern. In fact, the European Food Safety Authority (EFSA 2009) recommends sampling cattle between 1 April and 1 October to make the monitoring program more cost-effective. By re-capturing and re-sampling, Scaife et al. (2006) found a STEC-positive rabbit in the summer that had been STEC-negative during the winter. Another factor to consider is age, since in wildlife, as found in cattle, young animals may be more prone to STEC infection. Age factors can be difficult to measure since hunting bags rarely include enough young animals to obtain consistent conclusions. Several studies evaluating age as a risk factor have not found statistically significant differences (Bardiau et al. 2010; Díaz-Sánchez et al. 2013). Also, the existence of "supershedders" in wildlife remains unknown and is relevant to understanding the dynamics of this pathogen at the livestock–wildlife interface. Since "supershedders" in cattle populations have been associated with increased STEC prevalence and excretion within pens (Cobbold et al. 2007), "supershedders" in the wild would be very important in maintaining this pathogen in the environment.

STEC can enter the game meat chain(Pierard et al. 1997), and human infections associated with consumption of game meat are probably underestimated (Miko et al. 2009). In 2012, STEC O113:H8 in frozen deer meat caused an outbreak of three cases in Austria (AGES 2012). STEC has been isolated from wild boar meat and meat products from Spain (Díaz-Sánchez et al. 2012a), and STEC with the same PFGE profile have been found in feces and carcasses of wild boar and red deer, indicating that cross-contamination occurs during processing (Díaz-Sánchez et al. 2013). Some available data have revealed that wild boars and other wild animals play a role in human infection; Sánchez et al. (2010) found indistinguishable PFGE types in *E. coli* O157:H7 isolates from a wild boar and a human patient. Likewise, Mora et al. (2012) reported similarities between STEC from wildlife and humans. However, the subtypes of STEC strains that were isolated from fecal samples from wild ruminants in Switzerland (Hofer et al. 2012) did not have the patterns of typical strains that are highly pathogenic to humans. Additional characterization of virulence factors and the human clinical relevance of wildlife-derived strains not previously associated with human illness is needed. Regarding game birds, STEC has been isolated from raw meat from pheasant, red-legged partridge (*Alectoris rufa*), common quail (*Coturnix coturnix*), and other birds (Pierard et al. 1997).

Campylobacter spp.

Infections by *Campylobacter* pose a serious public health problem, as *Campylobacter* enteritis has been the most frequently reported zoonotic disease in humans in the EU since 2005, with 214,268 confirmed cases in 2012. *Campylobacter* infections in the EU human population have increased significantly in recent years (2008–2012), with a clear seasonal trend (EFSA and ECDC 2014). Thermophilic *Campylobacter*, specifically *C.jejuni* and *C.coli*, are the most common species associated with diarrheal disease in humans in the developed world, accounting for 80–90 and 5–10 % of *Campylobacter* infections, respectively (Fitzgerald et al. 2008). Nevertheless, other *Campylobacter* spp. are also associated with gastrointestinal and extra-gastrointestinal infections in animals and humans. "Emerging" species, specifically *C. concisus* and *C. upsaliensis*, have been regularly isolated from patients with gastroenteritis in Europe. However, some *Campylobacter* species are uncommon and are not considered as emergent since they are newly identified or little is known about their pathogenic potential (Man 2011). Frequently, the microbiological methodology used to isolate *Campylobacter* spp. can introduce culture bias thereby skewing the actual contribution of different species and sources. In the last decade, considerable effort has been devoted to improving the protocols for detecting this pathogen, as bacteriological culture of *Campylobacter* spp. can be a challenge due to the fragility of these microbes (Baylis et al. 2000; European Commission 2007).

Transmission pathways for *Campylobacter* include direct and indirect contact with infected animals (normally healthy carriers), people, and the environment. The vast majority of human cases of *Campylobacter* enteritis are sporadic and linked to the food chain, resulting from handling or consumption of raw or undercooked contaminated poultry products. However, the prevalence of *Campylobacter* carriage by broilers has remained largely at the same level while the number of human cases in the EU increased considerably during the same time period. Consequently, it has been suggested that other risk factors not related to chicken are involved in the increasing incidence of *Campylobacter* enteritis in humans. Examples include meat products from other species, ingestion of untreated drinking water or milk, consumption of contaminated fruits, vegetables, fish or fishery products and, less frequently, recreational activities in aquatic environments. Although large outbreaks of *Campylobacter* enteritis are uncommon, some notable outbreaks have been reported within the EU, and typically involved broiler meat or unpasteurized milk (EFSA and ECDC 2014; Schönberg-Norio et al. 2004). Furthermore, waterborne transmission has been associated with *Campylobacter* outbreaks because of contamination from sewage or heavy rainfall (Pitkanen 2013). Braeye et al. (2014) described a large community outbreak of gastroenteritis linked to the consumption of drinking water contaminated by river water. In addition, a substantial risk of zoonotic transmission could be associated with pets, livestock, and wild animals, as they can play a role in the contamination of food products, or transmission through direct contact with an animal fecally shedding the pathogen. In this context, wild bird droppings have been recognized as a significant environmental source of *Campylobacter* spp. infection for humans and animals, as well as the consumption and handling of game-derived products. However, the number of wildlife species serving as reservoirs of *Campylobacter* spp. is still unknown (Humphrey et al. 2007; Epps et al. 2013; Waldenström et al. 2002).

Wild birds, are considered, as with poultry, to be natural reservoirs of *Campylobacter* spp. Healthy wild birds can be a source for human or livestock infection, although their role is likely minor due to host specificity (Colles et al. 2008; Griekspoor et al. 2013). In Sweden, Waldenström et al. (2002) detected *C. jejuni*, *C. lari*, and *C. coli* in a wide range of wild and migrating birds, suggesting a potential role as vectors in long-distance transmission to livestock or humans. In agreement with this finding, Wahlström et al. (2003) frequently detected *Campylobacter* spp. in Swedish wild birds (Canada geese (*Branta Canadensis*) and seagulls (*Larus argentatus*, *L. canus*, and *L. marinus*)). In North-Western Italy, Robino et al. (2010) determined that hooded crows (*Corvus cornix*) were highly sensitive to *Campylobacter* infection. In Germany, studies by Atanassova and Ring (1999) revealed that wild pheasants shed both *C. coli* and *C. jejuni* (25.9 %). In Spain, free-living waterfowl were recognized as a source of *Campylobacter*, especially *C. coli* which was detected in an area densely populated by wild birds (Antilles et al. 2015). Other migrating birds, such as common quails, may also be environmental carriers serving as a source of infection for other birds, livestock, and humans (Dipineto et al. 2014). In addition, a higher occurrence of the pathogen was observed in artificial environments compared with populations in natural environments. Specifically, Nebola et al. (2007) determined in the Czech Republic a higher prevalence of *Campylobacter* spp. in pheasants originating from farms with intensive production (70.2 %) than in wild pheasants (27.5 %). Moreover, genetic diversity of*Campylobacter* strains isolated from farmed animals was greater than the diversity of strains isolated from wildlife, which might be due to close contact between animals or possibly contamination by farm workers. Díaz-Sánchez et al. (2012b) highlighted the potential risk in Spain of *Campylobacter* transmission to natural populations of partridges from farmed and restocked red-legged partridges.

Wahlström et al. (2003) detected *Campylobacter* in the majority of mammalian wildlife species analyzed in Sweden. The prevalence was considerable (>10 %) for wild boars; low for roe deer, mountain and European hares (*Lepus timidus* and *L. europeaus*) and moose (*Alces alces*), and absent for red deer and fallow deer. In Norway and Finland, Kemper et al (2006) found 1 of 2500 reindeer (*Rangifer tarandus*) was positive for *C. hyointestinalis* and Lillehaug et al (2005) in a Norwegian study observed that only 1 of 324 wild cervids was positive for *C. jejuni*. In central and southern Spain, Díaz-Sánchez et al. (2013) determined there was no relation between *Campylobacter* spp. in large game animals and the presence of livestock in hunting estates despite the high prevalence of *Campylobacter* found in wild boars (66 %). However, in northeastern Spain a lower prevalence of *Campylobacter* spp. was found in wild boars (10 % *C. lanienae*, 1.3 % *C. coli*), as well as a potential for cross-over from free-range livestock to wild boars and vice versa (Navarro-Gonzalez et al. 2014b). Such differences may be due to unique conditions in the hunting estates, such as estate fencing, high density of game species and different farming or management conditions that may also be affecting the presence and species of *Campylobacter*. In southern Spain, *Campylobacter* spp. was detected in wild artiodactyls (wild boar, red deer, and mouflon), although the findings suggested only wild boars constituted an important reservoir of infection (Carbonero et al. 2014). Atanassova et al. (2008) and Paulsen et al. (2003) isolated in Germany *Campylobacter* from wild boar meat (2.1 %) and roe deer meat (3 %) indicating its potential to enter the food chain through these vehicles. In contrast, Wacheck et al. (2010) did not recover *Campylobacter* from any of 153 wild boar samples analyzed in Switzerland.

In general, it is difficult to understand all aspects of *Campylobacter* epidemiology due to its being a multi-host pathogen and variable prevalence. Furthermore, seasonality, especially in temperate regions, may affect the recovery of *Campylobacter* (EFSA and ECDC 2014; Humphrey et al. 2007; Strachan et al. 2013). In wildlife, the remoteness of the study areas and the difficulty of processing samples quickly may influence *Campylobacter* recovery, which tends to be laborious and delicate. In fact, the literature reveals there is substantial variability in the presence of the bacteria in different animal species and locations. Other factors such as diet, habitat preferences, or migration patterns are likely very important variables in explaining the prevalence of *Campylobacter* species among host taxa (Griekspoor et al. 2013; Navarro-Gonzalez et al. 2014b; Waldenström et al. 2010).

Listeria monocytogenes

Despite the relatively low number of cases of listeriosis compared with other foodborne diseases, *Listeria monocytogenes* infection is a major health concern due to the severity of the illness and the high case-fatality rate reported (17.8 % in EU in 2012). Normally, infection is acquired through the consumption of contaminated food, especially dairy products, cooked poultry and meat, as well as ready-to-eat vegetables and seafood in which *Listeria* is able to multiply during cold-storage (EFSA and ECDC 2014).

Source tracking of *L. monocytogenes* often remains challenging because of its ubiquity, whereby the bacteria are widely distributed in nature and highly adapted to soil, water, and vegetation (Linke et al. 2014). Due to the ability of *Listeria* to survive for long periods in the environment, as well as in feces of asymptomatic carrier animals, transmission between farms and the wild habitats could introduce the pathogen into the food chain (Lyautey et al. 2007). Since 1999, relatively little has been reported regarding disease in wildlife, but listeriosis was described in European hares and fallow deer (Wuthe and Schonberg 1999; Tham et al. 1999). In Germany, Schwaiger et al. (2005) isolated *L. monocytogenes* from brain samples of wild ruminants. Additionally, Wacheck et al. (2010) recovered *L. monocytogenes* from wild boar tonsils and fecal samples in Switzerland. In Russia, Yegorova et al. (2012) found *L. monocytogenes* in aquatic organisms and in wild artiodactyls, and Zaytseva et al. (2007) detected the pathogen in wild animals, including small rodents and marine organisms. Conversely, Aschfalk et al. (2003) did not isolate this pathogen from reindeer in Norway, nor did Obwegeser et al. (2012) from fecal samples of red deer, roe deer, Alpine chamois, and Alpine ibex.

Consumption of game meat is considered to be a potential source of listeriosis in humans. In Germany, Atanassova et al. (2008) and Vargas et al. (2013) both detected *L. monocytogenes* in meat from wild boars, roe deer, and red deer. In Italy, Avagnina et al. (2012) recovered the bacteria from the carcasses of game ungulates. Hence, proper food safety practices for game meat, especially during the later stages of production (game chambers, cold storage rooms, processing factories), are of great importance due to the ability of *Listeria* to grow at low temperatures (Atanassova et al. 2008).

Antimicrobial Resistance

Antimicrobial resistance (resistance to antibiotic compounds) has been partly associated with the use and misuse of antimicrobial agents in food animal production, and a lower occurrence of resistant bacteria has been observed in extensive or organic farming systems when compared with intensive livestock production systems (e.g., Alvarez-Fernandez et al. 2012; Berge et al. 2010; Blake et al. 2003; Heuer et al. 2002). Divergent results suggest that antimicrobial resistance may also depend on the hosts, bacteria, and type of antibiotic studied.

Bacteria resistant to antimicrobials have been isolated from a large variety of wildlife species throughout Europe, including game ungulates and birds that might enter in the food chain. Wild boars in particular have been reported to carry antimicrobial-resistant bacteria to a greater extent than other wild species, which has been partly attributed to their omnivorous habits. Thus, many investigators have suggested wild boars to be reservoirs, and even sentinels or indicators of the antimicrobial resistant microbes circulating in their environment (Mokracka et al. 2012; Literak et al. 2010; Poeta et al. 2009).

As with other pathogens, the occurrence of antimicrobial-resistant enteric bacteria in free-roaming populations could result in wildlife serving as reservoirs and vectors for the introduction of antimicrobial-resistant bacteria to farm animals or vice versa (Smith et al. 2014). Several studies have revealed similarities in the patterns of antimicrobial resistance in bacterial isolates from livestock and small fauna (e.g., rodents, insects, and birds) obtained in the farm environment (Kozak et al. 2009; Literak et al. 2009; Rybarikova et al; 2010). Interestingly, both red-legged partridges sampled on farms and frequently restocked populations had a higher prevalence of resistant *E. coli* than partridges from natural populations where no farmed partridges had been released in the previous 5 years (Díaz-Sánchez et al. 2012b).

However, it is unclear if game ungulates co-habiting with livestock share antimicrobial-resistance profiles. For example, Navarro-Gonzalez et al. (2013a) found a high variability in the resistance patterns of *E. coli* isolated from sympatric cattle, wild boars, and Iberian ibex. Some results were concerning such as the presence of bacteria resistant to fluoroquinolones and third-generation cephalosporins, although the same resistance profile was rarely detected more than once. Similar findings were obtained from the same animals when testing antimicrobial resistance in *Salmonella enterica* (Navarro-Gonzalez et al. 2012), indicating the complexity of identifying the origin of antimicrobial-resistant bacteria carried by wildlife. Other findings of concern include the detection of enterobacteria resistant to beta-lactams (Literak et al. 2010; Guerra et al. 2014; Smith et al. 2014) or methicillin-resistant *Staphylococcus aureus* (Porrero et al. 2013; Loncaric et al. 2014; Gómez et al. 2014) in wild mammals and birds. Wild birds might have an important role in the spread of antimicrobial-resistant bacteria, hence special attention should be paid to migratory birds that can carry resistant bacteria over long distances.

The presence of antimicrobial-resistant microbes in the wildlife–livestock–human interface is an emerging concern in light of the increasing likelihood of contact occurring between wildlife, domestic animals, and humans. Increased interactions are anticipated in suburban and peri-urban environments, where some wildlife populations are proliferating. Interestingly, Navarro-Gonzalez et al. (2013b) isolated a linezolid-resistant *Enterococcus faecalis* from an urban wild boar in Barcelona (Spain). Linezolid is a synthetic antimicrobial agent reserved for human use.

The European Food Safety Authority (EFSA and ECDC 2013b) reported that the spread of antimicrobial-resistant enteric bacteria such as *Campylobacter*, *Salmonella* and some strains of *Escherichia coli* can occur through food vehicles. However, further research is needed to understand the transmission of antimicrobial-resistant bacteria between humans, livestock, and wildlife, and their relevance in the food chain.

Viruses

Hepatitis E Virus

Hepatitis E virus (HEV), a single-stranded 7.2-kb RNA virus in the family *Hepeviridae*, usually causes a self-limiting infection, but may develop into a fatal, acute icteric hepatitis in humans worldwide. HEV is considered the main enterically transmitted non-A, non-B hepatitis virus causing human hepatitis in the world. The World Health Organization (WHO) estimates there are 20 million cases of hepatitis E (HE) annually worldwide, with 3 million cases of acute hepatitis and approximately 56,600 deaths (http://www.who.int/mediacentre/factsheets/fs280/en/). To date, four genotypes of mammalian HEV are recognized and new genotypes have been proposed but are not yet classified. Genotypes 1 and 2 are largely isolated from human beings although they have been also found in animals, whereas genotypes 3 and 4 are reported mainly in animals worldwide. Evidence supports genotype differences in the potential for virulence in humans. Within the European context, HEV causes mainly sporadic human outbreaks of HE, large epidemics are infrequent. Epidemiologically, consumption of raw, smoked or undercooked products of animal, livestock, and wildlife origin are responsible for a high percentage of cases (Li et al. 2005; Kim et al. 2011). However, HEV may be shed in the feces of domestic and wild swine (Wiratsudakul et al. 2012; Nardini et al. 2014) and perhaps in feces of several other wild animal hosts (Jirintai et al. 2014), therefore suggesting that indirect food (e.g., contaminated fruit or vegetables) or waterborne transmission may also occur (Ayral et al. 2015). To date, no large outbreaks of human HE originating from animal-to-human transmission have been reported in developed countries, perhaps due to the low probability of direct fecal-oral human-to-human transmission. The most recent report from EFSA on trends and sources of zoonoses in the EU (EFSA and ECDC 2014) does not provide information on HEV infection. In contrast, there is an increasing incidence of sporadic human HE cases with suspected zoonotic links (Krumbholz et al. 2014). There is evidence that several domestic—cattle, sheep, goats, horses, dogs, rabbits—and wild—wild boars, red deer, sika deer (*Cervus nippon*), roe deer, rats (*Rattus rattus* and *R. norvegicus*), and even mollusks—have been exposed and/or infected by HEV (reviewed by Widén et al. 2012), indicating that HEV is a multi-host *Hepevirus*. From a European perspective, the major risk factor for human HEV infection from animal origin is direct consumption of improperly cooked game meat. High exposure and infection rates have been observed in wild boars and red deer in different European countries (de Deus et al. 2008; Boadella et al. 2010; Widén et al. 2012; Larska et al. 2015), which suggests both species may represent a significant risk for human infection in Europe. However, the role of wild ungulates may differ geographically because risk factors for HEV maintenance varies between different epidemiologic scenarios, hence influencing variation in transmission risks to target hosts (Haydon et al. 2002). As an example, an extensive study carried out in wild ungulates in Poland revealed no evidence of HEV exposure in wild ruminant species, whereas 44 % of coexisting wild boars were seropositive (Larska et al. 2015). Exposure to zoonotic pathogens from wildlife, either through direct contact with animals or their products, or indirectly through environmental contamination, presents an increasing concern for public health authorities worldwide (Ruiz-Fons 2015). Changes in human life styles, space use or wildlife management, among other factors, influence the risk of human exposure to wildlife pathogens and the chance of introducing pathogens into naïve territories (Gortázar et al. 2014). Hepatitis E virus is currently distributed almost worldwide, and increasing interactions between wildlife, livestock, and humans probably accounts for a high risk of HEV transmission, both in developing and industrialized countries. Several ecological, epidemiologic, and pathological aspects of the wildlife cycle of HEV remain unknown despite concerns that RNA viruses could be responsible for future emerging human pandemics (Woolhouse et al. 2005). Multi-host RNA viruses such as HEV may diversify into new viral strains with different tropism and virulence for humans. Therefore, evidence points to the need of further investigation of HEV–host relationships in different eco-epidemiologic scenarios to prevent the chance of emergence of new virulent strains in humans (see Gortázar et al. 2014). Additionally, spatial and temporal monitoring of HEV in the primary wildlife hosts should be implemented as a major preventive tool. Public and animal health authorities should jointly provide research funding to address this important gap in epidemiologic knowledge.

Parasites

Echinococcus multilocularis and *Echinococcus granulosus*

Echinococcosis (hydatid disease) is a neglected zoonotic infection caused by the larval stage of *Echinococcus* tapeworms. Despite the fact that several species are present in the EU (e.g., *Echinococcus equinus*, *Echinococcus ortleppi*, and *Echinococcus* spp.), *Echinococcus multilocularis* and *Echinococcus granulosus* are responsible for all human and animal cases (ECDC 2013; Otero-Abad and Torgerson 2013). Humans become infected through the accidental intake of tapeworm eggs, commonly through contact with infected animals (definitive hosts) or their environment, which become contaminated with their feces containing *Echinococcus* eggs. Normally, the infection appears as cystic echinococcosis, which is associated with *E. granulosus*. This species is predominately observed in the Mediterranean area and presents a domestic life-cycle involving dogs as the final host and livestock (sheep) as intermediate hosts (Deplazes et al. 2011; Romig et al. 2006). Hence, close contact between humans and domestic animals increases the risk of infection. However, wild canids could also represent an important reservoir of *E. granulosus* (Carmena and Cardona 2014; Otero-Abad and Torgerson 2013), as reported in Eastern Finland (Hirvelä-Koski et al. 2003) and Bulgaria (Breyer et al. 2004). In addition, wild boars infected with *Echinococcus granulosus* cysts have been found in Spain (Martín-Hernando et al. 2008) and Romania (Onac et al. 2013). Therefore, wild boars may be an intermediate host for infection of dogs and wolves through consumption of carcasses and hunting remains. The second form of disease is alveolar echinoccoccosis, which is caused by *E. multilocularis* and historically has a higher prevalence in Central Europe (France, Germany, and Switzerland) (Carmena and Cardona 2014). This species is transmitted in a predominantly sylvatic cycle in which mainly red foxes, but also raccoon dogs (*Nyctereutes procyonoides*), wolves (*Canis lupus*), and jackals (*Canis aureus*), are the final hosts (Carmena and Cardona 2014; Medlock and Leach 2009; Learmount et al. 2012; Süld et al. 2014). Furthermore, domestic dogs and cats can be sporadically infected, so they may play a role in transmission to humans due to close contact, especially dogs. Currently, alveolar echinoccoccosis is considered an emerging disease in Europe, with cases being reported in most countries, with the exception of the UK, Ireland (Medlock and Leach 2009), Malta and Finland (EFSA 2013). Notably, a synanthropic transmission cycle has been associated with an increase in fox densities since the 1990s due to vaccination campaigns against rabies. Hence, there is a greater likelihood of contact between foxes and humans in urban and suburban areas (Otero-Abad and Torgerson 2013; Romig et al. 2006). Agricultural practices such as permanent grasslands (Carmena and Cardona 2014), introduction of infected animals such as beavers into *Echinococcus*-free areas (Kosmider et al. 2013), or consumption of raw vegetables and wild berries (Wahlström et al. 2012) have been correlated with the increase observed in parasite transmission between animals and humans.

Toxoplasmagondii

Toxoplasma gondii is an apicomplexan protozoan whose infection causes from mild clinical disease to reproductive disorders in humans and animals. *T. gondii* presents one of the widest host spectrums among the parasitic protozoa because it is able to infect almost every species of warm-blooded animals (Dubey and Beattie 1988). Currently, *T. gondii* is distributed worldwide, in part due to its wide host range, high transmission plasticity, and the introduction of the protozoan by movement of infected domestic cats into naïve regions. Definitive hosts for *T. gondii* are felids—both domestic and wild—which shed viable and environmentally resistant oocysts into the environment. Additionally, *T. gondii* remains infective in the form of bradyzoites in muscular tissues of intermediate hosts that serve as sources of infection of definitive and other intermediate hosts (predators). Fatal infections are occasionally observed in wildlife (Fernández-Aguilar et al. 2013). Intermediate hosts, those that represent some of the most relevant sources of human infection, may be infected by *T. gondii* by a variety of pathways depending on their life strategies. Predators (both birds and mammals) become infected mostly by consumption of infected prey and less frequently by exposure to oocysts shed by felids. In contrast, herbivores become exposed to *T. gondii* mostly as a consequence of consumption of water and food (plants) contaminated with oocysts shed by felids. Nonetheless, omnivores such as the wild boar may become infected by *T. gondii* both by ingestion of contaminated meat from intermediate hosts and by exposure to infective oocysts, which may explain the high prevalence of *T. gondii* infection observed in European wild boar populations in comparison to other intermediate hosts such as ruminants (Gauss et al. 2005; Beral et al. 2012; Paştiu et al. 2013; Coelho et al. 2014; Ferroglio et al. 2014). *T. gondii* has been reported in several other wildlife species in the European continent: red deer, roe deer, fallow deer, reindeer, moose, chamois (*Rupicapra* spp.), Iberian ibex, mouflons, Armenian mouflons (*Ovis gmelini*), red foxes, common genets (*Genetta genetta*), European badgers (*Meles meles*), Iberian wolves (*Canis lupus signatus*), and many different wild bird species. Additionally, *T. gondii* is also present in domestic ungulates such as pigs, cattle and small ruminants; the latter accounting for the highest prevalence among European domestic ungulates (EFSA and ECDC 2014). It is estimated that approximately 50–80 % of Europeans are infected by *T. gondii* (EFSA and ECDC 2014); 1259 human toxoplasmosis cases were reported in 2009 in EU (Lahuerta et al. 2011). A large number of asymptomatic cases likely occur annually in the continent. The main risk factors for acquiring human infections in Europe include consumption of improperly cooked infected wild game meat, mainly wild boar; however, the risk associated with exposure to infected domestic cats should not be dismissed. In northern European countries, risk may additionally come from ingestion of raw or undercooked carnivore game meat (brown bear (*Ursus arctos*), seals, and walrus (*Odobenus rosmarus*)). The current status of *T. gondii* in European wildlife, especially in omnivorous and carnivorous species that may be able to maintain the circulation of *T. gondii* without the intervention of felids, suggests that prevention of infection through consumer education about food handling practices is perhaps a better, quicker, and cheaper strategy than any potential control measure in wildlife.

Trichinella spp.

Trichinosis is a human foodborne disease caused by parasitic nematodes of the genus *Trichinella*, and is reported to affect approximately 2500 people worldwide annually (Murrell and Pozio 2011). However, the incidence of human trichinosis is likely higher because underreporting frequently occurs in developing countries due to the absence of pathognomonic clinical signs and limited access to health care. This may explain why the European region has a higher contribution to the global burden of human trichinosis (Devleesschauwer et al. 2015). Human cases originate from consumption of raw or undercooked meat from infected hosts. From a European perspective, four *Trichinella* species are of concern: *T. spiralis*, *T. britovi*, *T. native*, and *T. pseudospiralis*. More than 150 mammal, bird, and reptile species can serve as hosts for *Trichinella* spp. (Pozio 2005). Although certain host specificities have been described, the encapsulated clade infects only mammals, whereas the non-encapsulated clade infects mammals, birds, and reptiles (Pozio and Zarlenga 2013). Moreover, different *Trichinella* species can infect the same host species, e.g., *T. spiralis*, *T. britovi*, *T. nativa*, and *T. pseudospiralis* can infect raccoon dogs (Pannwitz et al. 2010) and swine (EFSA and ECDC 2014). In contrast, *T. spiralis* and *T. britovi* primarily infect swine and carnivores, whereas *T. nativa* and *T. pseudospiralis* infect mainly carnivores. In 2012, all EU member states reported *Trichinella* spp. in animals and all except Denmark reported infection in humans. In 2012, the incidence of trichinosis increased 12.3 % compared to 2011 (EFSA and ECDC 2014), and most cases occurred in Eastern European countries. The sources of *Trichinella* spp. associated with human infection in EU differ geographically; whereas backyard domestic transmission constitutes an important source in Eastern Europe (e.g., Romania), consumption of improperly cooked wild boar meat is the main vehicle in other European regions (e.g., Spain). However, as the control of *Trichinella* infection progresses in European domestic pig populations, wildlife may become the main source for trichinosis outbreaks in Europe in the near future.

Education on the risks associated with consumption of raw or undercooked meat from wildlife would be effective for preventing trichinosis outbreaks.

Other Zoonotic Pathogens Shared with Wildlife

Some pathogens are not predominantly foodborne, but are good examples of disease transmission at the wildlife–livestock interface in Europe and can be of concern for human and animal health, food production, and wildlife conservation. Below, we briefly explain the current situation of brucellosis and bovine tuberculosis in Europe, two classical zoonoses shared between wildlife and livestock. Q fever (*Coxiella burnetti*) is also worth mentioning here since its status in wildlife is neglected. Finally, tularemia is considered an emerging infection in several parts of Europe.

Brucellosis

Six of the ten currently recognized species of the genus *Brucella* are pathogenic for humans and are shared with wildlife: *B. abortus*, *B. canis*, *B. ceti*, *B. melitensis*, *B. pinnipedialis*, and *B. suis* (Galińska and Zagórski 2013). Most reported cases of human brucellosis in the EU are caused by *B. melitensis*, either from occupational exposure to infected domestic ruminants (non-foodborne) or to consumption (foodborne) of unpasteurized milk or milk products (EFSA and ECDC 2013a). *Brucella melitensis* is present in domestic small ruminants in Mediterranean EU member states and has been found at low prevalences in European wild ruminant populations (e.g., in the Balkans, Spain, or Portugal). This epidemiologic status is, according to Godfroid et al. (2013), the reason for a low number of *B. melitensis* infections in wildlife, for example, the Alpine chamois and Alpine ibex in Italy and France, and the Iberian ibex in the Iberian Peninsula. It is hypothesized that wild ruminant cases are the result of exposure to infected domestic ruminants, and therefore wild ruminants do not constitute a reservoir for *B. melitensis*. The low prevalences of brucellosis occurring in many European wild ruminant populations (Muñoz et al. 2010; Astorga Márquez et al. 2014) may support this postulate. However, a recent outbreak of brucellosis in cattle caused by *B. melitensis* in the French Alps was transmitted by infected Alpine ibex (Mick et al. 2014), and this species is considered to be a true *B. melitensis* reservoir. Both routes of infection may be correct since the unique ecological traits of some local wildlife populations may contribute to the life cycle of *B. melitensis* in these animals (Mick et al. 2014). According to these authors, this bovine outbreak and a human case could have originated from the Alpine ibex population. However, from a public health perspective, the risk for EU citizens of acquiring *B. melitensis* from wildlife is likely low.

B. suis may constitute, from a worldwide perspective, the most relevant non-foodborne cause of brucellosis in humans. Four of the five *B. suis* biovars (1–4) are pathogenic for humans. *B. suis* biovars 1–3 are distributed worldwide in domestic and wild swine populations, with biovar 2 being the predominant biovar in the wild boar in Europe. This biovar has been very rarely reported in humans despite the high individual and population prevalence reported in European wild boars (Muñoz et al. 2010), and its presence in European hares. No human cases of brucellosis caused by biovar 4 have been reported in Europe. Reindeer populations in Norway were found to be free from *B. suis* (Asbakk et al. 1999), but no information from Sweden, Finland, and European Russia is currently available.

The prevalence of *B. abortus* in European wildlife is very low (Muñoz et al. 2010). This suggests that the risk of human infection with *B. abortus* from wildlife in Europe is low. Currently, it is very difficult to estimate the risk for Europeans to be infected by *B. ceti* and *B. pinnipedialis* since only a few cases of human brucellosis in the world have been reported and these were associated with the marine environment (Hernández-Mora et al. 2013). In the European Atlantic, Mediterranean, and North seas, several species of cetaceans have been found to be infected with *B. ceti* (Guzmán-Verri et al. 2012).

Bovine Tuberculosis

Human and animal tuberculosis is caused by acid-fast, gram-positive bacteria of the *Mycobacterium tuberculosis* complex (MTC). The MTC includes, among other members, the main causal agents of animal tuberculosis, *Mycobacterium bovis* and *Mycobacterium caprae* (Rodríguez et al. 2011; Bezos et al. 2014). More than the 95 % of human tuberculosis cases in developed countries (e.g., the USA and the EU) are caused by within-human transmission of *M. tuberculosis* and only a low percentage is caused by *M. bovis* and other tuberculous mycobacteria. Human cases of tuberculosis caused by *M. bovis* in the EU represent less than 1 %, with only 125 confirmed cases in 2012 (EFSA and ECDC 2014). Transmission of *M. bovis* from animals to humans occurs mainly by consumption of contaminated cattle meat or unpasteurized milk/milk products and by direct exposure to infected cattle. Infection from wildlife may occur mainly by direct exposure to wildlife or their products; hunters, wildlife veterinarians, researchers, gamekeepers, and people eating game meat may be at higher risk of acquiring tuberculosis from wildlife. The clinical outcome of human infection by *M. bovis* is similar to that caused by infection with *M. tuberculosis* (Pérez-Lago et al. 2014).

In Europe, wildlife reservoirs of *M. bovis* include wild ungulates (mainly European wild boars, red deer, and European bison—*Bison bonasus*) and the Eurasian badger. The role these species play in the maintenance and transmission of *M. bovis*/*M. caprae* varies according to variation in host demography and management (Muñoz-Mendoza et al. 2013). Particular demographic and management traits of wild ungulates in the southwestern Iberian Peninsula drive *M. bovis* prevalence and transmission risks (Vicente et al. 2006). Prevalence of tuberculosis-like lesions in wild boars in some central Iberian populations reaches 100 % and has remained at this rate for many years (Vicente et al. 2013). A high prevalence in wildlife increases the risk of transmission of *M. bovis* to domestic species (Martínez-López et al. 2014). The epidemiologic scenario is slightly different in Central European countries where wild boar and red deer densities and management schemes differ from those observed in Iberia. This translates into lower prevalences of *M. bovis* infection in wild ungulates (Muñoz-Mendoza et al. 2013). However, recent evidence points to an increasing relevance of wild ungulates—especially wild boars—in transmission to cattle in Central Europe (Richomme et al. 2013). This may translate to an increase in the risk of transmission from wildlife to humans—both directly from wildlife and indirectly through cattle—in central Europe. The role of the Eurasian badger as a *M. bovis* reservoir in Central Europe and the British Isles adds complexity to the epidemiologic scenario of bovine tuberculosis in continental and Atlantic European regions.

Q Fever

Q fever is caused by *Coxiella burnetii*, a pathogen with worldwide distribution. A wide variety of vertebrate species, including mammals, birds and reptiles, and some invertebrates, e.g., ticks, are susceptible to *C. burnetii* infection. Therefore, *C. burnetii* presents one of the widest host ranges of known pathogenic bacteria.

Q fever is endemic in the EU, where it circulates in domestic ruminants (Maurin and Raoult 1999). Transmission to humans occurs mainly by contaminated aerosols originating in domestic ruminant farms; however, a small percentage of human Q fever cases have non-domestic ruminant origins such as exposure to infected pets (cats and dogs) or are of unknown origin—perhaps associated with wildlife.

C.burnetii is not considered to be largely associated with European wildlife. However, the few studies conducted on *C. burnetii* in wildlife revealed the potential persistence of this zoonotic pathogen in nondomestic environments (Ruiz-Fons 2012). The wide host range of *C. burnetii* enables its establishment in wildlife in which it is subsequently maintained and thereafter enters into the domestic environment. The current epidemiology of *C. burnetii* in Europe indicates that domestic ruminants have medium-to-high farm seroprevalences (Ruiz-Fons et al. 2010; Piñero et al. 2014), hence placing emphasis on this reservoir and not on wildlife. However, as long as the disease is controlled in domestic ruminants (through vaccination or population control), the potential reservoir role of wildlife may become more evident. Recent studies in wild and farmed red deer and European rabbits in Iberia (González-Barrio et al. 2015; F. Ruiz-Fons, unpublished) have revealed that *C. burnetii* circulates in these two species with a wide geographic distribution and at medium-to-high prevalence. Pathological studies also indicate that both species are able to shed *C. burnetii* into the environment (González-Barrio et al. 2013; F. Ruiz-Fons, unpublished data), which may be a source of human infection. The current European situation suggests that *C. burnetii* transmission through wildlife may be an important threat to human health.

Tularemia

Francisella tularensis is the causative agent of tularemia, a zoonotic disease that affects humans, farm animals, and wildlife. Currently, four subspecies with differing degrees of virulence are recognized, but in Europe only the moderately virulent *F. tularensis* subsp. *holarctica* is present, and covers almost the entire European continent, with the exception of the British Isles and Iceland (reviewed by Gyuranecz 2012).

Similar to *C. burnetii*, *F. tularensis* is a multi-host pathogen that has been found in mammals, birds, amphibians, fish, and invertebrates (e.g., freshwater crayfish and ticks). Rodents (several species) and lagomorphs (mainly hares) are major hosts for *F. tularensis* in Europe. Infection by *F. tularensis* is enzootic in two interconnected cycles: (1) a terrestrial cycle involving rodents, hares (mainly the European hare but also the mountain hare and the Iberian hare—*Lepus granatensis*) and biological vectors (ticks); and (2) an aquatic cycle involving mainly the European water vole (*Arvicola amphibius*) and *Microtus* spp., among other host species.

Eight hundred and fifty cases of tularemia in humans were confirmed in Europe in 2008 (ECDC 2010), with most in Scandinavian countries. Humans are exposed to *F. tularensis* largely by handling infected wild animals, for example, hunters skinning hare carcasses. A portion of the cases arise from exposure to contaminated food or water, tick bites, and even through exposure to contaminated aerosols. The risk for humans is directly linked to the prevalence of *F. tularensis* in wildlife and the prevalence in wildlife may be linked to host density, among other factors. As an example, tularemia outbreaks in common voles (*Microtus arvalis*) in the North Spanish Plateau are thought to be linked to a cyclic demographic explosion of this species (Luque-Larena et al. 2013). Outbreaks in voles are associated with the increase in the number of cases in Iberian hares, which in turn increase the risk for hunters and is related to the peak in the number of reported cases. The mechanisms by which *F. tularensis* is maintained between epizooties are not fully elucidated. Gyuranecz et al. (2011) identified European brown hares and ticks as being responsible for the persistence of *F. tularensis* subsp. *holarctica* between epizooties in Hungary. However, in regions in which European brown hares are absent but other hare species are present, it is currently unknown if the role played by these hare species is the same as that of the European hare or involves other host species. Elucidating the factors that influence this enzootic situation within low host density scenarios is essential to propose preventive measures to mitigate this disease.

Conclusions

Each wildlife–livestock–human interface is unique in regard to its relationship with foodborne pathogens since local ecological and human differences determine the likelihood of interactions between hosts, transmission of pathogens, and contamination of food. The difficulty in elucidating the dynamics of diseases shared between wildlife and livestock is partly related to the challenge in obtaining an adequate number of biological samples of sufficient quality, and to limitations related to available methods used in the isolation and detection of pathogens. However, these difficulties may be overcome, or at least compensated for, by concurrently sampling livestock and wildlife (and humans, whenever possible), using the proper molecular tools and conducting appropriate statistical analyses.

It cannot be disregarded that some foodborne pathogens can also cause disease in wildlife, though their implication in wildlife conservation is less known than their effect on human and animal health. Wildlife population declines caused by disease are difficult to prove, but in conjunction with other contributing factors, infections of wildlife by foodborne pathogens could be an underappreciated cause of illness in vulnerable populations.

For the most part, food safety risks from wildlife in Europe, although sometimes underestimated, may not be the most important contributor to foodborne illness compared with other sources. However, current changes in wildlife population trends and their management, in food production systems and in consumer preferences, all should be closely monitored to be able to detect changes in food risks from wildlife.

References

AGES (Austrian Agency for Health and Food Safety) (2012) Report on zoonoses and zoonotic agents in Austria, 2012. http://www.ages.at/uploads/media/Report_on_Zoonoses_and_zoonotic_Agents_in_Austria__2012.pdf

Alvarez-Fernandez E, Dominguez-Rodriguez J, Capita R, Alonso-Calleja C (2012) Influence of housing systems on microbial load and antimicrobial resistance patterns of *Escherichia coli* isolates from eggs produced for human consumption. J Food Protect 75:847–853

Antilles N, Sanglas A, Cerdà-Cuéllar M (2015) Free-living waterfowl as a source of zoonotic bacteria in a dense wild bird population area in Northeastern Spain. Transbound Emerg Dis 62:516–521 doi:10.1111/tbed.12169

Apollonio M, Andersen R, Putman R (2010) European ungulates and their management in the 21st century. Cambridge University Press, New York, NY

Asbakk K, Gall D, Stuen S (1999) A screening ELISA for brucellosis in reindeer. Zentralbl Veterinarmed B 46:649–657

Aschfalk A, Kemper N, Holler C (2003) Bacteria of pathogenic importance in faeces from cadavers of free-ranging or corralled semi-domesticated reindeer in northern Norway. Vet Res Commun 27:93–100

Astorga Márquez RJ, Carvajal A, Maldonado A et al (2014) Influence of cohabitation between domestic goat (*Capra aegagrus hircus*) and Iberian ibex (*Capra pyrenaica hispanica*) on seroprevalence of infectious diseases. Eur J Wildl Res 60:387–390

Atanassova V, Ring C (1999) Prevalence of *Campylobacter* spp. in poultry and poultry meat in Germany. Int J Food Microbiol 51:187–190

Atanassova V, Apelt J, Reich F, Klein G (2008) Microbiological quality of freshly shot game in Germany. Meat Sci 78:414–419

Avagnina A, Nucera D, Grassi MA et al (2012) The microbiological conditions of carcasses from large game animals in Italy. Meat Sci 91:266–271

Ayral F, Artois J, Zilber AL et al (2015) The relationship between socioeconomic indices and potentially zoonotic pathogens carried by wild Norway rats: a survey in Rhône, France (2010–2012). Epidemiol Infect 143:586–599

Bardiau M, Gregoire F, Muylaert A et al (2010) Enteropathogenic (EPEC), enterohaemorragic (EHEC) and verotoxigenic (VTEC) *Escherichia coli* in wild cervids. J Appl Microbiol 109:2214–2222

Baylis CL, MacPhee S, Martin KW et al (2000) Comparison of three enrichment media for the isolation of *Campylobacter* spp. from foods. J Appl Microbiol 89:884–891

Beral M, Rossi S, Aubert D et al (2012) Environmental factors associated with the seroprevalence of *Toxoplasma gondii* in wild boars (*Sus scrofa*), France. Ecohealth 9:303–309

Berge AC, Hancock DD, Sischo WM, Besser TE (2010) Geographic, farm, and animal factors associated with multiple antimicrobial resistance in fecal *Escherichia coli* isolates from cattle in the western United States. J Am Vet Med A 12:1338–1344

Bezos J, Alvarez J, Romero B et al (2014) Bovine tuberculosis: historical perspective. Res Vet Sci Suppl S3–4

Blake DP, Humphry RW, Scott KP et al (2003) Influence of tetracycline exposure on tetracycline resistance and the carriage of tetracycline resistance genes within commensal *Escherichia coli* populations. J Appl Microbiol 94:1087–1097

Boadella M, Casas M, Martín M et al (2010) Increasing contact with hepatitis E virus in red deer, Spain. Emerg Infect Dis 16:1994–1996

Bradley CA, Altizer S (2007) Urbanization and the ecology of wildlife diseases. Trends Ecol Evol 2:95–102

Braeye T, de Schrijver K, Wollants E et al (2014) A large community outbreak of gastroenteritis associated with consumption of drinking water contaminated by river water, Belgium, 2010. Epidemiol Infect 143:711–719

Brainerd S (2007) European Charter on hunting and biodiversity. Convention on the conservation of European wildlife and natural habitats. http://fp7hunt.net/Portals/HUNT/Hunting_Charter.pdf

Breyer I, Georgieva D, Kurdova R et al (2004) *Echinococcus granulosus* strain typing in Bulgaria: the G1 genotype is predominant in intermediate and definitive wild hosts. Parasitol Res 93:127–130

Cahill S, Llimona F, Cabañeros L, Calomardo F (2012) Characteristics of wild boar (*Sus scrofa*) habituation to urban areas in the Collserola Natural Park (Barcelona) and comparison with other locations. Anim Biodivers Conserv 35:221–233

Caprioli A, Donelli G, Falbo V et al (1991) Antimicrobial resistance and production of toxins in *Escherichia coli* strains from wild ruminants and the Alpine Marmot. J Wildl Dis 27:324–327

Carbonero A, Paniagua J, Torralbo A et al (2014) *Campylobacter* infection in wild artiodactyl species from southern Spain: occurrence, risk factors and antimicrobial susceptibility. Comp Immunol Microbiol Infect Dis 37:115–121

Carmena D, Cardona GA (2014) Echinococcosis in wild carnivorous species: epidemiology, genotypic diversity, and implications for veterinary public health. Vet Parasitol 202:69–94

Cenci-Goga BT, Rossitto PV, Sechi P et al (2012) Effect of selected dairy starter cultures on microbiological, chemical and sensory characteristics of swine and venison (*Dama dama*) nitrite-free dry-cured sausages. Meat Sci 90:599–606

Coelho C, Vieira-Pinto M, Faria AS et al (2014) Serological evidence of *Toxoplasma gondii* in hunted wild boar from Portugal. Vet Parasitol 202:310–312

Chiari M, Zanoni M, Tagliabue S et al (2013) *Salmonella* serotypes in wild boars (*Sus scrofa*) hunted in northern Italy. Acta Vet Scand 55:42

Cobbold RN, Hancock DD, Rice DH et al (2007) Rectoanal junction colonization of feedlot cattle by Escherichia coli O157: H7 and its association with supershedders and excretion dynamics. Appl Environ Microbiol 73:1563–1568

Colles FM, Dingle KE, Cody AJ, Maiden MCJ (2008) Comparison of *Campylobacter* populations in wild geese with those in starlings and free-range poultry on the same farm. Appl Environ Microbiol 74:3583–3590 de Deus N, Peralta B, Pina S et al (2008) Epidemiological study of hepatitis E virus infection in European wild boars (*Sus scrofa*) in Spain. Vet Microbiol 129:163–170

De Jong H, Ekdahl M (1965) Salmonellosis in calves—the effect of dose rate and other factors on transmission. N Z Vet J 13:59–64

Deplazes P, van Knapen F, Schweiger A et al (2011) Role of pet dogs and cats in the transmission of helminthic zoonoses in Europe, with a focus on echinococcosis and toxocarosis. Vet Parasitol 182:41–53

Devleesschauwer B, Praet N, Speybroeck N et al (2015) The low global burden of trichinellosis: evidence and implications. Int J Parasitol 45:95–99

Díaz-Sánchez S, Sánchez S, Sánchez M et al (2012a) Detection and characterization of Shiga toxin-producing *Escherichia coli* in game meat and ready-to-eat meat products. Int J Food Microbiol 160:179–182

Díaz-Sánchez S, Moriones AM, Casas F et al (2012b) Prevalence of *Escherichia coli*, *Salmonella* spp and *Campylobacter* spp in the intestinal flora of farm-reared, restocked and wild red-legged partridges (*Alectoris rufa*): is restocking using farm-reared birds a risk? Eur J Wildl Res 58:99–105

Díaz-Sánchez S, Sánchez S, Herrera-Leon S et al (2013) Prevalence of Shiga toxin-producing *Escherichia coli*, *Salmonella* spp. and *Campylobacter* spp. in large game animals intended for consumption: relationship with management practices and livestock influence. Vet Microbiol 163:274–281

Dipineto L, Russo TP, Gargiulo A et al (2014) Prevalence of enteropathogenic bacteria in common quail (*Coturnix coturnix*). Avian Pathol 23:1–10

Dubey JP, Beattie CP (1988) Toxoplasmosis of animals and man. CRC Press, Boca Raton, FL

ECDC (European Centre for Disease Prevention and Control) (2010) Annual epidemiological report on communicable diseases in Europe 2010. ECDC, Stockholm, p 2010

ECDC (European Centre for Disease Prevention and Control) (2013) Reporting on 2011 surveillance data and 2012 epidemic intelligence data. Annual epidemiological report. ECDC, Stockholm, p 2013

EFSA (European Food Safety Authority) (2009) Technical specifications for the monitoring and reporting of verotoxigenic *Escherichia coli* (VTEC) on animals and food (VTEC surveys on animals and food). EFSA J 7:1366

EFSA (European Food Safety Authority) (2013) Assessment of *Echinococcus multilocularis* surveillance reports submitted 2013 in the context of Commission Regulation (EU) No 1152/2011. EFSA J 11:3465

EFSA (European Food Safety Authority) and ECDC (European Centre for Disease Prevention and Control) (2013a) The European Union summary report on trends and sources of zoonoses, zoonotic agents and food-borne outbreaks in 2011. EFSA J 11:3129

EFSA (European Food Safety Authority) and ECDC (European Centre for Disease Prevention and Control) (2013b) The European Union summary report on antimicrobial resistance in zoonotic and indicator bacteria from humans, animals and food in 2011. EFSA J 11:3196

EFSA (European Food Safety Authority) and ECDC (European Centre for Disease Prevention and Control) (2014) The European Union summary report on trends and sources of zoonoses, zoonotic agents and food-borne outbreaks in 2012. EFSA J 12:3547

Ejidokun OO, Walsh A, Barnett J et al (2006) Human vero cytotoxigenic *Escherichia coli* (VTEC) O157 infection linked to birds. Epidemiol Infect 134:421–423

Epps SV, Harvey RB, Hume ME et al (2013) Foodborne *Campylobacter*: infections, metabolism, pathogenesis and reservoirs. Int J Environ Res Public Health 10:6292–6304

European Commission (2007) 2007/516/EC: commission decision of 19 July 2007concerning a financial contribution from the community towards a survey on the prevalence and antimicrobial resistance of *Campylobacter* spp. in broiler flocks and on the prevalence of *Campylobacter* spp. and *Salmonella* spp. in broiler carcasses to be carried out in the Member States. Off J Eur Union 50:L190

Fedorka-Cray P, Kelley L, Stabel T et al (1995) Alternate routes of invasion may affect pathogenesis of *Salmonella* Typhimurium in swine. Infect Immun 63:2658–2664

Fernández-Aguilar X, Ajzenberg D, Cabezón O, Martínez-López A, Darwich L, Dubey JP, Almería S (2013) Fatal toxoplasmosis associated with an atypical *Toxoplasma gondii* strain in a Bennett's wallaby (*Macropus rufogriseus*) in Spain. Vet Parasitol 196:523–527
Ferroglio E, Bosio F, Trisciuoglio A, Zanet S (2014) *Toxoplasma gondii* in sympatric wild herbivores and carnivores: epidemiology of infection in the Western Alps. Parasit Vector 7:196
Freidl G, Stalder G, Kostic T et al (2011) Verocytotoxin-producing *Escherichia coli* in Chamois (*Rupicapra rupicapra*) and cattle in Austria. J Wildl Dis 47:704–708
Fitzgerald C, Whichard J, Nachamkin I (2008) Diagnosis and antimicrobial susceptibility of *Campylobacter* species. In: Nachamkin I, Szymanski CM, Blaser MJ (eds) *Campylobacter*, 3rd edn. ASM Press, Washington, DC, pp 227–243
Gaffuri A, Holmes JP (2012) *Salmonella* infections. In: Gavier-Widen D, Meredith A, Duff JP (eds) Infectious diseases of wild mammals and birds in Europe, 1st edn. Wiley-Blackwell, Chichester, pp 386–397
Galińska EM, Zagórski J (2013) Brucellosis in humans – etiology, diagnostics, clinical forms. Ann Agric Environ Med 20:233–238
García-Sánchez A, Sánchez S, Rubio R et al (2007) Presence of Shiga-toxin producing *E. coli* O157:H7 in a survey of wild artyodactils. Vet Microbiol 121:377
Gauss CBL, Dubey JP, Vidal D et al (2005) Seroprevalence of *Toxoplasma gondii* in wild pigs (*Sus scrofa*) from Spain. Vet Parasitol 131:151–156
Glawischnig W, Khaschabi D, Schopf K, Schonbauer M (2000) An outbreak of *Salmonella* Dublin in chamois (*Rupicapra rupicapra*). Wien Tierarztl Monatsschr 87:21–25
Godfrey ER, Randolph SE (2011) Economic downturn results in tick-borne disease upsurge. Parasit Vector 4:35
Godfroid J, Garin-Bastuji B, Saegerman C, Blasco JM (2013) Brucellosis in terrestrial wildlife. Rev Sci Tech OIE 32:27–42
Gómez P, González-Barrio D, Benito D et al (2014) Detection of methicillin-resistant *Staphylococcus aureus* (MRSA) carrying the mecC gene in wild small mammals in Spain. J Antimicrob Chemother 69:2061–2064
González-Barrio D, Almería S, Caro MR et al (2013) *Coxiella burnetii* shedding by farmed red deer (*Cervus elaphus*). Transbound Emerg Dis. doi:10.1111/tbed.12179
González-Barrio D, Maio E, Vieira-Pinto M et al (2015) European rabbits as reservoir for *Coxiella burnetii*. Emerg Infect Dis 21:1055–1058
Gortázar C, Acevedo P, Ruiz-Fons F, Vicente J (2006) Disease risks and overabundance of game species. Eur J Wildl Res 52:81–87
Gortázar C, Ferroglio E, Höfle U et al (2007) Diseases shared between wildlife and livestock: a European perspective. Eur J Wildl Res 53:241–256
Gortázar C, Reperant LA, Kuiken T et al (2014) Crossing the interspecies barrier: opening the door to zoonotic pathogens. PLoS Pathogens 10, e1004129
Griekspoor P, Colles FM, McCarthy ND et al (2013) Marked host specificity and lack of phylogeographic population structure of *Campylobacter jejuni* in wild birds. Mol Ecol 22:1463–1472
Guerra B, Fischer J, Helmuth R (2014) An emerging public health problem: acquired carbapenemase-producing microorganisms are present in food-producing animals, their environment, companion animals and wild birds. Vet Microbiol 171:290–297
Gummow B (2010) Challenges posed by new and re-emerging infectious diseases in livestock production, wildlife and humans. Livest Sci 130:41–46
Guzmán-Verri C, González-Barrientos R, Hernández-Mora G et al (2012) *Brucella ceti* and brucellosis in cetaceans. Front Cell Infect Microbiol 2:3
Gyuranecz M, Rigó K, Dán A et al (2011) Investigation of the ecology of *Francisella tularensis* during an inter-epizootic period. Vector Borne Zoonotic Dis 11:8
Gyuranecz M (2012) Tularaemia. In: Gavier-Widen D, Meredith A, Duff JP (eds) Infectious diseases of wild mammals and birds in Europe, 1st edn. Wiley-Blackwell, Chichester, pp 303–309
Hälli O, Ala-Kurikka E, Nokireki T et al (2012) Prevalence of and risk factors associated with viral and bacterial pathogens in farmed European wild boar. Vet J 194:98–101

Haydon DT, Cleaveland S, Taylor LH, Laurenson MK (2002) Identifying reservoirs of infection: a conceptual and practical challenge. Emerg Infect Dis 8:1468–1473

Hernández-Mora G, Palacios-Alfaro JD, González-Barrientes R (2013) Wildlife reservoirs of brucellosis: *Brucella* in aquatic environments. Rev Sci Tech OIE 32:89–103

Heuer OE, Pedersen K, Andersen JS, Madsen M (2002) Vancomycin-resistant enterococci (VRE) in broiler flocks 5 years after the avoparcin ban. Microb Drug Resist 8:133–138

Hilbert F, Smulders FJM, Chopra-Dewasthaly R, Paulsen P (2012) *Salmonella* in the wildlife-human interface. Food Res Int 45:603–608

Hirvelä-Koski V, Haukisalmi V, Kilpelä SS et al (2003) *Echinococcus granulosus* in Finland. Vet Parasitol 111:175–192

Hofer E, Cernela N, Stephan R (2012) Shiga toxin subtypes associated with Shiga toxin-producing *Escherichia coli* strains isolated from red deer, roe deer, chamois, and ibex. Foodborne Pathog Dis 9:792–795

Hoffman LC, Wiklund E (2006) Game and venison – meat for the modern consumer. Meat Sci 74:197–208

Horigan V, Davies RH, Kelly LA et al (2014) A qualitative risk assessment of the microbiological risks to consumers from the production and consumption of uneviscerated and eviscerated small game birds in the UK. Food Control 45:127–137

Humphrey T, O'Brien S, Madsen M (2007) Campylobacters as zoonotic pathogens: a food production perspective. Int J Food Microbiol 117:237–257

Jirintai S, Tanggis M, Suparyatmo JB et al (2014) Rat hepatitis E virus derived from wild rats (*Rattus rattus*) propagates efficiently in human hepatoma cell lines. Virus Res 185:92–102

Kemper N, Aschfalk A, Holler C (2006) *Campylobacter* spp., *Enterococcus* spp., *Escherichia coli*, *Salmonella* spp., *Yersinia* spp., and *Cryptosporidium* oocysts in semi-domesticated reindeer (*Rangifer tarandus tarandus*) in Northern Finland and Norway. Acta Vet Scand 48:7

Kim YM, Jeong S, Kim JY et al (2011) The first case of genotype 4 hepatitis E related to wild boar in South Korea. J Clin Virol 50:253–256

Kosmider R, Paterson A, Voas A et al (2013) *Echinococcus multilocularis* introduction and establishment in wildlife via imported beavers. Vet Rec 172:606

Kozak GK, Boerlin P, Janecko N et al (2009) Antimicrobial resistance in *Escherichia coli* isolates from swine and wild small mammals in the proximity of swine farms and in natural environments in Ontario, Canada. Appl Environ Microbiol 75:559–566

Krumbholz A, Joel S, Dremsek P et al (2014) Seroprevalence of hepatitis E virus (HEV) in humans living in high pig density areas of Germany. Med Microbiol Immunol 203:273–282

Lahuerta A, Westrell T, Takkinen J et al (2011) Zoonoses in the European Union: Origin, distribution and dynamics – the EFSA-ECDC summary report 2009. Euro Surveill 16:13

Larska M, Krzysiak MK, Jabłonski A et al (2015) Hepatitis E virus antibody prevalence in wildlife in Poland. Zoonoses Public Health 62:105–110

Learmount J, Zimmer IA, Conyers C et al (2012) A diagnostic study of *Echinococcus multilocularis* in red foxes (*Vulpes vulpes*) from Great Britain. Vet Parasitol 190:447–453

Li T, Chijiwa K, Sera N et al (2005) Hepatitis E virus transmission from wild boar meat. Emerg Infect Dis 11:1958–1960

Lillehaug A, Bergsjo B, Schau J et al (2005) *Campylobacter* spp., *Salmonella* spp., verocytotoxic *Escherichia coli*, and antibiotic resistance in indicator organisms in wild cervids. Acta Vet Scand 46:23–32

Linke K, Ruckerl I, Brugger K et al (2014) Reservoirs of *Listeria* species in three environmental ecosystems. Appl Environ Microbiol 80:5583–5592

Literak I, Dolejska M, Rybarikova J et al (2009) Highly variable patterns of antimicrobial resistance in commensal *Escherichia coli* isolates from pigs, sympatric rodents, and flies. Microb Drug Resist 15:229–237

Literak I, Dolejska M, Radimersky T et al (2010) Antimicrobial-resistant faecal *Escherichia coli* in wild mammals in central Europe: multiresistant *Escherichia coli* producing extended-spectrum beta-lactamases in wild boars. J Appl Microbiol 108:702–1711

Loncaric I, Kübber-Heiss A, Posautz A et al (2014) MecC- and mecA-positive methicillin-resistant *Staphylococcus aureus* (MRSA) isolated from livestock sharing habitat with wildlife previously tested positive for mecC-positive MRSA. Vet Dermatol 25:146–148

Luque-Larena JJ, Mougeot F, Viñuela J et al (2013) Recent large-scale range expansion and outbreaks of the common vole (*Microtus arvalis*) in NW Spain. Basic Appl Ecol 14:432–441

Lyautey E, Hartmann A, Pagotto F et al (2007) Characteristics and frequency of detection of fecal *Listeria monocytogenes* shed by livestock, wildlife, and humans. Can J Microbiol 53:1158–1167

Man SM (2011) The clinical importance of emerging *Campylobacter* species. Nat Rev Gastroenterol 8:669–685

Martin C, Pastoret P, Brochier B et al (2011) A survey of the transmission of infectious diseases/infections between wild and domestic ungulates in Europe. Vet Res 42:70

Martín-Hernando MP, González LM, Ruiz-Fons F et al (2008) Massive presence of *Echinococcus granulosus* (Cestoda, Taeniidae) cysts in a wild boar (*Sus scrofa*) from Spain. Parasitol Res 103:705–707

Martínez R, García A, Blanco JE (2011) Occurrence of verocytotoxin-producing *Escherichia coli* in the faeces of free-ranging lagomorphs in southwest Spain. Eur J Wildl Res 57:187–189

Martínez-López B, Barasona JA, Gortázar C et al (2014) Farm-level risk factors for the occurrence, new infection or persistence of tuberculosis in cattle herds from South-Central Spain. Prev Vet Med 116:268–278

Massei G, Roy S, Bunting R (2011) Too many hogs? A review of methods to mitigate impact by wild boar and feral hogs. Hum Wildl Interact 5:79–99

Maurin M, Raoult D (1999) Q fever. Clin Microbiol Rev 12:518–533

Medlock JM, Leach S (2009) *Echinococcus multilocularis* and possible cycles in UK wildlife. Vet Rec 164:789–790

Membré J, Laroche M, Magras C (2011) Assessment of levels of bacterial contamination of large wild game meat in Europe. Food Microbiol 28:1072–1079

Mentaberre G, Porrero MC, Navarro-Gonzalez N et al (2013) Cattle drive *Salmonella* infection in the wildlife-livestock interface. Zoonoses Public Health 60:510–518

Methner U, Heller M, Bocklisch H (2010) *Salmonella enterica* subspecies *enterica* serovar Choleraesuis in a wild boar population in Germany. Eur J Wildl Res 56:493–502

Mick V, Le Carrou G, Corde Y et al (2014) *Brucella melitensis* in France: persistence in wildlife and probable spillover from Alpine Ibex to domestic animals. PLoS One 9, e94168

Miko A, Pries K, Haby S et al (2009) Assessment of Shiga-Toxin producing *Escherichia coli* isolates from wildlife meat as potential pathogens for humans. Appl Environ Microbiol 75:6462–6470

Mokracka J, Koczura R, Kaznowski A (2012) Transferable integrons of Gram-negative bacteria isolated from the gut of a wild boar in the buffer zone of a national park. Ann Microbiol 62:877–880

Mora A, Lopez C, Dhabi G et al (2012) Seropathotypes, phylogroups, Stx subtypes, and intimin types of wildlife-carried, Shiga toxin-producing *Escherichia coli* strains with the same characteristics as human-pathogenic isolates. Appl Environ Microbiol 78:2578–2585

Muñoz PM, Boadella M, Arnal M et al (2010) Spatial distribution and risk factors of Brucellosis in Iberian wild ungulates. BMC Infect Dis 10:46

Muñoz-Mendoza M, Marreros N, Boadella M et al (2013) Wild boar tuberculosis in Iberian Atlantic Spain: a different picture from Mediterranean habitats. BMC Vet Res 9:176

Murrell KD, Pozio E (2011) Worldwide occurrence and impact of human trichinellosis, 1986–2009. Emerg Infect Dis 17:2194–2202

Mustin K, Newey S, Knott J et al (2011) Biodiversity impacts of game bird hunting and associated management practices in Europe and North America. RSPB report. http://www.hutton.ac.uk/sites/default/files/files/RSPB_ReportFINAL_Covers.pdf. Accessed 22 Nov 2014

Nebola M, Borilova G, Steinhauserova I (2007) *Campylobacter* subtypes in pheasants (*Phasianus colchicus* spp. *torquatus*) in the Czech Republic. Vet Med (Praha) 52:496–501

Nardini R, Verin R, Mazzei M et al (2014) Hepatitis E virus-related liver alterations and viral antigen localization in European wild boar (*Sus scrofa*). Eur J Wildl Res 60:835–838

Navarro-Gonzalez N, Mentaberre G, Porrero MC et al (2012) Effect of cattle on *Salmonella* carriage, diversity and antimicrobial resistance in free-ranging wild boar (*Sus scrofa*) in northeastern Spain. PLoS One 7, e51614

Navarro-Gonzalez N, Porrero MC, Mentaberre G et al (2013a) Antimicrobial resistance in indicator *Escherichia coli* from free-ranging livestock and sympatric wild ungulates in a natural environment (NE Spain). Appl Environ Microbiol 79:6184–6186

Navarro-Gonzalez N, Casas-Diaz E, Porrero MC et al (2013b) Food-borne zoonotic pathogens and antimicrobial resistance of indicator bacteria in urban wild boars (*Sus scrofa*) in Barcelona, Spain. Vet Microbiol 167:686–689

Navarro-Gonzalez N, Velarde R, Porrero MC et al (2014a) Lack of evidence of spill-over of *Salmonella enterica* between cattle and sympatric Iberian ibex (*Capra pyrenaica*) from a protected area in Catalonia, NE Spain. Transbound Emerg Dis 61:378–384

Navarro-Gonzalez N, Ugarte-Ruiz M, Porrero MC et al (2014b) *Campylobacter* shared between free-ranging cattle and sympatric wild ungulates in a natural environment (NE Spain). Ecohealth 11:333–342

Navarro-Gonzalez N, Porrero MC, Mentaberre G et al (2015) Escherichia coli O157:H7 in wild boars (Sus scrofa) and Iberian ibex (Capra pyrenaica) sharing pastures with free-ranging livestock in a natural environment in Spain. Vet Quart 25:1–5

Nielsen EM, Skov MN, Madsen JJ et al (2004) Verocytotoxin-producing *Escherichia coli* in wild birds and rodents in close proximity to farms. Appl Environ Microbiol 70:6944–6947

Obwegeser T, Stephan R, Hofer E, Zweifel C (2012) Shedding of foodborne pathogens and microbial carcass contamination in hunted wild ruminants. Vet Microbiol 159:149–154

Onac D, Gyorke A, Oltean M et al (2013) First detection of *Echinococcus granulosus* G1 and G7 in wild boars (*Sus scrofa*) and red deer (*Cervus elaphus*) in Romania using PCR and PCR-RFLP techniques. Vet Parasitol 193:289–291

Otero-Abad B, Torgerson PR (2013) A systematic review of the epidemiology of echinococcosis in domestic and wild animals. PLoS Negl Trop Dis 7, e2249

Paulsen P, Hilbert F, Winkelmayer R et al (2003) Zur tierarztlichen Fleischuntersuchung von Wild, dargestellt an der Untersuchung von Rehen in Wildfleischbearbeitungsbetrieben (Veterinary meat inspection of wildlife, examinations of roe deer in a cutting plant.). Arch Lebensmitt Hyg 54:137–140

Paulsen P, Smulders FJM, Hilbert F (2012) *Salmonella* in meat from hunted game: a Central European perspective. Food Res Int 45:609–616

Pannwitz G, Mayer-Scholl A, Balicka-Ramisz A, Nöckler K (2010) Increased prevalence of *Trichinella* spp., Northeastern Germany, 2008. Emerg Infect Dis 16:936–942

Paştiu AI, Györke A, Blaga R et al (2013) In Romania, exposure to *Toxoplasma gondii* occurs twice as often in swine raised for familial consumption as in hunted wild boar, but occurs rarely, if ever, among fattening pigs raised in confinement. Parasitol Res 112:2403–2407

Pérez-Lago L, Navarro Y, García-de-Viedma D (2014) Current knowledge and pending challenges in zoonosis caused by *Mycobacterium bovis*: a review. Res Vet Sci 97:S94–S100

Pierard D, Van Damme L, Moriau L et al (1997) Virulence factors of verocytotoxin-producing *Escherichia coli* isolated from raw meat. Appl Environ Microbiol 63:4585–4587

Piñero A, Ruiz-Fons F, Hurtado A et al (2014) Changes in the dynamics of *Coxiella burnetii* infection in dairy cattle: an approach to match field data with the epidemiological cycle of *C. burnetii* in endemic herds. J Dairy Sci 97:2718–2730

Pioz M, Loison A, Gibert P et al (2008) Antibodies against *Salmonella* is associated with reduced reproductive success in female alpine chamois (*Rupicapra rupicapra*). Can J Zool 86:1111–1120

Pitkanen T (2013) Review of *Campylobacter* spp. in drinking and environmental waters. J Microbiol Methods 95:39–47

Poeta P, Radhouani H, Pinto L et al (2009) Wild boars as reservoirs of extended-spectrum beta-lactamase (ESBL) producing *Escherichia coli* of different phylogenetic groups. J Basic Microbiol 49:584–588

Porrero C, Mentaberre G, Sánchez S et al (2013) Methicillin resistant *Staphylococcus aureus* (MRSA) carriage in different free-living wild animal species in Spain. Vet J 198:127–130

Pozio E (2005) The broad spectrum of *Trichinella* hosts: from cold- to warm blooded animals. Vet Parasitol 132:3–11

Pozio E, Zarlenga DS (2013) New pieces of the *Trichinella* puzzle. Int J Parasitol 43:983–997

Pritchard GC, Williamson S, Carson T et al (2001) Wild rabbits – a novel vector for verocytotoxigenic *Escherichia coli* O157. Vet Rec 149:567

Ramanzin M, Amici A, Casoli C et al (2010) Meat from wild ungulates: ensuring quality and hygiene of an increasing resource. Ital J Anim Sci 9, e61

Richomme C, Boadella M, Courcoul A et al (2013) Exposure of wild boar to *Mycobacterium tuberculosis* complex in France since 2000 is consistent with the distribution of bovine tuberculosis outbreaks in cattle. PLoS One 8, e77842

Robino P, Tomassone L, Tramuta C et al (2010) Prevalence of *Campylobacter jejuni*, *Campylobacter coli* and enteric *Helicobacter* in domestic and free living birds in North-Western Italy. Schweiz Arch Tierheilkd 152:425–431

Rodríguez S, Bezos J, Romero B et al (2011) *Mycobacterium caprae* infection in livestock and wildlife, Spain. Emerg Infect Dis 17:532–535

Romig T, Dinkel A, Mackenstedt U (2006) The present situation of echinococcosis in Europe. Parasitol Int 55:S187–S191

Ruiz-Fons F, Astobiza I, Barandika JF (2010) Seroepidemiological study of Q fever in domestic ruminants in semi-extensive grazing systems. BMC Vet Res 6:3

Ruiz-Fons F (2012) *Coxiella burnetii* infection. In: Gavier-Widen D, Meredith A, Duff JP (eds) Infectious diseases of wild mammals and birds in Europe, 1st edn. Wiley-Blackwell, Chichester, pp 409–412

Ruiz-Fons F (2015) A review of the current status of relevant zoonotic pathogens in wild swine (*Sus scrofa*) populations: changes modulating the risk of transmission to humans. Transbound Emerg Dis (in press) doi: 10.1111/tbed.12369

Rybarikova J, Dolejska M, Materna D et al (2010) Phenotypic and genotypic characteristics of antimicrobial resistant *Escherichia coli* isolated from symbovine flies, cattle and sympatric insectivorous house martins from a farm in the Czech Republic (2006–2007). Res Vet Sci 89:179–183

Saito M, Koike F, Momose H et al (2012) Forecasting the range expansion of a recolonising wild boar *Sus scrofa* population. Wildl Biol 18:383–392

Sánchez S, García-Sánchez A, Martínez R et al (2009) Detection and characterization of Shiga toxin-producing *Escherichia coli* other than *Escherichia coli* O157:H7 in wild ruminants. Vet J 180:384–388

Sánchez S, Martínez R, García A et al (2010) Detection and characterisation of O157:H7 and non-O157 Shiga toxin-producing *Escherichia coli* in wild boars. Vet Microbiol 143:420–423

Scaife HR, Cowan D, Finney J et al (2006) Wild rabbits (*Oryctolagus cuniculus*) as potential carriers of verocytotoxin-producing *Escherichia coli*. Vet Rec 159:175–178

Schönberg-Norio D, Johanna Takkinen J, Hänninen ML et al (2004) Swimming and *Campylobacter* Infections. Emerg Infect Dis 10:1471–1477

Schulp CJE, Thuiller W, Verburg PH (2014) Wild food in Europe: a synthesis of knowledge and data of terrestrial wild food as an ecosystem service. Ecol Econ 105:292–305

Schwaiger K, Stierstorfer B, Schmahl W et al (2005) Survey on bacterial CNS infections in roe deer (*Capreolus capreolus*), red deer (*Cervus elaphus*) and chamois (*Rupicapra rupicapra*) in Bavaria. Berl Munch Tierarztl Wochenschr 118:45–51

Smith S, Wang J, Fanning S et al (2014) Antimicrobial resistant bacteria in wild mammals and birds: a coincidence or cause for concern? Ir Vet J 67:8

Strachan NJ, Rotariu O, Smith-Palmer A et al (2013) Identifying the seasonal origins of human campylobacteriosis. Epidemiol Infect 141:1267–1275

Süld K, Valdmann H, Laurimaa L et al (2014) An invasive vector of zoonotic disease sustained by anthropogenic resources: the raccoon dog in Northern Europe. PLoS One 9, e96358

Tham W, Bannerman E, Bille J et al (1999) *Listeria monocytogenes* subtypes associated with mortality among fallow deer (*Dama dama*). J Zoo Wildl Med 30:545–549

Vargas RHM, Reich F, Klein G et al (2013) Bacterial contamination and antibiotic resistance of isolates from packaged game meat. Fleischwirtschaft 93:179–182

Vicente J, Höfle U, Garrido JM et al (2006) Wild boar and red deer display high prevalences of tuberculosis-like lesions in Spain. Vet Res 37:107–119

Vicente J, Barasona JA, Acevedo P et al (2013) Temporal trend of tuberculosis in wild ungulates from Mediterranean Spain. Transbound Emerg Dis 60:S92–S103

Vieira-Pinto M, Morais L, Caleja C et al (2011) *Salmonella* sp. in game (*Sus scrofa* and *Oryctolagus cuniculus*). Foodborne Pathog Dis 8:739–740

Wacheck S, Fredriksson-Ahomaa M, Konig M et al (2010) Wild boars as an important reservoir for foodborne pathogens. Foodborne Pathog Dis 7:307–312

Wahlström H, Tysen E, Engvall EO et al (2003) Survey of *Campylobacter* species, VTEC O157 and *Salmonella* species in Swedish wildlife. Vet Rec 153:74–80

Wahlström H, Lindberg A, Lindh J et al (2012) Investigations and actions taken during 2011 due to the first finding of Echinococcus multilocularis in Sweden. Euro Surveill 17:20215

Waldenström J, Broman T, Carlsson I et al (2002) Prevalence of *Campylobacter jejuni*, *Campylobacter lari* and *Campylobacter coli* in different ecological guilds and taxa of migrating birds. Appl Environ Microbiol 68:5911–5917

Waldenström J, Axelsson-Olsson D, Olsen B et al (2010) *Campylobacter jejuni* colonization in wild birds: results from an infection experiment. PLoS One 5, e9082

Widén F, Meredith A, Weissenböck H et al (2012) Other virus infections. In: Gavier-Widen D, Meredith A, Duff JP (eds) Infectious diseases of wild mammals and birds in Europe, 1st edn. Wiley-Blackwell, Chichester, pp 249–262

Wiratsudakul A, Sariya L, Prompiram P et al (2012) Detection and phylogenetic characterization of hepatitis E virus genotype 3 in a captive wild boar in Thailand. J Zoo Wildl Med 43:640–644

Woolhouse MEJ, Haydon DT, Antia R (2005) Emerging pathogens: the epidemiology and evolution of species jumps. Trends Ecol Evol 20:238–244

Wuthe HH, Schonberg A (1999) Listeriosis in the European brown hare in northern Germany. Berl Munch Tierarztl Wochenschr 112:98–99

Yegorova I, Selyaninov J, Fertickov V (2012) *Listeria* in the Wildlife of Russia. In: Romano A, Giordano CF (eds) Listeria infections: epidemiology, pathogenesis and treatment. Nova, Hauppage, NY, pp 167–176

Zaytseva E, Ermolaeva S, Somov GP (2007) Low genetic diversity and epidemiological significance of *Listeria monocytogenes* isolated from wild animals in the far east of Russia. Infect Genet Evol 7:736–742

Chapter 4
Microbiological Hazards of Wild Birds and Free-Range Chickens

Susan Sanchez, Monique França, and Nicole M. Nemeth

Abstract Zoonotic diseases are those that can be transmitted to people from animals. Back yard chickens are a reservoir of multiple zoonotic agents that can be transmitted to people directly through bird handling, eating meat or eggs, or indirectly by infecting pets that can then bring the pathogens into the home and put them in contact with other members of the family. Wild birds do also carry zoonotic pathogens that can infect us directly or indirectly by first infecting our back yard chickens, pets or other livestock such as pigs. With the increase in hobby-back yard chicken farms and wild bird fanciers it is important for the public to understand the risks associated with this practices. In this chapter we review the microbiological hazards of wild birds and back yard chickens to human health. From well known enteric pathogens such as Salmonella to vector borne viruses, this chapter hopes to reflect the microbiological risks of keeping chickens and being in contact with wild birds as a hobby or business. Wild birds and domestic poultry can transmit viral diseases to people. In fact, waterfowl have long been considered the natural reservoirs of Influenza A viruses (IAV). The large population of these natural reservoirs and the ability of IAV to undergo genetic mutations and reassortment make Influenza A a noneradicable zoonosis. Wild birds are also the natural reservoirs of Newcastle disease viruses (NDV) which can transmit to domestic poultry and cause outbreaks of high mortality after infection with virulent strains. Humans can become infected with these viruses after direct contact with infected birds and usually develop a self-limiting conjunctivitis.

Keywords Arbovirus • Avian influenza • Backyard chicken • Bacteria • Campylobacter • Campylobacteriosis • Hobby farm • One Health • Organic agriculture • Poultry • Salmonella • Salmonellosis • Wild bird • Zoonosis

---

S. Sanchez (✉) • M. França
College of Veterinary Medicine, The University of Georgia, Athens, GA, USA
e-mail: ssanchez@uga.edu

N.M. Nemeth
Ontario Veterinary College, University of Guelph, Guelph, ON, Canada

M. Jay-Russell, M.P. Doyle (eds.), *Food Safety Risks from Wildlife*,
Food Microbiology and Food Safety, DOI 10.1007/978-3-319-24442-6_4

Introduction

One Health Issues: New Food Fads, Habitat Encroachment, and Unrealized Risks

Several decades ago we lived with our animals in our homes and in our cities. As the industrial revolution progressed, the basis for infectious diseases and their transmission was elucidated, such that fewer food-producing animals were housed in cities and farmers struggled to produce enough animals to satisfy the demand for more meat that was consumed in these new metropolis. To increase productivity and efficiency farmers moved to preferentially grow on type of livestock depending on the local taste, type of farmalnd, water availability, transport distances, demand and cost. Farms that only produce one commodity are now the norm. Meat is relatively inexpensive in developed countries due to mass production, resulting in profit margins that are relatively small and rely on large numbers sold, low production costs per unit, and reduced numbers of condemned carcasses at slaughter and of losses at the farm. This is even more of a factor in integrated chicken production, where avoiding disease through biosecurity measures has become the standard in most countries. Birds are housed in large buildings where entry and exit are highly controlled, exposure to wild birds and rodents is eliminated or limited, and cost-effective mass vaccinations are administered to flocks. Diseases that historically had adverse effects on entire flocks, although not eliminated, have been greatly reduced, and the cost of meat production per pound has been greatly reduced. This progress in food production has had the unintended but beneficial consequence of reducing for the general population direct contact with live chickens. Furthermore, the risk of consumers acquiring diseases from foods has decreased so much over the years that acquiring illnesses from foods is no longer accepted. The presence of zoonotic bacteria (disease-causing bacteria that can be transmitted from animals to people) not pathogenic to birds is controlled in poultry across the food chain from the farm, at the processing plant to reduce the contamination of the carcasses, and at the supermarket at retail (Fortin 2013; USDA). Pathogen testing is used as an assessment of the efficacy of control measures that are in place to mitigate contamination on processed poultry. This testing is not limited to the presence or absence of pathogens, but also addresses the presence of antimicrobial resistance to clinically critical antimicrobials used to treat human diseases caused by these zoonotic agents (USDA 2014).

Over the past few years, there has been a movement toward buying food locally to promote the local economy, to enjoy better tasting meat and produce, and for a perceived reduced risk of microbial and chemical hazards. This is also part of the recently realized interest in food that has been produced organically, and where animals are given ample space to carry out their normal behaviors. Consumers are increasingly seeking food produced without additives such as use antibiotics or hormones, especially those used for growth promotion. Organic food is sold at supermarkets for a higher price, and is a common feature at farmers' markets and local stores. Many studies have addressed pathogen contamination, such as *Salmonella*, of chickens raised in an organic farm environment, to determine if these birds are less contaminated than their conventionally grown counterparts. Results have been inconsistent and sometimes contradictory, but most have revealed that the organically grown birds have a higher prevalence of *Salmonella*, even if they are perceived as tasting better (Bailey and Cosby 2005).

The changes in how food is perceived in developed parts of the world are still evolving; however, there is a recent movement to grow chickens and produce eggs in the backyard. This has been facilitated by the dissemination of information regarding the maintenance of animals and where to buy them via the Internet. In the United States, keeping certain types of animals in urban areas is regulated by local ordinances to avoid nuisance odors and noises. These ordinances are being challenged by many homeowners who want to grow their own food (Kaiser 2013).

Another ongoing risk is the breeding and possession of fighting cocks, which is considered a sport in many countries including Mexico, Central and South America. This practice has been banned for many years in the United States, but such activities continue to occur illegally. There is a risk that the movement of these birds across country borders and states may lead to transmission of pathogens to not only the people that keep these birds, but also to poultry grown in the vicinity.

The association of birds with humans is not limited to the use of birds for food. Birds have been bred and raised for their beauty, singing, flying, and homing abilities. Birds migrate annually, some over long distances, and they carry with them pathogens (West Nile virus, Avian Influenza virus, *Salmonella*, *Campylobacter jejuni*, and others). Climate change is affecting the geographical distribution of some of these pathogens and their vectors. Migratory birds can transport such pathogens, and together with anthropogenic changes, some zoonotic pathogens are becoming an increased risk for human disease (Fuller et al. 2012). Geese, ibises, and other wild birds are attracted to parks, golf courses, and many recreational areas where there is a safe environment and an abundance of human food (Cole et al. 2005). Here, the large accumulation of birds and inadequate sanitation of water and walking pathways put birds' excrement and people in close contact. Observing and photographing wild birds are a hobby of many avid birdwatchers, with 41.3 million people in the United States participating in such activities. Bird feeders in backyards lend themselves to this past time, and many wild birds are provided food and water in backyard feeders year around, thereby increasing the traffic of birds near homes. This pastime brings people frequently in contact with feeders and water bowls that are contaminated with bird feces when they are cleaned or replenished. This puts people in direct contact with pathogens such as *Salmonella* and *Campylobacter*. Many studies, mostly conducted in the UK, have revealed that the strains of *Salmonella* that are prevalent in backyard birds are also of the same genetic fingerprints as those most commonly isolated from people in the same region (Lawson et al. 2010, 2011, 2014), suggesting that people are being infected by enteric pathogens from contact with birds or their feeders. Cats are predators when outdoors and hunt wild birds at their feeders. They catch and eat birds that may carry enteric pathogenic bacteria that colonize cats without causing clinical symptoms. However, cats can shed the bacteria in feces for a few weeks or even months. If indoor–outdoor cats, they can transmit the pathogens to kitchen counters, floors, and other locations in the home, subsequently magnifying the birdfeeder-pathogen effect (Tauni and Osterlund 2000; Kock 2012). The connection between wild birds (as a pathogen source) and sick people may not be apparent, but is real. Bird feeders act to attract many birds that would not otherwise be in direct or indirect contact with each other, thereby enabling bacteria and viruses to spread across birds of the same and different species (Robinson et al. 2010). Not all wild animal species, including birds, are resistant to the same diseases, therefore some are more likely to become sick and amplify the pathogen than others. Furthermore, people are now expanding their living environments into previously sparsely inhabited forested and farm lands where the sylvatic cycle of diseases among wild birds via mosquitoes is common. Hence, people sitting in their backyards now can come into contact with mosquitoes that carry bird-borne pathogens disease.

This chapter will focus on the potential microbiological challenges associated with the practices described above.

Pathogens

There are many microorganisms carried by birds and poultry that are zoonotic pathogens. The two that most commonly cause disease in people are bacteria, namely *Salmonella* and *Campylobacter*. However, the potential for direct transmission of many other microbes from wild birds or backyard poultry to people has not been extensively studied, even though bacterial pathogen contamination of backyard-produced meat and eggs is clearly an issue (Pollock et al. 2012).

Bacteria

Salmonella

Non-typhoidal salmonellosis is one of the leading causes of acute bacterial gastroenteritis not only in the United States, but also worldwide (Scallan et al. 2011; de Jong and Ekdahl 2006). In the United States, *Salmonella* is estimated to sicken more than a million people annually. *Salmonella* is a zoonotic pathogen that can be part of the normal gut microflora of many different healthy animals, including backyard chickens and wild birds (Fig. 4.1) (Hernandez et al. 2012; Horton et al. 2013; Langholz 2013) (Sanchez et al. 2002). Therefore, this bacterium can readily contaminate raw foods during the slaughter process of backyard poultry, and also contaminate fresh produce, directly from animal contact or when manure is used as fertilizer. Direct contact with birds, housing, feeders, and their surroundings can infect adults and children. Although little chicks and ducklings are cute and many

Fig. 4.1 Contact with birds is not usually thought of as a major health risk. Backyard chickens and wild birds can harbor microbes pathogenic to people, in particular to the very young and the elderly. These pathogens can be transmitted directly or indirectly by colonizing pets, which in turn can bring these pathogens directly into the owner's homes. Birds, and other infected wildlife in contact with the birds can contaminate rivers, lakes, wells, and fresh produce. Pathogens now carried by these hosts can travel to other surrounding areas and contaminate parks and public spaces (Illustration by Will McAbee and Brad Gilleland)

children handle them as pets in the home, outbreaks of salmonellosis have been associated with such behavior (CDC 2014e).

Salmonella is a genus of gram-negative bacteria in the Family *Enterobacteriaceae* (gut dwelling bacteria transmitted by the oral–fecal route). The genus*Salmonella* is divided to the species *Salmonella bongori* and *Salmonella enterica. Salmonella* comprises more than 2500 distinct serotypes. However, those associated with most foodborne outbreaks in the United States are consistently caused by only three top serotypes: Enteritidis, Typhimurium, and Newport (Crim et al. 2014). These same serotypes are among the top ten serotypes normally isolated from non-human sources.

With the exception of *Salmonella* Pullorum and Gallinarum, *Salmonella* does not typically cause disease in poultry, and most birds are asymptomatic carriers of *Salmonella*, with the exception of some very young stressed animals. Some serotypes, especially Enteritidis, can be transmitted vertically inside of hens' eggs and these bacteria can be isolated from freshly hatched chicks at 1 day of age (Liljebjelke et al. 2005).

Public Health Concerns for Salmonella

The US Centers for Disease Control and Prevention investigate annually many *Salmonella* outbreaks in the United States. CDC estimates that over 47 million people in the United States become ill annually with a foodborne illness, of which 11 % are caused by *Salmonella* from all sources. Salmonellosis is the leading cause of hospitalizations by foodborne pathogens, which occurs in 35 % of the cases. Most of these infections are food associated, but not limited to consumption of meat and poultry products and eggs. Fresh produce, ice cream, peanut butter, spices, and pet reptiles are examples of other vehicles. In 2011, an outbreak was traced to live birds, specifically ducklings and chicks (CDC 2011b). *Salmonella* Altona and Johannesburg were the serotypes involved; neither serotype being among the top serotypes isolated most years from human or non-human sources. This outbreak in relation to the total number of cases reported annually is small, with 96 illnesses in 20 different states being traced back to the same hatchery. Mail-order chicks and ducklings were the sources of this outbreak. The drivers for human infections were the new availability of Internet order live birds, and the increase in popularity of the backyard bird phenomena.

From 2011 to 2014, there have been eight separate outbreaks of salmonellosis of single or multiple serotypes, involving baby birds and ducklings shipped from several hatcheries across the country (Table 4.1). Reported outbreaks of salmonellosis, as with cases of gastroenteritis in general, are the tip of the iceberg. Contact with animals and animal products has been a risk factor for *Salmonella* infections for farm workers, and is now also a risk for those that buy and keep backyard chickens and other pet poultry (Habing et al. 2014). Education on the risk of handling and raising birds for egg production or as a hobby has been initiated as a collaboration by two federal agencies, i.e., USDA and CDC (http://www.cdc.gov/Features/SalmonellaPoultry/) and the chick producers. This information has been produced in many formats for different audiences and all warn of the risks to both children and adults (http://www.cdc.gov/zoonotic/gi/animals.html).

In wild birds, *Salmonella*, and most often *Salmonella* Typhimurium, can present as a non-enteric lethal disease threat (Tizard 2004). When salmonellosis occurs, it results in large seasonal outbreaks both in aquatic waterfowl and songbirds, and these infections have been well documented in North America and Europe (Friend et al. 1987; Daoust et al. 2000; Friend and McLean 2002; Giovannini et al. 2013). Epizootics due to salmonellosis are remarkable due to the large number of birds involved, normally totaling more than tens of thousands per exposure when they occur. Backyard feeders act as a catalyst of pathogen transmission and a potential source of zoonotic disease. *Salmonella* Typhimurium has a wide host range in birds, and the advent of new molecular typing methods has revealed that a very narrow range of *Salmonella* Typhimurium strains have been associated with multiple outbreaks, suggesting specific host preferences (Alley et al. 2002; Rabsch et al. 2002; Hernandez et al. 2012). Furthermore, wild birds can be vehicles of other livestock-associated *Salmonella* strains such as those found near pig farms which

Table 4.1 Salmonellosis outbreaks in the United States associated with live birds, chicks, and ducklings between 2011 and 2014

	# Affected	# States involved	% Children	% Hospitalized	*Salmonella* serotypes
2014					
	363	43		35	*S.* Infantis *S.* Newport *S.* Hadar
2013					
#1	356	39	57	17	*S.* Typhimurium
#2	158	30	41	28	*S.* Infantis *S.* Lille *S.* Newport *S.* Mbandaka
2012					
#1	46	11	30	28	*S.* Hadar
#2	93	23	38	23	*S.* Montevideo
#3	195	27	33	34	*S.* Infantis *S.* Lille *S.* Newport
2011					
#1	68	20	31	31	*S.* Altona
#2	28	15	75	41	*S.* Johannesburg

Four hatcheries across the country were involved. No live birds were associated with salmonellosis outbreaks in the previous 5 years. CDC (2014e)

carry the same phage types and pulsed-field gel electrophoresis (PFGE) patterns as those isolated from pigs in the vicinity (Andres et al. 2013).

Transmission of *Salmonella* Typhimurium from wild garden birds to people has been described in Europe and New Zealand (Penfold et al. 1979; Tauni and Osterlund 2000; Alley et al. 2002). Later studies have further revealed that both in the USA and Great Britain specific strains, as determined by PFGE and multiple locus variable-number-tandem-repeat analysis (MLVA), were the same as those isolated from ill people within the same geographical region during the same time period (Lawson et al. 2011, 2014; Hernandez et al. 2012). It appears from these findings that the continuing increase in the popularity of bird watching and the number of bird feeders over the past decade has put more and more people at risk of contracting salmonellosis from wild birds or their droppings, directly or indirectly, through contamination of food in homes, as no contamination of other feed stuffs has been documented (Kapperud et al. 1998; Davies et al. 2009).

Salmonella is an enteric pathogen in people that is commensal in the gastrointestinal tract of most poultry, but causes disease in songbirds and can also be carried, although only temporarily, by gulls depending on their proximity to sewage or sewage waste sites (Refsum et al. 2002; Palmgren et al. 2006; Hughes et al. 2008; Kinzelman et al. 2008; Rodriguez et al. 2012). An epidemiological association has been determined between produce grown in the Eastern shore of the USA, specifically tomatoes, and contamination with *Salmonella* Newport and gulls carrying *Salmonella* in the same geographical area. The isolates from the tomatoes and several from the gulls had the same PFGE patterns. Furthermore, the isolation of such strains can persist over a 3-month sampling period. This persistence could enable the spread of *Salmonella* from landfills to gulls to produce and eventually sicken people (Gruszynski et al. 2014).

Pigeons are ubiquitous birds in cities where they are often a public nuisance. They are not afraid to raid food from tables, and land on people that may have food, although they are not as aggressive as seagulls. *Salmonella* prevalence in this group of birds is low, from 0.8 to 3.7 %, depending on location and country (Teske et al. 2013; Gargiulo et al. 2014). No studies thus far have associated human infections of salmonellosis with *Salmonella* isolates from pigeons either directly or through epidemiological evidence.

Wild birds carry *Salmonella* and are direct sources of sporadic salmonellosis in humans via oral–fecal transmission, or indirectly though the contamination of food, water (including wells), and through pets such as cats. These pets hunt frequently around bird feeders and homes getting the slow, sick, and baby birds that can carry *Salmonella*, thereby infecting the cat even if only transiently. Owners can acquire *Salmonella* from their cats, when these animals shed *Salmonella* in the home environment. Transmission of *Salmonella* during a spring birdfeeder cleanup can occur as can transmission at a seaside café outing during a warm afternoon as a seagull tries to take fries from a table of unsuspecting teens while pigeons search for crumbs on the floor and on nearby tables. Clearly, there are many opportunities for transmission of *Salmonella* from wild birds as exemplified in these scenarios. If the disease is mild, those infected typically do not seek medical attention; however, if the symptoms are severe and there is a visit to the doctor, there is a probability that fecal samples will be obtained for culture. Even if *Salmonella* is recovered, it may not be linked to other cases, therefore, it would be considered a sporadic case of unknown origin.

In summary, backyard chickens and wild birds are sources of salmonellosis (CDC 2011b, 2014e). This is a well-known fact and greater efforts by several federal agencies and commodity groups are being made to inform the public of the potential dangers associated with bird-related hobbies. As interest and social movements evolve to bring us closer to the source of our food by producing our own food, more needs to be done to distribute information regarding the potential health risks and mitigation strategies to prevent illnesses associated with keeping poultry in backyards and those associated with bird feeding and bird watching.

Antimicrobial Resistance

Antimicrobial resistance is one of the major One Health problems of the twenty-first century. Zoonotic pathogens such as *Salmonella* and *Campylobacter* from livestock and poultry production farms are monitored by the industry and through national federal programs such as the National Antimicrobial Resistance Monitoring system (NARMS). The NARMS program samples at slaughterhouses and at the retail level to in part provide an assessment of the efficacy of contamination reduction measures from the processing plant to retail. In the United States in 2011, the NARMS study revealed that the top serotypes for human salmonellosis were Enteritidis, Typhimurium, Newport, I 4, [5],12:i:- and Infantis. Of these, Typhimurium and Newport had the highest levels of multidrug resistance (USDA 2014). For animals, the 2011 NARMS study revealed that Kentucky, Enteritidis, Heidelberg, Typhimurium var 5-, Infantis and Typhimurium were the most commonly isolated *Salmonella* serotypes from chickens. Their multidrug resistance levels were much less than those determined for human isolates, with the highest resistance present in a single isolate to be ≥3 antimicrobial classes in 7.9 % of the chicken isolates of *S.* Typhimurium. Resistance to ACSSUTAxCx was 0.4 %, which was present in Newport and Typhimurium isolates, including var 5- (USDA 2014).

Backyard chickens are purchased from online sites or local farm stores; nevertheless, these chickens can be traced back to hatcheries and can harbor the same *Salmonella* serotypes with the same antibiotic resistance profile as those that are intended for human consumption. This presents a risk for those that raise them, including the most susceptible groups, i.e., children and the elderly. Habing et al. (2014) described a very detailed study in which shipments of hatching poultry were assayed for *Salmonella* over a 1-year period (2013). These young birds were of the type that can be purchased by those starting or re-supplying their backyard flocks. Results revealed that 10.5 % of the *Salmonella* serotypes isolated matched those associated with multi-state outbreaks. Nineteen percent of these *Salmonella* isolates were resistant to more than one antimicrobial, but just one of them was resistant to multiple antibiotics, including ceftriaxone, which is the drug of choice for children for treating acute salmonellosis. This isolate was serotype Kentucky that carried the gene bla_{CMY}. This same serotype is widely circulating in Europe and carries multiple resistance genes for cephalosporins and β-lactamases (Boyle et al. 2010; Le Hello et al. 2013; Harrois et al. 2014). One particular clone, ST198, has been determined for the first time to carry an IncA/C conjugative plasmid that harbors the relatively rare bla CTX_{-M-25} gene that confers resistance to β-lactamases (Wasyl et al. 2015).

Many antimicrobial resistance genes present in *Salmonella* are located in mobile elements such as plasmids and integrons and can move in and out of certain serovars of *Salmonella* into one another and from other *Enterobacteriaceae* with relative ease.

Geographical distribution of antimicrobial resistance in *Salmonella* is tightly associated with the serotype, and differences in antimicrobial susceptibility of *Salmonella* in wild birds in different parts of the world can be striking. A recent study revealed that there is a 67 % greater likelihood that *Salmonella* is carrying resistance to one or more antibiotics if the crows are from Europe than if they are from the USA. Interestingly, the serovars isolated were different in both locations (Janecko et al. 2014). A study in Spain revealed an association between livestock and resistance in *Salmonella* isolates from wild birds living within 200 m of pig farms. *Salmonella* Typhimurium-specific phage type U311 and serotype 1,4,5,12:i:- isolated from birds, pigs, and the environment were found to be resistant to three or more antibiotics (Andres et al. 2013). *Salmonella* isolated from wild birds (Kite, *Milvus migrans*) in Germany were recently described as producing New Delhi Metalobectalactamse 1 (NDM-1), a newer class B carbapenemase, that renders all carbapenems (imipenem, meropenem, doripenem, and ertapenem) useless (Fischer et al. 2013). Furthermore, this enzyme confers resistance to penicillins, cephalosporins, and carbapenems. These β-lactam drugs are considered of critical importance in the treatment of people with healthcare-associated infections and severe community-acquired infections by *Enterobacteriaceae*. Resistance to carbapenems has been rare until 10 years ago, and is produced by $bla_{NMD\text{-}1}$ that can be found in plasmids and the chromosome of many gram-negative bacteria, including *Pseudomonas aeruginosa*, *Klebsiella pneumoniae*, *Escherichia coli*, *Vibrio cholerae*, and *Acinetobacter* spp. This gene has no specificity for any bacterial species or bacterial clone, being readily transferred with speed and agility. Its origin has been traced to the Indian subcontinent as far back as 2008. Since then, it has moved to several hot spots in the Balkans and the Middle East where it is very prevalent (Patel and Bonomo 2013). The first report of $bla_{NMD\text{-}1}$ in *Salmonella* in Germany was in an isolate of serotype Corvalis. This is an unusual serotype in Europe, but is reported with increasing frequency in other parts of the world, including southeast Asia where it has been isolated from pigs and pork, and is emerging in North Africa where it has been isolated from many different animal species (Fischer et al. 2013). The kite from which this *Salmonella* isolate was obtained is a migratory bird that spends its winters in Northern Africa and comes to Germany for the summer months via the east coast of the Black Sea. It has been postulated that the $bla_{NMD\text{-}1}$ gene-carrying *Salmonella* isolate from this bird was picked up during transit either at the beginning of the trip or during rest stops via the Balkans (Fischer et al. 2013). Wild birds, including migratory birds, cannot only transfer pathogens across regions, countries, and across continents, but also can disperse with them antimicrobial resistance. Wild birds are a direct source of *Salmonella* contamination for produce, water, and other animals, including companion animals. Therefore, the risk of infection by *Salmonella* for people from wild birds can directly occur from handling birds or their feces, indirectly by eating contaminated produce, or via their pets when they bring *Salmonella* into the home.

Campylobacter

Campylobacteriosis is estimated to be the third leading cause of bacterial gastrointestinal disease in the USA after *Salmonella* and *Clostridium perfringens* (Scallan et al. 2011). In Europe, *Campylobacter* is the leading cause of bacterial gastroenteritis (de Wit et al. 2000), as it is for most of the rest of the world (WHO 2013a). Members of the *Campylobacter* genus are gram-negative, motile, thin, curved rods that require for growth a microaerophilic environment. Within the genus are sixteen species most commonly pathogenic to people including the thermophilic species, i.e., *C. jejuni* and *C. coli*. Both *C. jejuni* and *C. coli* are part of the normal flora of wild and domestic birds (Hughes et al. 2009; Teske et al. 2013), including poultry where they do not cause overt disease, except for hepatitis in ostriches (Stephens et al. 1998). Recent studies by Waldenstrom et al. (2010) have revealed that strains of *C. jejuni* isolated from birds are better adapted to colonize wild birds than *C. jejuni* isolates originally isolated from people, and that the bird isolates colonize transiently, i.e., for about a week, whereas the human isolate failed to colonize at all. More research on strain diversity and their ability to colonize wild birds and domestic poultry is needed to better elucidate the epidemiology of *C. jejuni*.

Public Health Concerns

The actual incidence of campylobacteriosis is not well known, due to the low frequency of doctor visits by sick patients, the few cultures requested by physicians for these ill patients, and the difficulty in culturing this microbe (Spencer et al. 2012). Although there are no definitive data, it is estimated that the incidence of illness is 4.4–9.3 per 1000 individuals around the world. Gastroenteritis symptoms are acute, and most patients readily recover without the need for treatment. Nevertheless, the possibility of sequelae to infection is a real concern. These sequelae include Guillian–Barré syndrome (GBS), reactive arthritis (RA), and irritable bowel syndrome (IBS) (WHO 2013a). These are autoimmune diseases that can be triggered by *Campylobacter* infection and appear many days after the diarrheic episode. In the case of GBS, the host immune system attacks the nerves, causing tingling and numbing of the extremities that may progress to total paralysis. Up to 40 % of all cases of GBS in the developed world are attributed to a previous infection with *Campylobacter* (Baker et al. 2012). The other two syndromes have recently been recognized and are more difficult to diagnose; nevertheless, there is evidence indicating that RA is diagnosed in 1–5 % of patients that have been infected with *Campylobacter*, and IBS in 36 % of patients that had had clinical campylobacteriosis 1–2 years prior to presentation of IBS (WHO 2013a).

Thermophilic campylobacters are present in large numbers (ca. 10^6 CFU/g) in the GI tract of chickens, and contaminated chicken meat and eggs are major routes by which they are spread to people. Backyard chickens are sometimes raised for food, but mostly for egg production. Egg handling and litter removal are principal sources for cross-contamination of foodstuffs. Wild birds are also a vehicle for transmitting *Campylobacter* to people during the replenishment of birdfeeders and bird water baths. Runoff from chicken houses is known to contaminate waterways, and wild birds (i.e., seagulls, geese, and ducks) directly contaminate waterways and beaches (Kinzelman et al. 2008; Van Dyke et al. 2010). *Campylobacter* carriage and shedding by seagulls has been correlated with livestock, and food refuse contamination (Ramos et al. 2010). *Campylobacter*-infected wild birds can also contaminate home-reared chickens. Although this is possible, an in-depth study by Colles et al. (2008b) failed to obtain the same genotypes of *Campylobacter* in wild birds, i.e., geese and starlings, and free-range chickens during the grow-out period, unless colonization of the wild birds was transient and the genotypes of isolates were constantly replaced after a short colonization period and were only colonized by strains that are host adapted (Waldenstrom et al. 2010). Some isolates of *C. jejuni* in the UK had the same sequence types as those from people suffering from campylobacteriosis in the same geographical area, as determined by multilocus sequence typing (MLST) of seven different genes. This association revealed that wild birds could be a source of *C. jejuni* for people (Colles et al. 2008). One recent study conducted in Alaska revealed a link between isolates of *C. jejuni* from people, those obtained from fresh peas, and those obtained from Sandhill cranes that were in the proximity of the pea-growing areas (Kwan et al. 2014).

Antimicrobial Resistance

In the United States in 2012, greater than 86 % of the species of *Campylobacter* isolated from symptomatic people were *C. jejuni* and about 10 % were *C. coli* (Crim et al. 2014; Johnson et al. 2014); CDC 2014b). Antibiotic susceptibility testing of *Campylobacter* isolates obtained in 10 different states that are part of CDC's FoodNet (CDC 2014a, b, c, d, e, f, g, h, i) is conducted by the NARMS program. Resistance value interpretations for this genus have changed with time from the use of clinical to epidemiological cut-off values (ECOFF) values. Based on these epidemiologically more relevant values, the percentage of *Campylobacter* isolates resistant to ciprofloxacin, gentamycin, and erythromycin in 2012 were 25, 0, and 2 % for *C. jejuni*, and 34, 6, and 9 %, respectively, for *C. coli* (CDC 2014a, b, c, d, e, f, g, h, i; CLSI 2010–2011). During 2011, of 577 *Campylobacter* isolates tested for antimicrobial resistance by the NARMS animal program, 60 % were identified as C. *jejuni*, and 40 % as *C. coli*. Using CLSI (Clinical and Laboratory Standards Institute, M100 documents) breakpoints, 19, 0.3, and 0.6 % of the *C. jejuni* isolates and 28, 6, and 3 % of the *C. coli* were resistant to ciprofloxacin, gentamycin, and erythromycin, respectively (USDA 2014; CLSI 2010–2011). Overall, the chicken isolates were less resistant than the human isolates. Scant information is available on the antibiotic susceptibility of campylobacters isolated from backyard chickens, but it is reasonable to believe that if these birds are purchased online or from a farm store, the *Campylobacter* isolates will have similar antibiotic-resistance profiles as those present in other chicks from the same hatchery of origin (Anderson et al. 2012). For *C. jejuni* isolates obtained from free-range chickens in Spain from 2003 to 2005, 58 % were resistant to ciprofloxacin while none were resistant to erythromycin. Gentamycin resistance was not tested in this study (Oporto et al. 2009). Bester and Essack (2012) determined that in South Africa, free-range-produced broilers had less ciprofloxacin and erythromycin resistance than their conventionally produced counterparts.

Many studies have been conducted on the prevalence of *Campylobacter* in wild birds, but very few have addressed the issue of antimicrobial resistance. This is in part because of the complexity of the test protocol. Waldenstrom et al. (2005) in Sweden tested thrushes, shorebirds, and raptors for *Campylobacter* and determined their antibiotic susceptibility profiles. Ciprofloxacin resistance was detected in only 0.7 of the isolates, and all were from raptors (Waldenstrom et al. 2005). The overall resistance of campylobacters was low compared to previous studies of wild birds, but very similar to that of isolates from poultry in the same region during that same time period; however, the wild birds were migratory and could have obtained resistant isolates in other parts of the world. A recent study on the role of wild birds as a source of *Campylobacter* contamination of fresh peas associated with a foodborne outbreak in Alaska revealed that of all the isolates tested, only one was resistant to ciprofloxacin (Kwan et al. 2014). Hence, although wild birds can be directly or indirectly a source of *Campylobacter* infections, for the most part, *Campylobacter* isolates from wild birds are likely to be responsive to first-line antibiotics.

Campylobacter in poultry is known to be transmitted vertically as well as horizontally (Idris et al. 2006). Vertical transmission is an important factor to consider in the case of backyard chickens, as they can harbor strains of *Campylobacter* present in the progeny flock from the farm where they originated, including strains with antibiotic resistance genes. In addition to wild birds, water and the environment may also be sources of antibiotic-resistant *Campylobacter*.

Other Bacteria

Other enteric bacteria that can cause human disease such as *Yersinia enterocolitica*, *Listeria monocytogenes*, pathogenic *E. coli*, and *Clostridium perfringens* can be found in both wild birds and backyard chickens. But, in general, these other bacterial species are not prevalent in birds (Fukushima and Gomyoda 1991; Capita et al. 2002; Asaoka et al. 2004; Kobayashi et al. 2009; Cortés et al. 2010; Catherine Racicot et al. 2012; Mete et al. 2013).

Viruses

Influenza A

Influenza A viruses (IAV) are negative-sense, single-stranded, and segmented RNA viruses from the *Orthomyxoviridae* family. IAV are classified into subtypes based on the surface glycoproteins hemagglutinin (HA) and neuraminidase (NA). There are currently 18 HA and 11 NA subtypes of IAV. Wild waterfowl in the orders *Anseriformes* and *Charadriiformes* are the natural reservoirs of all the avian HA (1–16) and NA (1–9) subtypes of IAV, and are considered the source of most IAV that infect domestic poultry and mammalian species (Webster et al. 1992; Krauss et al. 2004; Stallknecht and Brown 2008). The H17N10 and the H18N11 viruses have only been detected in bats and have gene segments that are distinct from other known IAV (Tong et al. 2012, 2013). IAV cause acute and contagious respiratory viral disease in humans, birds, horses and have also caused sporadic disease in marine mammals, dogs, and cats among other mammalian species (Webster et al. 1992; Yoon et al. 2014). These viruses are prone to mutations (antigenic drift) and are also capable of reassortment (antigenic shift) due to the segmented nature of the virus genome (Murphy and Webster 1996). Antigenic drift and shift of the HA and/or NA of IAV can result in the generation of pandemic viruses against which the human population may be immunologically naïve (Murphy and Webster 1996).

Influenza infections in humans may occur in two epidemiological forms: epidemics and pandemics (Nicholson 1998). Epidemics of IAV in humans are currently caused by antigenic variants of H1N1, H3N2, and their reassortant H1N2 IAV. There were four influenza A pandemics in humans during the last 100 years: the H1N1 "Spanish flu" pandemic in 1918, the H2N2 "Asian flu" pandemic in 1957, the H3N2 "Hong Kong flu" pandemic in 1968, and the H1N1 "swine flu" pandemic in 2009 (Smith et al. 2009a). The H1N1 "Spanish flu" virus had all gene segments traced back to an avian source, whereas the H2N2 and the H3N2 pandemic viruses were reassortant viruses from circulating avian and human influenza viruses (Scholtissek et al. 1978; Kawaoka et al. 1989; Webster et al. 1992; Belshe 2005; Taubenberger et al. 2005). The A(H1N1)pdm09 virus emerged in North America in 2009, causing approximately 12,469 deaths in the U.S. and an estimated 280,000 global deaths (Dawood et al. 2009; Shrestha et al. 2011). This virus was later determined to be a quadruple reassortant from circulating Eurasian and North American lineages of swine, avian and human viruses, and the gene combination of this virus had never before been identified in swine (Dawood et al. 2009; Smith et al. 2009b; Trifonov et al. 2009). Swine are believed to be "mixing vessels" for IAV because they can support reassortment after infection with viruses from different hosts and can generate IAV with pandemic potential (Kida et al. 1994).

Swine influenza (SI), also known as "swine flu," is an acute respiratory disease of pigs that usually causes high morbidity and low mortality in infected animals (CDC 2012). In the United States, the triple reassortant H1N1, H3N2, and H1N2 viruses have caused swine influenza in pigs in recent years (CDC 2012). The swine influenza viruses (SIV) that infect humans are called "variant viruses" and are designated by adding the letter "v" after the virus subtype (CDC 2014h). The SIV variant viruses H1N1v, H3N2v, and H1N2v have caused human infections and illnesses (CDC 2014h). The triple reassortant H3N2v virus has caused most reported human infections with SIV; this virus is endemic in U.S. swine and turkey populations (Yassine et al. 2007; CDC 2014g). Human infections with swine influenza viruses have been associated with direct or indirect contact with swine herds and visits to agricultural fairs (CDC 2014h).

IAV circulating in birds are known as avian influenza (AI) viruses. The AI virus subtypes H5, H6, H7, H9, and H10 have been sporadically transmitted from birds to humans and have resulted in mild to severe disease (Shortridge 1992; Peiris et al. 2007; Shi et al. 2013). Most human infections have been caused by highly pathogenic avian influenza (HPAI) H5N1. More than 660 cases of HPAI H5N1 in humans have been reported by the World Health Organization (WHO) since 2003, and approximately 59 % of the reported cases died (WHO 2014a). Procedures to minimize infection in poultry and transmission of these viruses to humans include depopulation of infected poultry, vaccination, and implementation of biosecurity measures in poultry farms. Preventing exposure of poultry to wild birds is paramount, as some species can asymptomatically shed HPAI viruses.

The presence of multiple natural reservoirs for IAV as well as the ability of IAV to undergo genetic mutations and reassortment make Influenza A a noneradicable zoonosis (Yen and Webster 2009). Surveillance, biosecurity, and elimination are key strategies to prevent influenza A outbreaks in poultry, whereas prophylaxis and control of human infections rely on vaccination and antiviral treatment. Antivirals available to treat human influenza infections include the M2 ion channel blockers amantadine and rimantadine, and the neuraminidase (NA) inhibitors oseltamivir and zanamivir (Monto 2003). The NA inhibitors are recommended for treatment and prophylaxis of influenza A as the ion-channel blockers can cause rapid emergence of resistant variants (Hayden et al. 1989; CDC 2014c). However, oseltamivir-resistant viruses have been detected in patients infected with H5N1 and also during the 2008–2009 influenza season, when a high prevalence (98 %) of H1N1 human influenza viruses was found resistant to this antiviral (de Jong et al. 2005; Le et al. 2005; Yen and Webster 2009). Seasonal human influenza is prevented by vaccination with inactivated or live-attenuated influenza vaccines (LAIV) containing the influenza A viruses H1N1 and H3N2, as well as one or two influenza B viruses (CDC 2014d). A recombinant influenza vaccine is also available (CDC 2014d). The seasonal influenza viruses continually undergo antigenic drift leading to the emergence of new strains and the need to review and update vaccines on an annual basis. Vaccines for HPAI H5N1 have been developed and are being stockpiled by the U.S. federal government in case of a human outbreak with sustained human-to-human transmission (CDC 2014f).

Preventative measures for personnel involved with SI and AI outbreaks and recommended by the Centers for Disease Control and Prevention (CDC) include vaccination with the seasonal influenza vaccine, phophylactic use of antiviral drugs and the use of appropriate personnel protective equipment, including disposable gloves, coveralls, shoe covers and the use of masks or particulate respirators (N-95, N-99, and N-100) (CDC 2006, 2011a).

Public Health Concerns - Avian Influenza

More than 105 bird species belonging to 13 different avian orders have been found to carry. AI viruses (Olsen et al. 2006), but wild waterfowl in the orders *Anseriformes* and *Charadriiformes* are considered the natural reservoirs for these viruses (Stallknecht and Brown 2008). In *Anseriformes*, most AI virus isolations are from dabbling ducks of the *Anas* genus and Mallard is the species most frequently detected with these viruses (Stallknecht and Shane 1988). In wild waterfowl, transmission of low pathogenicity avian influenza (LPAI) viruses occurs frequently mainly by the fecal–oral route in shared and contaminated aquatic habitats (Stallknecht and Brown 2008). LPAI viruses replicate in the intestinal and respiratory tracts and can be detected from both ends of these aquatic birds during wild bird surveillance (Yoon et al. 2014). Re-infection with the same virus and co-infection with different viruses in the same season occur frequently, and these birds usually remain asymptomatic (Sharp et al. 1997). Wild waterfowl transmit LPAI viruses to domestic waterfowl and gallinaceous poultry species by direct contact, ingestion of contaminated water, and indirectly via contaminated fomites (Suarez 2008).

HPAI viruses in wild birds are believed to be spillover events from outbreaks in domestic poultry. The very first report of AI in wild birds occurred in 1961 during a mortality event of Common Terns (*Sterna hirundo*) caused by a HPAI H5N3 virus in South Africa (Becker 1966). Prior to 2002, infections with AI viruses were generally asymptomatic in wild bird species. This situation changed in 2002, when two waterfowl parks in Hong Kong experienced high mortality in captive and wild aquatic birds infected with HPAI H5N1 virus (Ellis et al. 2004). In 2005, the Asian lineage of HPAI H5N1 viruses caused a large die-off of migratory wild birds in Qinghai Lake, China (Liu et al. 2005) and an estimated 10 % decrease in the global population of the Bar-headed Goose (*Anser indicus*) (Olsen et al. 2006). During 2005 and 2006, HPAI H5N1 viruses spread westward to Eurasia and Africa, possibly due to movement of infected wild birds as well as trade of live poultry and poultry products.

Human infections with HPAI H5N1 generally occur as a result of direct contact with infected poultry and contaminated environment (Beigel et al. 2005). There are rare reports of AIV transmission from wild waterfowl to humans which have generally been associated with direct contact during hunting and bird handling (Gill et al. 2006; Gilsdorf et al. 2006; Shafir et al. 2012).

LPAI viruses are occasionally transmitted from wild and domestic waterfowl to gallinaceous poultry, and infections in these birds can be subclinical or result in reductions in egg production, respiratory symptoms, and an increase in mortality. LPAI outbreaks have been more frequently reported in turkeys, less frequently in laying chickens, and rarely in other poultry species (Swayne et al. 2013). Turkeys are more susceptible to LPAI infections than chickens (Capua and Terregino 2009). Of all 16 HA subtypes, only the H5 and H7 are currently capable of becoming highly pathogenic viruses after circulation in domestic poultry. Infections with HPAI H5N1 viruses have caused devastating outbreaks in poultry as well as human infections and fatalities since 1997 (de Jong et al. 1997; To et al. 2001), and endemic infections with these viruses are currently found in poultry in Bangladesh, China, Egypt, India, Indonesia, and Vietnam (CDC 2014a). The high human population density and the presence of backyard poultry and live bird markets in these countries provide the opportunity for interspecies transmission and outbreaks of HPAI H5N1 in humans (Shortridge 1992). There is still no evidence of sustained human-to-human transmission for HPAI H5N1, although limited human-to-human transmission is believed to have occurred within some family clusters (Ungchusak et al. 2005; Wang et al. 2008). There is currently a great concern that HPAI H5N1 viruses can cause the next human influenza pandemic, as a few mutations in the HA and PB2 genes can result in efficient airborne transmission in ferrets (Herfst et al. 2012; Imai et al. 2012). HPAI H5N8 viruses geneticallly related to the Eurasian lineage of HPAI H5N1 viruses (Goose Guangdong lineage) were first detected in

January 2014 in South Korea in association with outbreaks of morbidity and mortality in wild birds and poultry flocks (Jeong et al. 2014). These HPAI H5N8 viruses reassorted with other wild bird AI viruses and spread along migratory flyways in late 2014, causing outbreaks in domestic poultry flocks and wild bird deaths in several countries in Europe, Asia (Adlhoch et al. 2014) and the worst animal disease outbreak ever documented in the U.S. The U.S. outbreak affected 21 states and caused the loss of more than 48.8 million birds, mostly commercial table egg layers and turkeys, as of mid-June 2015 (USDA-APHIS, 2015). Backyard poultry facilities have also been detected positive with these viruses. Wild bird migration along the Pacific, Central and Mississippi flyways is believed to be the source of HPAI H5N8 and H5N2 in different states; however, biosecurity lapses due to sharing of contaminated equipment and trucks between farms have also been implicated and likely contributed to the spread of these viruses in some cases. The risk of human infection with the Eurasian/American HPAI H5N8 and H5N2 viruses is considered low by public health agencies and these viruses have not been associated with human infections to date.

H7N9 AI virus has caused asymptomatic infections in poultry and a new human influenza outbreak in China since March 2013 (Gao et al. 2013; WHO 2013b). Phylogenetic analysis revealed that all genes of this virus were of avian origin (Gao et al. 2013). A total of 453 cases of human infection, including 175 deaths, have been caused by the H7N9 virus as of October 2, 2014 (WHO 2014b). This virus does not cause disease in poultry species and may be difficult to eradicate as it can be transmitted silently in these birds (Zhang et al. 2013). Human cases are characterized by a rapidly progressive pneumonia and the development of an acute respiratory distress syndrome leading to death in approximately 30 % of the cases (WHO 2013b; Yu et al. 2013). Transmission of H7N9 AI virus from poultry to humans is believed to result from direct contact with infected poultry and exposure to a contaminated environment, including visits to live bird markets (Chen et al. 2014; WHO 2014b). Limited human-to-human transmission has been suggested within family clusters in a few cases (Hu et al. 2014; Xiao et al. 2014). Some H7N9 isolates have acquired the ability to transmit via respiratory droplets among ferrets (ferrets are the laboratory animal model of choice for transmission experiments) (Richard et al. 2013), and there is a concern that this virus may reassort or acquire mutations that would allow sustained human-to-human transmission.

Human infections with H9, H10, and other H7 AI viruses have been sporadic. The H9N2 subtype of AI viruses is endemic in poultry in parts of Asia and the Middle East, and has caused sporadic human infections and mild influenza-like illness since 1999 (Peiris et al. 1999; Guo et al. 2007). H9N2 AI viruses with human-type receptor specificity have been detected in Asia (Matrosovich et al. 2001), and researchers speculate that these H9N2 viruses have the potential to evolve into a pandemic strain (Yen and Webster 2009; Imai and Kawaoka 2012). The H10N8, H10N7, H7N3, and H7N2 AI viruses have caused rare human infections characterized by conjunctivitis and influenza-like illnesses (Freidl et al. 2014). The H7N7 HPAI virus outbreak in The Netherlands, Belgium, and Germany in 2003 resulted in the culling of 30 million birds and at least 89 human cases (Fouchier et al. 2004; Stegeman et al. 2004). Most human infections with this virus resulted in conjunctivitis, but one fatal case of pneumonia and acute respiratory distress syndrome also occurred (Fouchier et al. 2004). This virus was most likely derived from a LPAI H7N7 virus that was detected in wild ducks that evolved to its highly pathogenic phenotype in domestic poultry (Fouchier et al. 2004).

The "Asian flu" pandemic of 1957 was caused by an H2N2 virus which has disappeared from the human population since the 1968 Hong Kong pandemic (Scholtissek et al. 1978), but this subtype still circulates in wild and domestic birds. Re-emergence of another H2N2 pandemic influenza is also currently considered a threat as individuals born after 1968 lack immunity to this virus. Additionally, the H2N2 avian viruses are very similar genetically and antigenically to the ancestral viruses associated with the 1957 pandemic (Schafer et al. 1993; Webster 1997).

Newcastle Disease Virus

Newcastle disease viruses (NDV), also known as avian paramyxovirus (APMV) serotype 1, are single-stranded, negative-sense RNA viruses of the *Paramyxoviridae* family and *Avulavirus* genus. NDV have been detected in at least 241 species of birds from 27 orders (Kaleta and Baldauf 1988), and the severity of the disease caused by these viruses varies with host species, virus virulence, age, immune status of the host, concomitant infections, and environmental stressors (OIE 2014). NDV have been grouped into five pathotypes based on the clinical signs seen in experimentally infected chickens: asymptomatic enteric, lentogenic or respiratory, mesogenic, neurotropic velogenic, and viscerotropic velogenic (OIE 2014). Viruses of the asymptomatic enteric pathotype have tropism for the gastrointestinal tract and cause subclinical infections (OIE 2014). The lentogenic or respiratory pathotype consists of mild or subclinical respiratory infection, whereas the mesogenic pathotype is characterized by respiratory signs, occasional nervous signs, and mortality in young birds (Miller and Koch 2013; OIE 2014). The velogenic pathotype typically causes high mortality in naïve chickens and are further divided into neurotropic velogenic and viscerotropic velogenic pathotypes. The neurotropic velogenic pathotype is characterized by respiratory and neurological signs, whereas the viscerotropic velogenic consists of hemorrhagic lesions in the gastrointestinal tract and often neurological signs (Miller and Koch 2013; OIE 2014). Newcastle disease (ND) is a highly infectious disease caused by virulent NDV (vNDV). These vNDV have an intracerebral pathogenicity index (ICPI) of 0.7 or greater in day-old chicks as well as the presence of multiple basic amino acids at the C-terminus of the fusion (F) protein cleavage site and a phenylalanine residue at position 117 (Miller and Koch 2013; OIE 2014). ND has caused devastating outbreaks in commercial and backyard poultry, and only a few areas of the world have not been affected by this disease (Miller and Koch 2013). ND limits the development of the poultry industry in many countries, significantly affecting the trade of poultry products and is currently a significant problem for poultry producers in the Middle East, Africa, and Asia (Miller and Koch 2013).

Wild waterfowl are natural reservoirs of Newcastle disease viruses (NDV) and most of the isolates obtained from these birds have been of the asymptomatic or lentogenic pathotypes for chickens. Natural infections of wild birds are apparently subclinical, but outbreaks of severe disease with high mortality have been reported in double-crested cormorants (*Phalacrocorax auritus*) in North America and Scotland (Blaxland 1951; Glaser et al. 1999; Kuiken et al. 1999; Rue et al. 2010). In 1992, there were large outbreaks in which more than 20.000 cormorants in the Northern Midwest of the USA and in Canada died as a result of infections with velogenic neurotropic NDV, which also spread to domestic turkeys (Glaser et al. 1999). Affected cormorants generally had neurological signs characterized by weakness, incoordination, ataxia, torticollis, paralysis of the wing and neck, head tremors, and blindness (Meteyer et al. 1997; Kuiken et al. 1998, 1999). High mortality rates occurred in juvenile birds, whereas adult birds were more resistant to disease and mortality (Glaser et al. 1999). Several other outbreaks of ND affecting cormorants have been reported in North America since then, which indicates that vNDV has become established in this species (Rue et al. 2010; Diel et al. 2012). Pigeons are natural reservoirs of "variant strains" of vNDV, also known as pigeon paramyxovirus type 1 (PPMV-1), which cause neurological signs and mortality in these birds, but variable outcomes in infected poultry (Alexander et al. 1984; Alexander 2000; Fuller et al. 2007; Miller and Koch 2013). Pigeons were responsible for a panzootic of PPMV-1 during the 1980s, and these viruses continue to circulate worldwide (Alexander 2000; Miller and Koch 2013). The potential spread of vNDV by pigeons into a country or disease-free area has led to strict regulations and mandatory vaccinations in racing and show birds (Miller and Koch 2013). World trade of pet and exotic birds can also be a source of vNDV introduction into a disease-free area, as previously reported in an outbreak involving six U.S. states in 1991 (Panigrahy et al. 1993). Infected cormorants, pigeons, and parrots can have asymptomatic infections with prolonged viral shedding (Erickson et al. 1977; Panigrahy et al. 1993; Kuiken et al. 1998). Asymptomatic infections with vNDV strains have been detected in other aquatic and terrestrial wild bird species, which may act as reservoirs for these viruses and potentially transmit ND to susceptible poultry flocks (Roy et al. 1998; Kaleta and Kummerfeld 2012).

NDV infections in chickens vary in their clinical presentation, which is dependent on the viral strain and host susceptibility. Chickens are the most susceptible species, followed in order by turkeys, pigeons, and ducks (Aldous et al. 2010). NDV may cause a peracute disease with few clinical signs and high sudden mortality, reaching up to 100 %. Birds that succumb to the acute disease caused by vNDV may present depression, respiratory signs, diarrhea, edema of the head and wattles, and neurological signs (Alexander 2000). Layers and breeders can have a drastic decline and even complete cessation of egg production (Alexander 2000). Well-vaccinated birds infected with vNDV may have few or no clinical signs of ND, but they can still shed viruses (Miller and Koch 2013). Transmission of NDV in poultry flocks can occur horizontally by direct contact with oropharyngeal secretions or feces, and indirectly via contaminated fomites, personnel, insects, and poultry products (Miller and Koch 2013). Implementation of strict biosecurity measures, including controlled movement of infected birds, poultry products, fomites and personnel, are critical to contain outbreaks of ND. Vaccination with live, killed, and recombinant vaccines is a complementing tool in the prevention and control of ND in poultry flocks and can restrict the devastating effects of the disease; however, it should not be considered as an alternative to good biosecurity practices (Alexander 2000).

Public Health Concerns

Human infections caused by NDV are mainly characterized by self-limiting conjunctivitis without corneal involvement that develops within 24 h of NDV exposure to the eye (Swayne and King 2003; OIE 2014). These infections have generally been reported in personnel handling infected birds, live viruses and vaccine strains, including laboratory workers, veterinary diagnosticians, workers in poultry processing plants and vaccination crews (Alexander 1995). Clinical signs mainly consist of unilateral or bilateral reddening, excessive lacrimation, edema of the eyelids and subconjunctival hemorrhage (OIE 2014), although chills, fever, headache and swelling of the preauricular lymph nodes have also been reported (Chang 1981; Swayne and King 2003). However, an immunocompromised patient developed pneumonia and died after infection with a pigeon-like APMV-1 (Goebel et al. 2007). There are no reports of human-to-human transmission for NDV to date (Miller and Koch 2013).

Viral Encephalomyelitides

Arthropod-borne viruses (i.e., arboviruses) are a diverse group of viruses that infect a variety of hosts over a broad geographical range. A number of these viruses can adversely affect the health of birds as well as humans. Transmission and maintenance cycles of arboviruses generally involve vertebrate hosts and hematophagous (i.e., blood-feeding) arthropod vectors (e.g., mosquitoes and less commonly ticks, biting midges, sand flies [*Culicoides*], and others) (McLean and Ubico 2007). Sylvatic transmission cycles for numerous arboviruses involve a variety of wildlife hosts, from rodents to birds, which are, in most cases, subclinically infected. However, in some cases, infection can cause disease in wildlife, humans, and domestic animals. The disease manifestations of arbovirus infections depend on host factors, including age, immune status, and genetic resistance (Daep et al. 2014). Humans and domestic mammals are usually dead-end hosts (i.e., do not play a role in virus transmission cycles) for most arboviruses that involve birds as amplifying hosts (Iversen 1994; Mackenzie et al. 2004). Most arboviral infections in humans are asymptomatic, but when disease manifests, symptoms range from generalized (e.g., fever and aches) to hemorrhagic or neurologic (e.g., meningitis, encephalitis, or acute flaccid paralysis) (Gaensbauer et al. 2014). The complexity of arbovirus transmission cycles is in part due to the biology and behavior of arthropod vectors, which have varied life histories, geographical ranges, host feeding preferences, climate-dependent development, and virus kinetics (Kenney and Brault 2014).

At least 77 arboviruses have been isolated from birds, and are distributed among five families: Bunyaviridae, Flaviviridae, Rhabdoviridae, Reoviridae, and Togaviridae (McLean and Ubico 2007). Arboviruses within the families Togaviridae, Flaviviridae, and Bunyaviridae can cause neurological disease in humans (Gaensbauer et al. 2014; Hubalek et al. 2014b). Further, only arboviruses within the families Togaviridae (genus *Alphavirus*) and Flaviviridae (genus *Flavivirus*) have been known to cause disease in poultry and game birds (Guy 2013). Transmission cycles of most of these viruses involve wild birds as virus-amplifying hosts, which are often subclinically infected, but in some cases undergo concurrent illness and death. Passerines (e.g., West Nile virus, St. Louis encephalitis virus, eastern equine encephalitis virus, western equine encephalitis virus, and Highlands J virus) and aquatic birds (e.g., Japanese encephalitis virus, Murray Valley encephalitis virus, and eastern equine encephalitis virus) are most commonly implicated in virus transmission (Iversen 1994; Endy and Nisalak 2002; Hubalek et al. 2014b; Nemeth and Oesterle 2014; Selvey et al. 2014). Wild birds and chickens have played an important role as sentinels of arbovirus transmission in public health programs (Komar 2003; Broom and Whelan 2005; McLean and Ubico 2007; Estep et al. 2013). The risk of human disease for most of these viruses is low, with the primary means of acquisition of human infections being via mosquito bites. Such an event is most likely to occur within areas of active or ongoing arbovirus transmission, where competent amplifying hosts, such as wild birds, and competent vectors, such as certain mosquito species, exist. Because transmission of these viruses depends on actively feeding mosquitoes, transmission of many arboviruses is highly seasonal, especially in temperate zones (McLean and Ubico 2007).

Notable arboviruses that involve avian amplifying hosts in transmission and can cause disease in humans and poultry and other domesticated fowl, including gamebirds, are West Nile virus (WNV), eastern equine encephalitis virus (EEEV), western equine encephalitis virus (WEEV), and Highlands J virus (HJV). St. Louis encephalitis (SLEV), Murray Valley (MVEV), and Japanese encephalitis viruses (JEV) are additional zoonotic arboviruses that circulate in wild birds (Mackenzie et al. 2004; Gaensbauer et al. 2014; Selvey et al. 2014). These viruses vary by geographical distribution, diversity of virus-amplifying host and mosquito vector species, and in their effects on birds and other potential hosts. Many are also recognized for their ability to cause neurological disease in domestic mammals, most notably, equids (Hubalek et al. 2014b; Long 2014). Fortunately, morbidity and mortality associated with most of these and other arboviral infections in poultry and other domesticated fowl are uncommon.

The primary transmission and infection route of arboviruses is via blood-feeding insect vectors, such as mosquitoes. For many arboviruses of both medical and veterinary importance, *Culex* spp. mosquitoes are the most commonly implicated vector. However, a broad diversity of mosquito species are involved in natural transmission of arboviruses, including *Aedes* spp., *Culiseta* spp., *Anopheles* spp., and others (Hassan et al. 2003; Mackenzie et al. 2004; Long 2014), and provide the potential for rapidly evolving viruses to adapt to new vector and host species (Stapleford et al. 2014). Virus inoculation of hosts can occur via mosquito bite or probing into subepidermal tissue. Initial virus replication is within Langerhans dendritic cells in the skin; these infected cells migrate to local lymph nodes and subsequently to secondary lymphoid organs, such as the spleen. Virus replication in these tissues is followed by its entry into the circulation (viremia), and systemic virus dissemination, including the central nervous system (Diamond et al. 2003).

Public Health Concerns

The major public health risk of arbovirus infection is via contact with mosquitoes and other arthropod vectors. Other routes of infection, such as ingestion, inhalation, and mucosal contact are exceedingly rare in humans. As such, arboviruses are generally not considered a risk for foodborne illness to humans. For most arboviruses, including WNV, infection in chickens does not usually lead to sufficient viremia titers to infect mosquitoes, and they readily seroconvert (Langevin et al. 2001; Komar 2003). However, some arboviruses can cause significant viremia titers and virus replication in tissues and bodily secretions of birds (e.g., WNV, EEEV, and WEEV), and in these cases, mucosal and conjunctival (e.g., oral and ocular) contact with raw muscle or other tissues and aerosolization of droplets should be minimized and protective measures taken while handling infected carcasses (e.g., use of disposable gloves, hand washing, disinfection of in-contact surfaces) (Guy 2013).

The geographical range of arboviruses is in part determined by the presence of competent vectors and amplifying hosts. Human exposures occur opportunistically in areas where these vectors and hosts overlap. Ecological factors such as climate and habitat also play a role in determining arbovirus transmission. Climate change in the form of global warming allows greater potential for arbovirus spread and establishment in novel geographical regions due to suitability of warmer temperatures for arthropod vectors, as well as lengthened transmission seasons. In addition, continued evolution of human land use practices and expanding populations, including higher densities in urban and suburban areas, expansion into less developed areas, including instigation of irrigation methods, and habitat alterations, may create opportunities for more frequent contact between humans and arthropods that are competent vectors for various arboviruses (Gallana et al. 2013).

The ability of arboviruses to spread great distances (e.g., transcontinental) in a short period of time is an important indicator of the ongoing threat of introduction of novel viruses or re-emergence of existing viruses. Recent examples of novel or emerging arboviruses with detrimental effects to public health or agriculture include the emergence of *Culicoides* spp.-transmitted Schmallenberg virus (family *Bunyaviridae*, genus *Orthobunyavirus*) (Balenghien et al. 2014), and expansion of *Culicoides* spp.-transmitted bluetongue virus (family *Reoviridae*, genus *Orbivirus*) serotypes in Europe (Mackenzie and Jeggo 2013), and locally acquired infections of mosquito-borne Chikungunya virus (family *Togaviridae*, genus *Alphavirus*) in humans in Florida, United States (Kendrick et al. 2014). JEV has also undergone recent transcontinental spread, most likely via wind-blown mosquitoes, from mainland Australia to Papua New Guinea and the Torres Strait (van den Hurk et al. 2009). Long-distance spread of these viruses can also occur via migrating viremic birds, travel or shipment of viremic domestic or wild animals, and undetected "hitchhiker" mosquitoes on shipping vessels or aircraft that can rapidly adapt to new geographical locales (Mackenzie et al. 2004; Nemeth et al. 2012; Daep et al. 2014). As with the recent emergence of WNV in the Western Hemisphere, zoonotic pathogens have become increasingly mobile and widespread, in part due to global human travel, and trade and commerce, including the sale and shipment of domesticated and wild-caught animals (Kuiken et al. 2003; Karesh et al. 2005; Marano et al. 2007). WNV is the most important modern example of the ongoing risks arboviruses pose to the health of humans and birds.

West Nile virus (WNV; family *Flaviviridae*, genus *Flavivirus*; Japanese encephalitis antigenic group) is the most geographically widespread flavivirus, occurring on all continents except Antarctica. Lineage 1 viruses are associated with fatal avian and human disease and continue to circulate in Europe, North and South America, Africa, and elsewhere; however, the range of lineage 2 viruses appears to be expanding, including strains capable of causing human and avian disease in Austria, Hungary, Greece, Italy, the Balkans, with the evolution of a possible novel lineage recently detected in Spain (Bakonyi et al. 2006; Jimenez-Clavero et al. 2008; Vazquez et al. 2010; Bagnarelli et al. 2011; Papa et al. 2011; Wodak et al. 2011; Valiakos et al. 2012). Lineage 1 WNV have caused recent outbreaks in Romania, northern Italy, and southern Russia; whereas the southward spread of WNV to Central and South America has not been associated with human or avian outbreaks, and the public and avian health impacts remain poorly understood in these expansive regions (Ulloa et al. 2009; Hubalek et al. 2014b). Further, the recent characterization of viruses within putative novel WNV lineages in Senegal and Denmark demonstrates the constant evolution of WNV on a wide geographical scale and the continuous risk posed by this virus (Fall et al. 2014; Pachler et al. 2014). Disease associated with WNV infection in humans varies from systemic febrile illness to neuroinvasive disease, including meningitis, encephalitis, or acute flaccid paralysis (Gaensbauer et al. 2014). However, human WNV infections are subclinical, and the case-fatality rate in humans is approximately 4–11 % (Komar 2003). The rate and incidence of WNV outbreaks will likely continue to be unpredictable (Zeller and Schuffenecker 2004; Pachler et al. 2014). Unprecedented fatalities in humans and birds in the past 10 years also suggest greater virulence of more recently isolated WNV strains, traits which have allowed it to exploit its host and facilitated its spread (Gubler 2007).

North American wild birds have suffered the most ill-fated outcome of the 1999 introduction of WNV to New York. More than 300 species of birds, representing over 80 families, are susceptible to WNV infection, although most avian infections are subclinical (Nemeth and Oesterle 2014). Numerous species within the family Corvidae are exquisitely sensitive to WNV-induced mortality, and additional passerine as well as raptor species (especially Falconiformes and Strigiformes) are also relatively susceptible to fatal infections. Birds that die acutely due to WNV have systemic infections with widespread hematogenous dissemination of infectious virus, which is also present in oral and cloacal secretions. For example, infectious WNV has been isolated from skeletal muscle of experimentally infected owls and crows, and heart muscle of numerous avian species included in surveillance (Nemeth et al. 2006b, 2007, 2011). In addition, experimentally infected 7-week-old chickens had infectious WNV in heart, kidney, spleen, intestine, and lung from 3 to 10 days post-inoculation; liver and brain were negative (skeletal muscle was not tested) (Senne et al. 2000). No virus was isolated from heart, brain, intestine, and kidney of turkey poults experimentally infected with WNV on 21 days post-inoculation (Swayne et al. 2001). Further, WNV viremia titers appear to be relatively low and are transient in some domestic fowl species (Komar et al. 2003; Nemeth and Bowen 2007). In birds that survive experimental inoculation, infection results in a transient period of viremia and viral shedding, ranging from 1 to 6 days post-inoculation, followed by seroconversion; infectious virus is usually undetectable in tissues by 1–2 weeks following inoculation, corresponding with seroconversion (Komar et al. 2003; Nemeth et al. 2006a). WNV-neutralizing antibodies likely persist and are protective for life in immunocompetent birds (Nemeth et al. 2008, 2009).

Poultry, backyard (i.e., domesticated) fowl, and related taxa have had varied susceptibility to WNV infection both by species, age, and likely additional host factors. For example, mature chickens (*Gallus domesticus*) are highly resistant to disease and readily seroconvert following infection (Nemeth and Bowen 2007). For this reason, they have been used as sentinels in public health for both WNV and related viruses such as SLEV and MVEV (Broom and Whelan 2005; Chaskopoulou et al. 2013; Estep et al. 2013; Selvey et al. 2014). Peak WNV viremia titers documented experimentally in chickens and turkeys are not within the range considered infectious to several mosquito species (Senne et al. 2000; Langevin et al. 2001; Swayne et al. 2001; Nemeth and Bowen 2007), and therefore they are not likely involved in WNV transmission cycles, lessening the risk to human handlers. However, WNV is rapidly fatal in very young chicks (i.e., 1-day-old), which experience high viremia titers (Nemeth and Bowen 2007). Experimentally induced lesions in 7-week-old chickens included myocardial necrosis, pneumonia, and nephritis, but no associated clinical disease (Senne et al. 2000). However, North American strains of WNV are highly pathogenic in some Galliformes, such as certain species of grouse (e.g., the greater sage grouse [*Centrocercus urophasianus*]) (Clark et al. 2006), leading to concerns of potential declines in population numbers in the western United States (Naugle et al. 2004). WNV-associated declines are suspected but not yet confirmed in additional grouse species, such as the ruffed grouse (*Bonasa umbellus*), in Pennsylvania (L. Williams, pers. comm.). In addition, Mediterranean WNV strains are highly pathogenic for red-legged partridges (*Alectoris rufa*) (Sotelo et al. 2011). Species within the family Phasianidae, such as the ring-necked pheasant (*Phasianus colchicus*) may be relatively more resistant to disease (Komar et al. 2003). Age-associated differences in WNV outcome are likely a factor for most avian species, with higher morbidity and mortality rates observed in young domestic geese and mallards (*Anas platyrhynchos*), both experimentally and naturally induced (Swayne et al. 2001; Austin et al. 2004; Cox et al. 2015), whereas no signs of clinical disease resulted from experimental WNV infection in mature mallards and Canada geese (*Branta canadensis*) (Komar et al. 2003).

Public Health Concerns

Mosquito-borne transmission is by far the most common means of WNV infection; however, potential alternate routes that may involve humans in contact with poultry or other birds include oral, percutaneous (i.e., needle- or scalpel-induced), and direct contact between conjunctival surfaces and infectious tissue or secretions. The latter two routes appear to be extremely rare, but have been documented in a person whose ocular conjunctiva contacted brain of an infected crow and in laboratory personnel working with highly infectious biological materials (Centers for Disease and Prevention 2002; Fonseca et al. 2005). Foodborne infections in humans via ingestion of poultry, game birds, or other birds are unlikely. This route of infection is difficult to document in nature, and therefore, the frequency of oral WNV transmission in nature involving birds and mammals is unknown. Oral infection has occurred experimentally when high viral-dose fluids or carcasses are fed or administered to birds of prey and passerines (Komar et al. 2003; Nemeth et al. 2006a), as well as cats (Austgen et al. 2004), hamsters (Sbrana et al. 2005), fox squirrels (Tiawsirisup et al. 2010), and alligators (Klenk et al. 2004). However, this route appears to be less efficient than mosquito-induced infections, and whether this route consistently produces the same level of disease severity or rates of mortality is not well understood.

Efforts toward preventing WNV infection in humans, domestic birds, and other animals are best focused on avoiding contact with mosquitoes. In large commercial poultry and other livestock facilities, this is likely achieved through standard biosafety protocols preventing entry of mosquitoes into animal holding areas. However, in more open-housing scenarios, including backyard poultry and gamebirds, as well as zoological avian collections and other captive birds, this is harder to accomplish. In these cases, tactics to decrease the likelihood of breeding mosquitoes or mosquito contact include the elimination of standing water, installation of screens or fine mesh over bird enclosures, and placement of aerosolized, bird-safe mosquito repellants around caging. Successful dampening of local or regional mosquito activities has been through aerial application of insecticides (Smallwood and Nakamoto 2009).

Currently, there are no WNV vaccines licensed for use in birds, although out of need, the use of available vaccination formulations (especially those approved for use in horses) has become routine in some zoological, private and educational collections, and has also been used as a management strategy for threatened or endangered avian species (Clark et al. 2006; Chang et al. 2007; Boyce et al. 2011; Glavis et al. 2011; Wheeler et al. 2011; Jarvi et al. 2013). A study with recombinant envelope (E) protein WNV vaccine compared routes of administration, antibody responses, and viremia titers in chickens (Fassbinder-Orth et al. 2009). However, in general, equid vaccines along with numerous proprietary or experimental vaccines have either not been adequately tested in birds (i.e., lack of challenge experiments) or have been inconsistent in their ability to induce seroconversion or protective long-lasting immunity in birds. Adverse effects of WNV vaccination in birds are rare and are generally limited to vaccine-site reactions (Gamino et al. 2012; Nemeth and Oesterle 2014). Additional efforts are needed to determine whether existing vaccines can aid in preventing infections and disease in a variety of avian species and to develop novel vaccine candidates specifically for birds.

Japanese encephalitis virus (JEV; family *Flaviviridae*, genus *Flavivirus*; Japanese encephalitis antigenic group) is an emerging, zoonotic, mosquito-borne virus. JEV is noteworthy because it has undergone significant geographical spread from southeastern Asia to Papua New Guinea, the northern Australian islands in the Torres Strait, and Pakistan, and is considered an impending threat of geographical spread from endemic to non-endemic areas (e.g., North America) (Mackenzie 2005; Nett et al. 2009). The geographical distribution of JEV currently includes Japan, eastern Russia, Korea, China, Taiwan, Thailand, Vietnam, Cambodia, Laos, Malaysia, Indonesia, Papua New Guinea, northern Australia, Guam, Philippines, and India (Mackenzie et al. 2007). In many of these regions, JEV is an important cause of viral encephalitis in humans, and disease cases are most numerous and severe in children. Symptoms include mental dullness, tremors, hypertonia, cranial nerve palsies, paralysis, motor deficits, and cognitive impairment (Mackenzie et al. 2004). Wild birds are primary amplifying hosts, with colonial water birds such as herons and egrets, and various passerines deemed important in transmission cycles, along with pigs (Nemeth et al. 2012; Hubalek et al. 2014b). The most recent assessment of JEV infection in birds examined the potential reservoir competence status of various North American species. Relatively high viremia titers were observed in ring-billed gulls (*Larus delawarensis*), house finches (*Haemorhous mexicanus*), common grackles (*Quiscalus quiscula*), rock pigeons (*Columba livia*), and great egrets (*Ardea alba*) as compared to other species examined, such as mallards. Viremia titers in gallinaceous birds (i.e., chickens and ring-necked pheasants) were low to undetectable (Nemeth et al. 2012). The minimum threshold of virus titer needed to infect mosquitoes has not been well defined for JEV, so it is difficult to assess the potential role of these birds in JEV transmission. However, none of 17 avian species infected with JEV had clinically evident disease, oral and cloacal shedding was rarely detected and at low titers, and no virus was detected in tissues. These data, combined with those of historic studies, reveal that overt JEV-associated clinical disease is infrequent in birds (Buescher et al. 1959; Nemeth et al. 2012). Earlier experimental infection studies in birds led to mortality of some Indian moorhens (*Gallinala chloropus parisfrons*), shovellers (*Spatula clypeata*) and mallards, and brain, spinal cord, spleen and liver tested positive in some of these individuals (Kitaoka et al. 1953). Overall, however, the likelihood of JEV transmission to humans via handling of birds or ingestion is minimal.

St. Louis encephalitis virus (SLEV; family *Flaviviridae*, genus *Flavivirus*; Japanese encephalitis antigenic group) is a mosquito-borne virus for which wild birds are the amplifying hosts. Geographically, SLEV covers large areas of North and South America from southern Canada to Argentina and the Caribbean Islands. SLEV is emerging in some regions of South America, such as central Argentina (Diaz et al. 2011). House sparrows (*Passer domesticus*), house finches, and other passerines, as well as mourning doves (*Zenaida macroura*) and pigeons, are considered amplifying hosts and infection is subclinical in these birds (Reisen et al. 2000; Reisen 2003; Diaz et al. 2011). Major SLEV epidemics involving human encephalitis cases have occurred in parts of the Midwestern and southern United States, with smaller outbreaks in California, New Jersey, and Ontario, Canada (Reisen 2003). Most human infections are subclinical; occasionally, mild transient malaise is experienced, and less commonly, encephalitis with neurological dysfunction and high fever. Experimental inoculation of young (<1 month) chickens and ducks produced viremia titers sufficient to infect mosquitoes but no clinical disease resulted; in adults, infection was similarly subclinical but viremia titers were undetectable. Infection in all ages of chickens readily produced long-lasting antibodies, for which these birds are often used as sentinels in surveillance, along with targeted wild bird species, such as mourning doves (Day and Stark 1999).

Murray Valley encephalitis virus (MVEV; family *Flaviviridae*, genus *Flavivirus*; Japanese encephalitis antigenic group) is the most important endemic arbovirus in Australia, and outbreaks have occurred in southeastern, western and central regions of the continent. Its distribution also includes Papua New Guinea and Indonesia. Recent MVEV activity in southeastern Australia following 37 years of apparent inactivity in this region demonstrates that this virus is capable of re-emergence, which has been associated with climatic factors favorable to mosquito breeding, such as high rainfall and flooding (Selvey et al. 2014). Similar to other flaviviruses in the JEV antigenic group, MVEV circulates between avian amplifying hosts (mainly water birds such as egrets) and mosquito vectors, and although most human infections are subclinical, non-specific illness (e.g., head ache, myalgia, rash) can lead to more severe and potentially fatal disease associated with encephalitis and fever (Hubalek et al. 2014b; Selvey et al. 2014). Chickens are not clinically affected by MVEV infection, and therefore, similar to SLEV, are used for early warning or sentinel surveillance programs throughout MVEV-endemic areas in Australia (Selvey et al. 2014).

Eastern equine encephalitis virus (EEEV; family *Togaviridae*, genus *Alphavirus*) is a mosquito-borne virus whose transmission involves wild bird virus-amplifying hosts (mainly passerines), and potentially reptiles, amphibians, and rodents to a lesser extent. The geographical range of EEEV includes North, Central and South America and the Caribbean, from Canada south to Argentina. Similar to WNV, numerous mosquito species are competent vectors, including *Aedes* spp., *Culex* spp., *Uranotaenia* spp., and others (Hubalek et al. 2014b). Epornitics due to EEEV have involved over 50 avian species, with passerines such as the blue jay (*Cyanocitta cristata*), European starling (*Sturnus vulgaris*), Carolina chickadee (*Poecile carolinensis*), northern mockingbird (*Mimus polyglottis*), and others as likely important virus-amplifying hosts in some regions such as the eastern United States (Crans et al. 1994; Komar et al. 1999; Estep et al. 2011; Hubalek et al. 2014b). In most cases, these avian hosts sustain subclinical EEEV infections (Guy 2013). Wetland birds such as herons and ducks also likely contribute to transmission cycles in EEEV-endemic areas (Kissling et al. 1954; Estep et al. 2011). Gamebird species are adversely affected by EEEV infection, with outbreaks having involved significant mortality in pen-raised ring-necked pheasants and chukar partridges (*Alectoris chukar*), with the highest mortality rates in immature birds (Bigler et al. 1976; Guy 2013). In addition, pigeons, ducks, turkeys, and African penguins (*Spheniscus demersus*) are also susceptible to EEEV-induced morbidity and mortality (Tuttle et al. 2005; Guy 2013). In the aforementioned species, disease generally involves neurological abnormalities (e.g., listlessness, paresis, paralysis, torticollis, and tremors). Severity of disease in chickens varies with age; chickens of ages 1–14 days are most susceptible (up to 80 %) and adults are generally refractory to infection (Guy et al. 1994b; Guy 2013). Emus (*Dromaius novaehollandiae*) are uniquely susceptible to EEEV-induced fatal gastrointestinal disease that includes vomiting and hemorrhagic diarrhea associated with acute hemorrhagic enterocolitis, with a nearly 50 % mortality rate in some flocks (Brown et al. 1993; Chenier et al. 2010). A similar disease manifestation was observed following experimental oral and needle-induced EEEV inoculation in ostriches (*Struthio camelus*) and turkey poults (Brown et al. 1993). EEEV is a significant cause of neurological disease in humans; infections can lead to paralysis, convulsions, coma, and death with a fatality rate of over 30 % (Gibbs and Tsai 1994; Gaensbauer et al. 2014). Because EEEV has been isolated from tissues (small intestine, liver, and brain) of infected ratite carcasses (Brown et al. 1993) and horizontal transmission occurred among turkeys, appropriate biosafety precautions should be taken when handling these animals.

Western equine encephalitis virus (WEEV; family *Togaviridae*, genus *Alphavirus*; Western equine encephalitis antigenic complex) is a mosquito-borne virus that, similar to EEEV, has a transmission cycle that involves a variety of mosquito species, including *Culex* spp., *Culiseta* spp., and *Aedes* spp., as well as vertebrate hosts. The latter includes wild birds as the primary virus-amplifying host, which mostly includes passerines such as the house sparrow, house finch, white-throated sparrow (*Zonotrichia albicollis*) and tricolored blackbird (*Agelaius tricolor*), with rodents and lagomorphs, and possibly amphibian and reptile species serving as secondary hosts in some geographical regions (Calisher 1994; McLean and Ubico 2007; Hubalek et al. 2014b). The historic range of WEEV includes North and South America, from western Canada south to Argentina (Hubalek et al. 2014b). WEEV-associated mortality rates in birds are generally lower than for EEEV; pheasants, chickens and pigeons are relatively resistant and the latter two species have been used as sentinels in serosurveillance (McLean and Ubico 2007; Hubalek et al. 2014b). The infrequency of WEEV-associated disease in birds is evident in the relatively few published reports, several of which describe outbreaks in pen-raised birds, including somnolence and paralysis in turkeys in Wisconsin (Woodring 1957) and chukar partridges in Florida (Ranck et al. 1965). WEEV infection was associated with decreased egg production in turkeys in California in the early 1990s (Cooper et al. 1997). Short-term viremia was confirmed following experimental inoculations of chickens with WEEV, with no corresponding illness described (Hammon and Reeves 1946). Humans are infected with WEEV by mosquito bite, and are dead-end hosts. Although most human WEEV infections are subclinical, permanent neurological disease is a possible outcome, and the case-fatality rate is approximately 3–7 % (Peters and Dalrymple 1990). However, the threat of WEEV infection has been reduced in some regions due to immunization of equines and ratites, water management, screened-in windows and porches, mosquito control, and public awareness leading to increased mosquito avoidance efforts (Reisen et al. 2008). Documented human infections have recently been in decline in North America; the last case was in 1994, with negative mosquito pools as of 2008 (Bergren et al. 2014).

Highlands J virus (HJV; family *Togaviridae*, genus *Alphavirus*; Western equine encephalitis antigenic complex) was initially believed to be a variant of WEEV, but is now considered a distinct virus (McLean and Ubico 2007). The geographical range (i.e., eastern United States) and mosquito vectors of HJV overlap with those of EEEV (Hubalek et al. 2014b). Wild birds are virus-amplifying hosts, with the northern cardinal (*Cardinalis cardinalis*), tufted titmouse (*Baeolophus bicolor*) and wood thrush (*Hylocichla mustelina*) as likely candidate species for involvement in transmission (McLean et al. 1985). Wild bird reservoirs appear resistant to disease and natural disease has been rarely documented in domestic birds (Hubalek et al. 2014b). HJV outbreaks in the 1990s involved turkeys in North Carolina that exhibited decreased activity and egg production, and chukar partridges that experienced depression, hind limb paresis, and head and neck tremors, with a 35 % mortality rate (Wages et al. 1993; Eleazer and Hill 1994). Disease was primarily viscerotropic in the chukars, and despite minimal brain lesions, virus was isolated from brain tissues (Eleazer and Hill 1994). A small percentage (7 %) of HJV-experimentally inoculated 2-week-old chickens succumbed to infection, one of which had myocardial necrosis (Guy et al. 1994b), whereas 27 % of inoculated 2-week-old turkeys died of HJV, and had multi-organ necrosis and lymphoid depletion (Guy et al. 1994a). Turkey breeders experimentally infected with HJV had early depression and decreased food intake and egg production for 1 week following infection. Virus was isolated from reproductive tissues (including ovary), heart, kidney, spleen, pancreas, liver, and less commonly from brain of these birds early after infection (Guy et al. 1994a). HJV has not been associated with human disease (Guy 2013; Hubalek et al. 2014b).

Usutu virus (USUV; family *Flaviviridae*, genus *Flavivirus*; Japanese encephalitis antigenic group) is an emerging mosquito-borne virus that likely originated in Africa and has recently spread to Europe (Austria, Hungary, Switzerland, Italy, Germany, Czechland, and Spain), where significant epiornitics have occurred (Hubalek et al. 2014a). Wild birds are the amplifying hosts of USUV, which has caused morbidity and mortality in some wild bird species in Europe, including raptors and passerines (especially *Turdus* spp.), with blackbirds (*Turdus merula*) suffering the highest mortality rates (up to 100 %) (Hassan et al. 2003; Bakonyi et al. 2007; Hubalek et al. 2014a). Disease in these birds can involve depression, ataxia, and seizures, and lesions can include encephalitis, myocarditis, and hepatic and neuronal necrosis (Steinmetz et al. 2011; Hubalek et al. 2014b). Experimental USUV infections in chickens revealed limited pathogenicity, with no clinical disease and seroconversion in only 1 of 10 infected chickens (Chvala et al. 2005). USUV-associated disease is rarely reported in humans but can lead to fever and rash, and may induce meningoencephalitis (Vilibic-Cavlek et al. 2014).

Fungi

Wild birds can be hosts to two zoonotic fungi that can persist in the environment. This environmental contamination can then indirectly infect people.

Cryptococcus is a capsulated yeast normally found in wild bird feces, most commonly in pigeons, and soil contaminated with bird guano. It causes severe infections in people such as pneumonia and meningitis. It can be disseminated and is mostly fatal in immunocompromised hosts.

Histoplasma capsulatum is a dimorphic fungus that is commonly found in wild bird feces. In people this fungus causes pneumonia and occasionally can cause disseminated disease.

Concluding Comments

Life is not without hazards and everything we do has an inherent risk, from walking down the street to eating food at a restaurant. In most countries in the developed world, food is normally not considered to be a hazard. Nevertheless, foodborne illnesses do occur. Over the last century, food has become safer, and often more convenient to prepare by the consumer. Examples include bagged fresh-cut lettuce and skinned turkey breast that is ready to cook. For those who are not familiar with farm production practices, animals become idyllic creatures not related to the meat pieces we purchase in a Styrofoam package in the supermarket. This disconnect, together with concerns over genetically modified animals, growth promoters, antibiotics and the presence of a variety of chemicals in foods, have initiated a movement to return to their roots to enjoy back yard-raised foods, such as fresh-laid eggs. In addition, many people are fascinated by nature, especially by birds. Hence, bird watching and feeding are huge pastimes that can congregate large populations of birds, creating sinks for pathogens. These trends have led to pathogen exposure in many people, including family, friends, and companion animals. Furthermore, wild birds and backyard chickens pose a risk for nearby production birds, which could adversely affect the local economy and the availability of meat and eggs.

References

Adlhoch C, Gossner C, Koch G et al (2014) Comparing introduction to Europe of highly pathogenic avian influenza viruses A(H5N8) in 2014 and A(H5N1) in 2005. Euro Surveill 19:20996

Alexander DJ, Parsons G, Marshall R (1984) Infections of fowls with Newcastle disease virus by food contamination with pigeon faeces. Vet Rec 115:601–602

Alexander DJ (1995) The epidemiology and control of avian influenza and Newcastle disease. J Comp Pathol 112:105–126

Aldous EW, Seekings JM, McNally A et al (2010) Infection dynamics of highly pathogenic avian influenza and virulent avian paramyxovirus type 1 viruses in chickens, turkeys and ducks. Avian Pathol 39:265–273

Alley MR, Connolly JH, Fenwick SG et al (2002) An epidemic of salmonellosis caused by Salmonella Typhimurium DT160 in wild birds and humans in New Zealand. N Z Vet J 50:170–176

Anderson J, Horn BJ, Gilpin BJ (2012) The prevalence and genetic diversity of Campylobacter spp. in domestic 'backyard' poultry in Canterbury, New Zealand. Zoonoses Public Health 59:52–60

Andres S, Vico JP, Garrido V et al (2013) Epidemiology of subclinical salmonellosis in wild birds from an area of high prevalence of pig salmonellosis: phenotypic and genetic profiles of Salmonella isolates. Zoonoses Public Health 60:355–365

Asaoka Y, Yanai T, Hirayama H et al (2004) Fatal necrotic enteritis associated with Clostridium perfringens in wild crows (Corvus macrorhynchos). Avian Pathol 33:19–24

Austgen LE, Bowen RA, Bunning ML et al (2004) Experimental infection of cats and dogs with West Nile virus. Emerg Infect Dis 10:82–86

Austin RJ, Whiting TL, Anderson RA et al (2004) An outbreak of West Nile virus-associated disease in domestic geese (*Anser anser domesticus*) upon initial introduction to a geographic region, with evidence of bird to bird transmission. Can Vet J 45:117–123

Bagnarelli P, Marinelli K, Trotta D et al (2011) Human case of autochthonous West Nile virus lineage 2 infection in Italy, September 2011. Euro Surveill 16:5–8

Bailey JS, Cosby DE (2005) Salmonella prevalence in free-range and certified organic chickens. J Food Prot 68:2451–2453

Baker MG, Kvalsvig A, Zhang J et al (2012) Declining Guillain-Barre syndrome after campylobacteriosis control, New Zealand, 1988–2010. Emerg Infect Dis 18:226–233

Bakonyi T, Ivanics T, Erdelyi K et al (2006) Lineage 1 and 2 strains of encephalitic West Nile virus, central Europe. Emerg Infect Dis 12:618–623

Bakonyi T, Erdelyi K, Ursu K et al (2007) Emergence of Usutu virus in Hungary. J Clin Microbiol 45:3870–3874

Balenghien T, Pages N, Goffredo M et al (2014) The emergence of Schmallenberg virus across Culicoides communities and ecosystems in Europe. Prevent Vet Med 116:360–369

Becker WB (1966) The isolation and classification of tern virus: influenza virus A/Tern/South Africa/1961. J Hyg 64:309–320

Beigel JHJ, Farrar AM, Han FG, Hayden R, Hyer MD, de Jong S, Lochindarat TK, Nguyen TH, Nguyen TH, Tran A, Nicoll S, Touch KY, Yuen AH (2005) Writing Committee of the World Health Organization Consultation on Human Influenza Avian influenza A (H5N1) infection in humans. N Engl J Med 353:1374–1385

Belshe RB (2005) The origins of pandemic influenza – lessons from the 1918 virus. N Engl J Med 353:2209–2211

Bergren NA, Auguste AJ, Forrester NL et al (2014) Western equine encephalitis virus: evolutionary analysis of a declining alphavirus based on complete genome sequences. J Virol 88: 9260–9267

Bester LA, Essack SY (2012) Observational study of the prevalence and antibiotic resistance of *Campylobacter* spp. from different poultry production systems in KwaZulu-Natal, South Africa. J Food Prot 75:154–159

Bigler WJ, Lassing EB, Buff EE et al (1976) Endemic eastern equine encephalomyelitis in Florida: a 20-year analysis, 1955–1974. Am J Trop Med Hyg 25:884–890

Blaxland JD (1951) Newcastle disease in shags and cormorants and its significance as a factor in the spread of this disease among domestic poultry. Vet Rec 63:731–733

Boyce WM, Vickers W, Morrison SA et al (2011) Surveillance for West Nile vicrus and vaccination of free-ranging island scrub-jays (*Aphelocoma insularis*) on Santa Cruz Island, California. Vector Borne Zoonotic Dis 11:1063–1068

Boyle F, Morris D, O'Connor J et al (2010) First report of extended-spectrum-beta-lactamase-producing Salmonella enterica serovar Kentucky isolated from poultry in Ireland. Antimicrob Agents Chemother 54:551–553

Broom AK, Whelan PI (2005) Sentinel chicken surveillance program in Australia, July 2003 to June 2004. Commun Dis Intell Q Rep 29:65–70

Brown TP, Roberts W, Page RK (1993) Acute hemorrhagic enterocolitis in ratites: isolation of eastern equine encephalomyelitis virus and reproduction of the disease in ostriches and turkey poults. Avian Dis 37:602–605

Buescher EL, Scherer WF, Rosenberg MZ et al (1959) Immunologic studies of Japanese encephalitis virus in Japan. III. Infection and antibody responses of birds. J Immunol 83:605–613

Calisher CH (1994) Medically important arboviruses of the United States and Canada. Clin Microbiol Rev 7:89–116

Capita R, Alonso-Calleja C, Prieto M et al (2002) Incidence and pathogenicity of Yersinia spp. isolates from poultry in Spain. Food Microbiol 19:295–301

Capua I, Terregino C (2009) Clinical traits and pathology of avian influenza infections, guidelines for farm visit and differential diagnosis. In: Capua I, Alexander DJ (eds) Avian influenza and newcastle disease – a field and laboratory manual, 3rd edn. Springer-Verlag, Milan, Italy, pp 45–71

Catherine Racicot B, Catharine P, Patrick B et al (2012) Chicken as reservoir for extraintestinal pathogenic *Escherichia coli* in humans, Canada. Emerg Infect Dis 18:415

CDC (2006) Interim guidance for protection of persons involved in US avian influenza outbreak disease control and eradication activities. http://www.cdc.gov/flu/avian/professional/pdf/protectionguid.pdf

CDC (2011a) CDC interim guidance for workers who are employed at commercial swine farms: preventing the spread of influenza A viruses. http://www.cdc.gov/flu/swineflu/guidance-commercial-pigs.htm

CDC (2011b) Multistate outbreak of human Salmonella Altona and Salmonella Johannesburg infections linked to chicks and ducklings. http://www.cdc.gov/salmonella/altona-baby-chicks/

CDC (2012) Swine influenza (influenza in swine). http://www.cdc.gov/flu/swineflu/influenza-in-swine.htm

CDC CfDCaP (2014a) Surveillance for foodborne disease outbreaks, United States, 2012, Annual report. US Department of Health and Human Services, CDC, Atlanta, GA

CDC (2014a) Avian influenza current situation. http://www.cdc.gov/flu/avianflu/avian-flu-summary.htm

CDC (2014c) Foodborne Diseases Active Surveillance Network (FoodNet): FoodNet surveillance report for 2012 (final report). U.S. Department of Health and Human Services, CDC, Atlanta, GA

CDC (2014c) Influenza antiviral medications: summary for clinicians. http://www.cdc.gov/flu/professionals/antivirals/summary-clinicians.htm

CDC (2014d) Key facts about seasonal flu vaccine. http://www.cdc.gov/flu/protect/keyfacts.htm

CDC (2014e) Multistate outbreak of human Salmonella infections linked to live poultry in backyard flocks (final update) posted 12:45pm, 21 Oct 2014, ET http://www.cdc.gov/salmonella/live-poultry-05-14/

CDC (2014f) Prevention and treatment of avian influenza A viruses in people. http://www.cdc.gov/flu/avianflu/prevention.htm

CDC (2014g) Reported Infections with variant influenza viruses in the United States since 2005. http://www.cdc.gov/flu/swineflu/variant-cases-us.htm

CDC (2014h) Variant (swine origin) influenza viruses in humans. http://www.cdc.gov/flu/swineflu/variant.htm

Centers for Disease C, Prevention (2002) Laboratory-acquired West Nile virus infections—United States, 2002. MMWR Morb Mortal Wkly Rep 51:1133–1135

Chang PW (1981) Newcastle disease. In: Beran GW (ed) CRC handbook series in zoonoses, First Edition, Section B: viral zoonoses. CRC, Boca Raton, FL, pp 261–264

Chang G-JJ, Davis BS, Stringfield C et al (2007) Prospective immunization of the endangered California condors (*Gymnogyps californianus*) protects this species from lethal West Nile virus infection. Vaccine 25:2325–2330

Chaskopoulou A, Dovas CI, Chaintoutis SC et al (2013) Detection and early warning of West Nile virus circulation in Central Macedonia, Greece, using sentinel chickens and mosquitoes. Vector Borne Zoonotic Dis 13:723–732

Chen Z, Li K, Luo L et al (2014) Detection of avian influenza A(H7N9) virus from live poultry markets in Guangzhou, China: a surveillance report. PLoS One 9, e107266

Chenier S, Cote G, Vanderstock J et al (2010) An eastern equine encephalomyelitis (EEE) outbreak in Quebec in the fall of 2008. Can Vet J 51:1011–1015
Chvala S, Bakonyi T, Hackl R et al (2005) Limited pathogenicity of Usutu virus for the domestic chicken (*Gallus domesticus*). Avian Pathol 34:392–395
Clark L, Hall J, McLean R et al (2006) Susceptibility of greater sage-grouse to experimental infection with West Nile virus. J Wildl Dis 42:14–22
CLSI (2010–2011) Performance standards for antimicrobial susceptibility testing: informational supplement. CLSI Document M100-S. Clinical and Laboratory Standards Institute, Wayne, PA
Cole D, Drum DJ, Stalknecht DE et al (2005) Free-living Canada geese and antimicrobial resistance. Emerg Infect Dis 11:935–938
Colles F, Dingle K, Cody A et al (2008) Comparison of Campylobacter populations in wild geese with those in starlings and free-range poultry on the same farm. Appl Environ Microbiol 74:3583–3590
Cooper GL, Medina HA, Woolcock PR et al (1997) Experimental infection of turkey poults with western equine encephalitis virus. Avian Dis 41:578–582
Cortés P, Blanc V, Mora A et al (2010) Isolation and characterization of potentially pathogenic antimicrobial-resistant Escherichia coli strains from chicken and pig farms in Spain. Appl Environ Microbiol 76:2799–2805
Cox SL, Campbell GD, Nemeth NM (2015) Outbreaks of West Nile virus in captive waterfowl in Ontario. Pathology, Canada Avian
Crans WJ, Caccamise DF, McNelly JR (1994) Eastern equine encephalomyelitis virus in relation to the avian community of a coastal cedar swamp. J Med Entomol 31:711–728
Crim SM, Iwamoto M, Huang JY et al (2014) Incidence and trends of infection with pathogens transmitted commonly through food—Foodborne Diseases Active Surveillance Network, 10 U.S. sites, 2006–2013. MMWR Morb Mortal Wkly Rep 63:328–332
Daep CA, Munoz-Jordan JL, Eugenin EA (2014) Flaviviruses, an expanding threat in public health: focus on dengue, West Nile, and Japanese encephalitis virus. J Neurovirol 20:539–560
Daoust PY, Busby DG, Ferns L et al (2000) Salmonellosis in songbirds in the Canadian Atlantic provinces during winter-summer 1997–98. Can Vet J Rev 41:54–59
Davies ZG, Fuller RA, Loram A et al (2009) A national scale inventory of resource provision for biodiversity within domestic gardens. Biol Conserv 142:761–771
Dawood FS, Jain S, Finelli L et al (2009) Emergence of a novel swine-origin influenza A (H1N1) virus in humans. N Engl J Med 360:2605–2615
Day JF, Stark LM (1999) Avian serology in a St. Louis encephalitis epicenter before, during, and after a widespread epidemic in south Florida, USA. J Med Entomol 36:614–624
de Jong B, Ekdahl K (2006) The comparative burden of salmonellosis in the European Union member states, associated and candidate countries. BMC Public Health 6:4
de Jong JC, Claas EC, Osterhaus AD et al (1997) A pandemic warning? Nature 389:554
de Jong MD, Tran TT, Truong HK et al (2005) Oseltamivir resistance during treatment of influenza A (H5N1) infection. N Engl J Med 353:2667–2672
de Wit MA, Hoogenboom-Verdegaal AM, Goosen ES et al (2000) A population-based longitudinal study on the incidence and disease burden of gastroenteritis and Campylobacter and Salmonella infection in four regions of The Netherlands. Eur J Epidemiol 16:713–718
Diamond MS, Shrestha B, Mehlhop E et al (2003) Innate and adaptive immune responses determine protection against disseminated infection by West Nile encephalitis virus. Viral Immunol 16:259–278
Diaz LA, Flores FS, Contigiani MS (2011) Viremia profiles and host competence index for West Nile virus (Flavivirus, Flaviviridae) in three autochthonous birds species from Argentina. J Ornithol 152:21–25
Eleazer TH, Hill JE (1994) Highlands J virus-associated mortality in chukar partridges. J Vet Diagn Invest 6:98–99
Ellis TM, Bousfield RB, Bissett LA et al (2004) Investigation of outbreaks of highly pathogenic H5N1 avian influenza in waterfowl and wild birds in Hong Kong in late 2002. Avian Pathol 33:492–505

Endy TP, Nisalak A (2002) Japanese encephalitis virus: ecology and epidemiology. In: Mackenzie JS, Barrett ADT, Deubel V (eds) Japanese encephalitis and West Nile viruses. pp 11–48

Erickson GA, Mare CJ, Gustafson GA et al (1977) Interactions between viscerotropic velogenic Newcastle disease virus and pet birds of six species. II. Viral evolution through bird passage. Avian Dis 21:655–669

Estep LK, McClure CJW, Burkett-Cadena ND et al (2011) A multi-year study of mosquito feeding patterns on avian hosts in a southeastern focus of eastern equine encephalitis virus. Am J Trop Med Hyg 84:718–726

Estep LK, McClure CJW, Vander Kelen P et al (2013) Risk of exposure to eastern equine encephalomyelitis virus increases with the density of northern cardinals. PLoS One 8, e57879

Fall G, Diallo M, Loucoubar C et al (2014) Vector competence of *Culex neavei* and *Culex quinquefasciatus* (Diptera: Culicidae) from Senegal for lineages 1, 2, Koutango and a putative new lineage of West Nile virus. Am J Trop Med Hyg 90:747–754

Fassbinder-Orth CA, Hofmeister EK, Weeks-Levy C et al (2009) Oral and parenteral immunization of chickens (*Gallus gallus*) against West Nile virus with recombinant envelope protein. Avian Dis 53:502–509

Fischer J, Schmoger S, Jahn S et al (2013) NDM-1 carbapenemase-producing Salmonella enterica subsp. enterica serovar Corvallis isolated from a wild bird in Germany. J Antimicrob Chemother 68:2954–2956

Fonseca K, Prince GD, Bratvold J et al (2005) West Nile virus infection and conjunctival exposure. Emerg Infect Dis 11:1648–1649

Fortin ND (2013) HACCP and Other Regulatory Approaches to Prevention of Foodborne Diseases. Foodborne Infections and Intoxications, 4th Edition:497–510

Fouchier RA, Schneeberger PM, Rozendaal FW et al (2004) Avian influenza A virus (H7N7) associated with human conjunctivitis and a fatal case of acute respiratory distress syndrome. Proc Natl Acad Sci U S A 101:1356–1361

Freidl GS, Meijer A, de Bruin E et al (2014) Influenza at the animal-human interface: a review of the literature for virological evidence of human infection with swine or avian influenza viruses other than A(H5N1). Euro Surveill 19(pii):20793

Friend M, McLean RG (2002) The role of native birds and other wildlife on the emergence of zoonotic diseases. Emergence of zoonotic diseases, workshop summary 52–58

Friend M, Laitman CJ, Kampen RS et al (1987) Field guide to wildlife diseases. U.S. Department of the Interior, Fish and Wildlife Service, Washington, DC

Fukushima H, Gomyoda M (1991) Intestinal carriage of Yersinia pseudotuberculosis by wild birds and mammals in Japan. Appl Environ Microbiol 57:1152–1155

Fuller T, Bensch S, Muller I et al (2012) The ecology of emerging infectious diseases in migratory birds: an assessment of the role of climate change and priorities for future research. Ecohealth 9:80–88

Gaensbauer JT, Lindsey NP, Messacar K et al (2014) Neuroinvasive arboviral disease in the United States: 2003 to 2012. Pediatrics 134:E642–E650

Gallana M, Ryser-Degiorgis M-P, Wahli T et al (2013) Climate change and infectious diseases of wildlife: altered interactions between pathogens, vectors and hosts. Curr Zool 59:427–437

Gamino V, Gutierrez-Guzman A-V, Fernandez-de-Mera IG et al (2012) Natural Bagaza virus infection in game birds in southern Spain. Vet Res 43:65

Gao R, Cao B, Hu Y et al (2013) Human infection with a novel avian-origin influenza A (H7N9) virus. N Engl J Med 368:1888–1897

Gargiulo A, Russo TP, Schettini R et al (2014) Occurrence of enteropathogenic bacteria in urban pigeons (Columba livia) in Italy. Vector Borne Zoonotic Dis 14:251–255

Gibbs EPJ, Tsai TF (1994) Eastern encephalitis. In: Beran GW (ed) Handbook of Zoonoses, section B: viral. CRC Press, Boca Raton, FL, pp 11–24

Gill JS, Webby R, Gilchrist MJR et al (2006) Avian Influenza among Waterfowl Hunters and Wildlife Professionals. Emerg Infect Dis 12:1284–1286

Gilsdorf A, Boxall N, Gasimov V et al (2006) Two clusters of human infection with influenza A/H5N1 virus in the Republic of Azerbaijan, February–March 2006. Euro Surveill 11:122–126

Giovannini S, Pewsner M, Hussy D et al (2013) Epidemic of salmonellosis in passerine birds in Switzerland with spillover to domestic cats. Vet Pathol 50:597–606

Glaser LC, Barker IK, Weseloh DV et al (1999) The 1992 epizootic of Newcastle disease in double-crested cormorants in North America. J Wildl Dis 35:319–330

Glavis J, Larsen RS, Lamberski N et al (2011) Evaluation of antibody response to vaccination against West Nile virus in thick-billed parrots (*Rhynchopsitta pachyrhyncha*). J Zool Wildl Med 42:495–498

Goebel SJ, Taylor J, Barr BC et al (2007) Isolation of avian paramyxovirus 1 from a patient with a lethal case of pneumonia. J Virol 81:12709–12714

Gruszynski K, Pao S, Kim C et al (2014) Evaluating gulls as potential vehicles of Salmonella enterica serotype Newport (JJPX01.0061) contamination of tomatoes grown on the eastern shore of Virginia. Appl Environ Microbiol 80:235–238

Gubler DJ (2007) The continuing spread of West Nile virus in the Western Hemisphere. Clin Infect Dis 45:1039–1046

Guo CT, Takahashi N, Yagi H et al (2007) The quail and chicken intestine have sialyl-galactose sugar chains responsible for the binding of influenza A viruses to human type receptors. Glycobiology 17:713–724

Guy JS (2013) Arbovirus infections. In: Swayne DE, Glisson JR (eds) Diseases of poultry, 13th edn. Wiley-Blackwell, Ames, Iowa, pp 465–512

Guy JS, Barnes HJ, Ficken MD et al (1994a) Decreased egg production in turkeys experimentally infected with eastern equine encephalitis virus or Highlands J virus. Avian Dis 38:563–571

Guy JS, Barnes HJ, Smith LG (1994b) Experimental infection of young broiler chickens with eastern equine encephalitis virus and Highlands J virus. Avian Dis 38:572–582

Habing GG, Kessler SE, Mollenkopf DF et al (2014) Distribution and diversity of salmonella strains in shipments of hatchling poultry, United States, 2013. Zoonoses Public Health, Washington, DC

Hammon WM, Reeves WC (1946) Western equine encephalomyelitis virus in the blood of experimentally inoculated chickens. J Exp Med 83:163–173

Harrois D, Breurec S, Seck A et al (2014) Prevalence and characterization of extended-spectrum beta-lactamase-producing clinical Salmonella enterica isolates in Dakar, Senegal, from 1999 to 2009. Clin Microbiol Infect 20:O109–O116

Hassan HK, Cupp EW, Hill GE et al (2003) Avian host preference by vectors of eastern equine encephalomyelitis virus. Am J Trop Med Hyg 69:641–647

Hayden FG, Belshe RB, Clover RD et al (1989) Emergence and apparent transmission of rimantadine-resistant influenza A virus in families. N Engl J Med 321:1696–1702

Herfst S, Schrauwen EJ, Linster M et al (2012) Airborne transmission of influenza A/H5N1 virus between ferrets. Science 336:1534–1541

Hernandez SM, Keel K, Sanchez S et al (2012) Epidemiology of a Salmonella enterica subsp. enterica serovar Typhimurium strain associated with a songbird outbreak. Appl Environ Microbiol 78:7290–7298

Horton RA, Wu G, Speed K et al (2013) Wild birds carry similar Salmonella enterica serovar Typhimurium strains to those found in domestic animals and livestock. Res Vet Sci 95:45–48

Hu J, Zhu Y, Zhao B et al (2014) Limited human-to-human transmission of avian influenza A(H7N9) virus, Shanghai, China, March to April 2013. Euro Surveill 19(pii):20838

Hubalek Z, Rudolf I, Capek M et al (2014a) Usutu Virus in blackbirds (*Turdus merula*), Czech Republic, 2011–2012. Transbound Emerg Dis 61:273–276

Hubalek Z, Rudolf I, Nowotny N (2014b) Arboviruses pathogenic for domestic and wild animals. In: Maramorosch K, Murphy FA (eds) Advances in virus research, vol 89. pp 201–275

Hughes LA, Shopland S, Wigley P et al (2008) Characterisation of Salmonella enterica serotype Typhimurium isolates from wild birds in northern England from 2005–2006. BMC Vet Res 4:4

Hughes LA, Bennett M, Coffey P et al (2009) Molecular epidemiology and characterization of Campylobacter spp. isolated from wild bird populations in northern England. Appl Environ Microbiol 75:3007–3015

Idris U, Lu J, Maier M et al (2006) Dissemination of fluoroquinolone-resistant Campylobacter spp. within an integrated commercial poultry production system. Appl Environ Microbiol 72:3441–3447

Imai M, Kawaoka Y (2012) The role of receptor binding specificity in interspecies transmission of influenza viruses. Curr Opin Virol 2:160–167

Imai M, Watanabe T, Hatta M et al (2012) Experimental adaptation of an influenza H5 HA confers respiratory droplet transmission to a reassortant H5 HA/H1N1 virus in ferrets. Nature 486:420–428

Iversen JO (1994) Western equine encephalomyelitis. In: Beran GW (ed) Handbook of Zoonoses, section B: viral. CRC Press, Boca Raton, FL, pp 25–31

Janecko N, Cizek A, Halova D et al (2014) Prevalence, characterization and antibiotic resistance of Salmonella isolates in large corvid species of Europe and North America between 2010 and 2013. Zoonoses Public Health 62:292–300

Jarvi SI, Hu D, Misajon K et al (2013) Vaccination of captive nēnē (*Brenta sandvicensis*) against West Nile virus using a protein-based vaccine (WN-80E). J Wildl Dis 49:152–156

Jeong J, Kang HM, Lee EK et al (2014) Highly pathogenic avian influenza virus (H5N8) in domestic poultry and its relationship with migratory birds in South Korea during 2014. Vet Micobiol 173:249–257

Jimenez-Clavero MA, Sotelo E, Fernandez-Pinero J et al (2008) West Nile virus in golden eagles, Spain, 2007. Emerg Infect Dis 14:1489–1491

Johnson NB, Hayes LD, Brown K et al (2014) CDC National Health Report: leading causes of morbidity and mortality and associated behavioral risk and protective factors-United States, 2005–2013. MMWR Surveill Summ 63:3–27

Kaiser J (2013) Meet your new neighbors: chickens are moving from the henhouse to the backyard and looking for veterinary care. J Am Vet Med Assoc 243:458–463

Kaleta EF, Baldauf C (1988) Newcastle disease in free-living and pet birds. In: Alexander DJ (eds.) Newcastle Disease, edition. Kluwer Academic Publishers, Boston, MA, p 197–246

Kapperud G, Stenwig H, Lassen J (1998) Epidemiology of Salmonella typhimurium O:4-12 infection in Norway: evidence of transmission from an avian wildlife reservoir. Am J Epidemiol 147:774–782

Karesh WB, Cook RA, Bennett EL et al (2005) Wildlife trade and global disease emergence. Emerg Infect Dis 11:1000–1002

Kawaoka Y, Krauss S, Webster RG (1989) Avian-to-human transmission of the PB1 gene of influenza A viruses in the 1957 and 1968 pandemics. J Virol 63:4603–4608

Kendrick K, Stanek D, Blackmore C (2014) Transmission of Chikungunya virus in the Ccntinental United States – Florida, 2014. Morb Mortal Wkly Rep 63:1137

Kenney JL, Brault AC (2014) The role of environmental, virological and vector interactions in dictating biological transmission of arthropod-borne viruses by mosquitoes. In: Maramorosch K, Murphy FA (eds) Advances in virus research, vol 89. pp 39–83

Kida H, Ito T, Yasuda J et al (1994) Potential for transmission of avian influenza viruses to pigs. J Gen Virol 75(Pt 9):2183–2188

Kinzelman J, McLellan SL, Amick A et al (2008) Identification of human enteric pathogens in gull feces at Southwestern Lake Michigan bathing beaches. Can J Microbiol 54:1006–1015

Kissling RE, Chamberlain RW, Sikes RK et al (1954) Studies on the North American arthropod-borne encephalitides. III Eastern equine encephalitis in wild birds. Am J Hyg 60:251–265

Kitaoka M, Okubo K, Miura T et al (1953) Relationship between Japanese B and Russian spring-summer encephalitis and birds. Jpn J Med Sci Biol 6:247–259

Klenk K, Snow J, Morgan K et al (2004) Alligators as West Nile virus amplifiers. Emerg Infect Dis 10:2150–2155

Kobayashi H, Kanazaki M, Hata E et al (2009) Prevalence and characteristics of eae- and stx-positive strains of Escherichia coli from wild birds in the immediate environment of Tokyo Bay. Appl Environ Microbiol 75:292–295

Kock R (2012) One Health and its importance to wildlife. Vet Rec 171:613–614

Komar N (2003) West Nile virus: epidemiology and ecology. Advances in virus research. Academic. pp 185–234

Komar N, Dohm DJ, Turell MJ et al (1999) Eastern equine encephalitis virus in birds: relative competence of European starlings (*Sturnus vulgaris*). Am J Trop Med Hyg 60:387–391

Komar N, Langevin S, Hinten S et al (2003) Experimental infection of North American birds with the New York 1999 strain of West Nile virus. Emerg Infect Dis 9:311–322

Krauss S, Walker D, Pryor SP et al (2004) Influenza A viruses of migrating wild aquatic birds in North America. Vector Borne Zoonotic Dis 4:177–189

Kuiken T, Heckert RA, Riva J et al (1998) Excretion of pathogenic Newcastle disease virus by double-crested cormorants (Phalacrocorax auritus) in absence of mortality or clinical signs of disease. Avian Pathol 27:541–546

Kuiken T, Fouchier R, Rimmelzwaan G et al (2003) Emerging viral infections in a rapidly changing world. Curr Opin Biotechnol 14:641–646

Kwan PSL, Xavier C, Santovenia M et al (2014) Multi-locus sequence typing confirms wild birds as the source of a Campylobacter outbreak associated with the consumption of raw peas. Appl Environ Microbiol 80(15):PMC4148789

Langevin SA, Bunning M, Davis B et al (2001) Experimental infection of chickens as candidate sentinels for West Nile virus. Emerg Infect Dis 7:726–729

Langholz JMJ-R (2013) The potential role of wildlife in pathogenic contamination of fresh produce. Hum Wildl Interact 7:140–157

Lawson B, Howard T, Kirkwood JK et al (2010) Epidemiology of salmonellosis in garden birds in England and Wales, 1993 to 2003. Ecohealth 7:294–306

Lawson B, Hughes LA, Peters T et al (2011) Pulsed-field gel electrophoresis supports the presence of host-adapted Salmonella enterica subsp. enterica serovar Typhimurium strains in the British garden bird population. Appl Environ Microbiol 77:8139–8144

Lawson B, de Pinna E, Horton RA et al (2014) Epidemiological evidence that garden birds are a source of human salmonellosis in England and Wales. PLoS One 9, e88968

Le Hello S, Harrois D, Bouchrif B et al (2013) Highly drug-resistant Salmonella enterica serotype Kentucky ST198-X1: a microbiological study. Lancet Infect Dis 13:672–679

Le QM, Kiso M, Someya K et al (2005) Avian flu: isolation of drug-resistant H5N1 virus. Nature 437:1108

Liljebjelke KA, Hofacre CL, Liu T et al (2005) Vertical and horizontal transmission of salmonella within integrated broiler production system. Foodborne Pathog Dis 2:90–102

Liu J, Xiao H, Lei F et al (2005) Highly pathogenic H5N1 influenza virus infection in migratory birds. Science 309:1206

Long MT (2014) West Nile virus and equine encephalitis viruses new perspectives. Vet Clin North Am Equine Pract 30:523

Mackenzie JS (2005) Emerging zoonotic encephalitis viruses: lessons from Southeast Asia and Oceania. J Neurovirol 11:434–440

Mackenzie JS, Jeggo M (2013) Reservoirs and vectors of emerging viruses. Curr Opin Virol 3:170–179

Mackenzie JS, Gubler DJ, Petersen LR (2004) Emerging flaviviruses: the spread and resurgence of Japanese encephalitis, West Nile and dengue viruses. Nat Med 10:S98–S109

Mackenzie JS, Williams DT, Smith DW (2007) Japanese encephalitis virus: the geographic distribution, incidence, and spread of a virus with a propensity to emerge in new areas. In: Tabor E (ed) Emerging viruses in human populations. pp 201–268

Marano N, Arguin PM, Pappaioanou M (2007) Impact of globalization and animal trade on infectious disease ecology. Emerg Infect Dis 13:1807–1809

Matrosovich MN, Krauss S, Webster RG (2001) H9N2 influenza A viruses from poultry in Asia have human virus-like receptor specificity. Virology 281:156–162

McLean R, Ubico S (2007) Arboviruses in birds. In: Thomas NJ, Hunter DB, Atkinson CT (eds) Infectious diseases of wild birds. Blackwell Pub, Ames, IA, pp 17–62

McLean RG, Frier G, Parham GL et al (1985) Investigations of vertebrate hosts of eastern equine encephalitis during an epizootic in Michigan, 1980. Am J Trop Med Hyg 34:1190–1202

Mete A, Giannitti F, Barr B et al (2013) Causes of mortality in backyard chickens in northern California: 2007–2011. Avian Dis 57:311–315

Meteyer CU, Docherty DE, Glaser LC et al (1997) Diagnostic findings in the 1992 epornitic of neurotropic velogenic Newcastle disease in double-crested cormorants from the upper midwestern United States. Avian Dis 41:171–180

Miller P, Koch G (2013) Newcastle Disease. In: Swayne DE, Glisson JR, McDougald LR, Nolan LK, Suarez DL, Nair V (eds.) Diseases of Poultry, 13th edition. John Wiley & Sons, Inc., Ames, IA p 89–107

Monto AS (2003) The role of antivirals in the control of influenza. Vaccine 21:1796–1800

Murphy BR, Webster RG (1996) Orthomyxoviruses. Fields virology. Lippincott-Raven, Philadelphia, PA, pp 1397–1445

Naugle DE, Aldridge CL, Walker BL et al (2004) West Nile virus: pending crisis for greater sage-grouse. Ecol Lett 7:704–713

Nemeth NM, Bowen RA (2007) Dynamics of passive immunity to West Nile virus in domestic chickens (*Gallus gallus domesticus*). Am J Trop Med Hyg 76:310–317

Nemeth NM, Oesterle PT (2014) West Nile virus from an avian conservation perspective. Int Zoo Yearbook 48:101–115

Nemeth N, Gould D, Bowen R et al (2006a) Natural and experimental West Nile virus infection in five raptor species. J Wildl Dis 42:1–13

Nemeth NM, Hahn DC, Gould DH et al (2006b) Experimental West Nile virus infection in Eastern Screech Owls (*Megascops asio*). Avian Dis 50:252–258

Nemeth NM, Beckett S, Edwards E et al (2007) Avian mortality surveillance for West Nile virus in Colorado. Am J Trop Med Hyg 76:431–437

Nemeth NM, Kratz GE, Bates R et al (2008) Naturally induced humoral immunity to West Nile virus infection in raptors. Ecohealth 5:298–304

Nemeth NM, Oesterle PT, Bowen RA (2009) Humoral immunity to West Nile virus Is long-lasting and protective in the house sparrow (*Passer domesticus*). Am J Trop Med Hyg 80:864–869

Nemeth NM, Thomsen BV, Spraker TR et al (2011) Clinical and pathologic responses of American crows (*Corvus brachyrhynchos*) and fish crows (*C. ossifragus*) to experimental West Nile virus infection. Vet Pathol 48:1061–1074

Nemeth N, Bosco-Lauth A, Oesterle P et al (2012) North American birds as potential amplifying hosts of Japanese encephalitis virus. Am J Trop Med Hyg 87:760–767

Nett RJ, Campbell GL, Reisen WK (2009) Potential for the emergence of Japanese encephalitis virus in California. Vector Borne Zoonotic Dis 9:511–517

Nicholson KG (1998) Human influenza. In: Nicholson KG, Webster RG, Hay AJ (eds) Textbook of influenza. Blackwell, London, pp 219–264

OIE (2014) Newcastle Disease. In: (eds.) Manual of Diagnostic Tests and Vaccines for Terrestrial Animals edition. p 555–573.

Olsen B, Munster VJ, Wallensten A et al (2006) Global patterns of influenza a virus in wild birds. Science 312:384–388

Oporto B, Juste RA, Hurtado A (2009) Phenotypic and genotypic antimicrobial resistance profiles of Campylobacter jejuni isolated from cattle, sheep, and free-range poultry faeces. Int J Microbiol 2009:8

Pachler K, Lebl K, Berer D et al (2014) Putative new West Nile virus lineage in *Uranotaenia unguiculata* mosquitoes, Austria, 2013. Emerg Infect Dis 20:2119–2122

Palmgren H, Aspan A, Broman T et al (2006) Salmonella in Black-headed gulls (Larus ridibundus); prevalence, genotypes and influence on Salmonella epidemiology. Epidemiol Infect 134:635–644

Panigrahy B, Senne DA, Pearson JE et al (1993) Occurrence of velogenic viscerotropic Newcastle disease in pet and exotic birds in 1991. Avian Dis 37:254–258

Papa A, Xanthopoulou K, Gewehr S et al (2011) Detection of West Nile virus lineage 2 in mosquitoes during a human outbreak in Greece. Clin Microbiol Infect 17:1176–1180

Patel G, Bonomo RA (2013) "Stormy waters ahead": global emergence of carbapenemases. Front Microbiol 4:48

Peiris M, Yuen KY, Leung CW et al (1999) Human infection with influenza H9N2. Lancet 354:916–917

Peiris JS, de Jong MD, Guan Y (2007) Avian influenza virus (H5N1): a threat to human health. Clin Microbiol Rev 20:243–267

Penfold JB, Amery HC, Peet PJ (1979) Gastroenteritis associated with wild birds in a hospital kitchen. Br Med J 2:802

Peters CJ, Dalrymple JM (1990) Alphaviruses. In: Fields BN, Knipe DM, Chanok RM, Hirsch MS, Melnick JL, Monath TP, Roizman B (eds) Virology, 2nd edn. Raven Press, Ltd., New York, NY, pp 713–762

Pollock SL, Stephen C, Skuridina N et al (2012) Raising chickens in city backyards: the public health role. J Community Health 37:734–742

Rabsch W, Andrews HL, Kingsley RA et al (2002) Salmonella enterica serotype Typhimurium and its host-adapted variants. Infect Immun 70:2249–2255

Ramos R, Cerda-Cuellar M, Ramirez F et al (2010) Influence of refuse sites on the prevalence of Campylobacter spp. and Salmonella serovars in seagulls. Appl Environ Microbiol 76:3052–3056

Ranck FM, Gainer JH, Hanley JE et al (1965) Natural outbreak of eastern and western encephalitis in pen-raised chukars in Florida. Avian Dis 9:8

Refsum T, Heir E, Kapperud G et al (2002) Molecular epidemiology of Salmonella enterica serovar typhimurium isolates determined by pulsed-field gel electrophoresis: comparison of isolates from avian wildlife, domestic animals, and the environment in Norway. Appl Environ Microbiol 68:5600–5606

Reisen WK (2003) Epidemiology of St. Louis encephalitis virus. Adv Virus Res 61:139–183

Reisen WK, Kramer LD, Chiles RE et al (2000) Response of house finches to infection with sympatric and allopatric strains of western equine encephalomyelitis and St. Louis encephalitis viruses from California. J Med Entomol 37:259–264

Reisen WK, Fang Y, Brault AC (2008) Limited interdecadal variation in mosquito (Diptera : Culicidae) and avian host competence for western equine encephalomyelitis virus (Togaviridae : Alphavirus). Am J Trop Med Hyg 78:681–686

Richard M, Schrauwen EJ, de Graaf M et al (2013) Limited airborne transmission of H7N9 influenza A virus between ferrets. Nature 501:560–563

Robinson RA, Lawson B, Toms MP et al (2010) Emerging infectious disease leads to rapid population declines of common British birds. PLoS One 5, e12215

Rodriguez F, Moreno J, Ortega R et al (2012) Evidence for Kelp Gulls (Larus dominicanus) and Franklin's Gulls (Leucophaeus pipixcan) as carriers of Salmonella by real-time polymerase chain reaction. J Wildl Dis 48:1105–1108

Sanchez S, Hofacre CL, Lee MD et al (2002) Animal sources of salmonellosis in humans. J Am Vet Med Assoc 221:492–497

Sbrana E, Tonry JH, Xiao SY et al (2005) Oral transmission of West Nile virus in a hamster model. Am J Trop Med Hyg 72:325–329

Scallan E, Griffin PM, Angulo FJ et al (2011) Foodborne illness acquired in the United States—unspecified agents. Emerg Infect Dis 17:16–22

Schafer JR, Kawaoka Y, Bean WJ et al (1993) Origin of the pandemic 1957 H2 influenza A virus and the persistence of its possible progenitors in the avian reservoir. Virology 194:781–788

Scholtissek C, Rohde W, Von Hoyningen V et al (1978) On the origin of the human influenza virus subtypes H2N2 and H3N2. Virology 87:13–20

Selvey LA, Dailey L, Lindsay M et al (2014) The changing epidemiology of Murray Valley encephalitis in Australia: the 2011 outbreak and a review of the literature. PLoS Negl Trop Dis 8(1), e2656

Senne DA, Pedersen JC, Hutto DL et al (2000) Pathogenicity of West Nile virus in chickens. Avian Dis 44:642–649

Shafir SC, Fuller T, Smith TB et al (2012) A national study of individuals who handle migratory birds for evidence of avian and swine-origin influenza virus infections. J Clin Virol 54:364–367

Sharp GB, Kawaoka Y, Jones DJ et al (1997) Coinfection of wild ducks by influenza A viruses: distribution patterns and biological significance. J Virol 71:6128–6135

Shi W, Shi Y, Wu Y et al (2013) Origin and molecular characterization of the human-infecting H6N1 influenza virus in Taiwan. Protein Cell 4:846–853

Shortridge KF (1992) Pandemic influenza: a zoonosis? Semin Respir Infect 7:11–25

Shrestha SS, Swerdlow DL, Borse RH et al (2011) Estimating the burden of 2009 pandemic influenza A (H1N1) in the United States (April 2009-April 2010). Clin Infect Dis 52(Suppl 1):S75–S82

Smallwood KS, Nakamoto B (2009) Impacts of the West Nile virus epizootic on the yellow billed magpie, American crow, and other birds in the Sacramento Valley, California. Condor 111:247–254

Smith GJ, Bahl J, Vijaykrishna D et al (2009a) Dating the emergence of pandemic influenza viruses. Proc Natl Acad Sci U S A 106:11709–11712

Smith GJ, Vijaykrishna D, Bahl J et al (2009b) Origins and evolutionary genomics of the 2009 swine-origin H1N1 influenza A epidemic. Nature 459:1122–1125

Sotelo E, Gutierrez-Guzman AV, del Amo J et al (2011) Pathogenicity of two recent Western Mediterranean West Nile virus isolates in a wild bird species indigenous to Southern Europe: the red-legged partridge. Vet Res 42:11

Spencer SE, Marshall J, Pirie R et al (2012) The spatial and temporal determinants of campylobacteriosis notifications in New Zealand, 2001–2007. Epidemiol Infect 140:1663–1677

Stallknecht DE, Brown JD (2008) Ecology of avian influenza in wild birds. In: Swayne DE (ed) Avian influenza, 1st edn. Blackwell Publishing, Ames, IA, pp 43–58

Stallknecht DE, Shane SM (1988) Host range of avian influenza virus in free-living birds. Vet Res Commun 12:125–141

Stapleford KA, Coffey LL, Lay S et al (2014) Emergence and transmission of arbovirus evolutionary intermediates with epidemic potential. Cell Host Microbe 15:706–716

Stegeman A, Bouma A, Elbers AR et al (2004) Avian influenza A virus (H7N7) epidemic in The Netherlands in 2003: course of the epidemic and effectiveness of control measures. J Infect Dis 190:2088–2095

Steinmetz HW, Bakonyi T, Weissenboeck H et al (2011) Emergence and establishment of Usutu virus infection in wild and captive avian species in and around Zurich, Switzerland-Genomic and pathologic comparison to other central European outbreaks. Vet Microbiol 148:207–212

Stephens CP, On SL, Gibson JA (1998) An outbreak of infectious hepatitis in commercially reared ostriches associated with *Campylobacter coli* and *Campylobacter jejuni*. Vet Microbiol 61:183–190

Suarez DL (2008) Influenza A virus. In: Swayne DE (ed) Avian influenza. Blackwell, Ames, IA, pp 3–22

Swayne DE, Beck JR, Smith CS et al (2001) Fatal encephalitis and myocarditis in young domestic geese (*Anser anser domesticus*) caused by West Nile virus. Emerg Infect Dis 7:751–753

Swayne DE, King DJ (2003) Avian influenza and Newcastle disease. J Am Vet Med Assoc 222:1534–1540

Swayne DE, Suarez DL, Sims LD (2013) Influenza. In: Swayne DE, Glisson JR, McDougald LR, Nolan LK, Suarez DL, Nair V (eds) Diseases of poultry, 13th edn. Wiley, Ames, IA, pp 191–218

Taubenberger JK, Reid AH, Lourens RM et al (2005) Characterization of the 1918 influenza virus polymerase genes. Nature 437:889–893

Tauni MA, Osterlund A (2000) Outbreak of Salmonella typhimurium in cats and humans associated with infection in wild birds. J Small Anim Pract 41:339–341

Teske L, Ryll M, Rubbenstroth D et al (2013) Epidemiological investigations on the possible risk of distribution of zoonotic bacteria through apparently healthy homing pigeons. Avian Pathol 42:397–407

Tiawsirisup S, Blitvich BJ, Tucker BJ et al (2010) Susceptibility of fox squirrels (*Sciurus niger*) to West Nile virus by oral exposure. Vector Borne Zoonotic Dis 10:207–209

Tizard I (2004) Salmonellosis in wild birds. Semin Avian Exot Pet Med 13:50–66

To KF, Chan PK, Chan KF et al (2001) Pathology of fatal human infection associated with avian influenza A H5N1 virus. J Med Virol 63:242–246

Tong S, Li Y, Rivailler P et al (2012) A distinct lineage of influenza A virus from bats. Proc Natl Acad Sci U S A 109:4269–4274

Tong S, Zhu X, Li Y et al (2013) New world bats harbor diverse influenza A viruses. PLoS Pathog 9, e1003657

Trifonov V, Khiabanian H, Rabadan R (2009) Geographic dependence, surveillance, and origins of the 2009 influenza A (H1N1) virus. N Engl J Med 361:115–119

Tuttle AD, Andreadis TG, Frasca S et al (2005) Eastern equine encephalitis in a flock of African penguins maintained at an aquarium. J Am Vet Med Assoc 226:2059–2062

Ulloa A, Hann Ferguson H, Mendez-Sanchez JD et al (2009) West Nile virus activity in mosquitoes and domestic animals in Chiapas, Mexico. Vector Borne Zoonotic Dis 9:555–560

Ungchusak K, Auewarakul P, Dowell SF et al (2005) Probable person-to-person transmission of avian influenza A (H5N1). N Engl J Med 352:333–340

USDA (2014) National antimicrobial resistance monitoring system for enteric bacteria (NARMS): 2011 NARMS animal arm annual report. U.S. Department of Agriculture, Agricultural Research Service, Athens, GAUSDA-APHIS (2005) Update on Avian Influenza Findings Poultry Findings Confirmed by USDA's National Veterinary Services Laboratories. https://www.aphis.usda.gov/wps/portal/aphis/ourfocus/animalhealth/sa_animal_disease_information/sa_avian_health/ct_avian_influenza_disease/

Valiakos G, Touloudi A, Athanasiou LV et al (2012) Serological and molecular investigation into the role of wild birds in the epidemiology of West Nile virus in Greece. Virol J 9:266

van den Hurk AF, Ritchie SA, Mackenzie JS (2009) Ecology and geographical expansion of Japanese encephalitis virus. Annu Rev Entomol 54:17–35

Van Dyke MI, Morton VK, McLellan NL et al (2010) The occurrence of *Campylobacter* in river water and waterfowl within a watershed in southern Ontario, Canada. J Appl Microbiol 109:1053–1066

Vazquez A, Paz Sanchez-Seco M, Ruiz S et al (2010) Putative new lineage of West Nile virus, Spain. Emerg Infect Dis 16:549–552

Vilibic-Cavlek T, Kaic B, Barbic L et al (2014) First evidence of simultaneous occurrence of West Nile virus and Usutu virus neuroinvasive disease in humans in Croatia during the 2013 outbreak. Infection 42:689–695

Wages DP, Ficken MD, Guy JS et al (1993) Egg-production drop in turkeys associated with alphaviruses: eastern equine encephalitis virus and Highlands J virus. Avian Dis 37:1163–1166

Waldenstrom J, Mevius D, Veldman K et al (2005) Antimicrobial resistance profiles of *Campylobacter jejuni* isolates from wild birds in Sweden. Appl Environ Microbiol 71:2438–2441

Waldenstrom J, Axelsson-Olsson D, Olsen B et al (2010) Campylobacter jejuni colonization in wild birds: results from an infection experiment. PLoS One 5, e9082

Wang H, Feng Z, Shu Y et al (2008) Probable limited person-to-person transmission of highly pathogenic avian influenza A (H5N1) virus in China. Lancet 371:1427–1434

Wasyl D, Kern-Zdanowicz I, Domanska-Blicharz K et al (2015) High-level fluoroquinolone resistant Salmonella enterica serovar Kentucky ST198 epidemic clone with IncA/C conjugative plasmid carrying bla (CTX-M-25)gene. Vet Microbiol 175(1):85–91

Webster RG (1997) Predictions for future human influenza pandemics. J Infect Dis 176(Suppl 1): S14–S19

Webster RG, Bean WJ, Gorman OT et al (1992) Evolution and ecology of influenza A viruses. Microbiol Rev 56:152–179

Wheeler SS, Langevin S, Woods L et al (2011) Efficacy of three vaccines in protecting western scrub-jays (*Aphelocoma californica*) from experimental infection with West Nile virus: implications for vaccination of island scrub-jays (*Aphelocoma insularis*). Vector Borne Zoonotic Dis 11:1069–1080

WHO (eds) (2013a) Global view of campylobacteriosis. http://appswhoint/iris/bitstream/10665/80751/1/9789241564601_engpdf

WHO (2013b) Overview of the emergence and characteristics of the avian influenza A(H7N9) virus. http://www.who.int/influenza/human_animal_interface/influenza_h7n9/WHO_H7N9_review_31May13.pdf?ua=1

WHO (2014a) Cumulative number of confirmed human cases of avian influenza A/(H5N1) reported to WHO: 2 Oct 2014. http://www.who.int/csr/disease/avian_influenza/country/cases_table_2009_04_23/en/index.html

WHO (2014b) WHO risk assessment of human infection with avian influenza A(H7N9) virus. http://www.who.int/influenza/human_animal_interface/influenza_h7n9/riskassessment_h7n9_2Oct14.pdf

Wodak E, Richter S, Bago Z et al (2011) Detection and molecular analysis of West Nile virus infections in birds of prey in the eastern part of Austria in 2008 and 2009. Vet Microbiol 149:358–366

Woodring FR (1957) Naturally occurring infection with equine encephalomyelitis virus in turkeys. J Am Vet Med Assoc 130:511–512

Xiao XC, Li KB, Chen ZQ et al (2014) Transmission of avian influenza A(H7N9) virus from father to child: a report of limited person-to-person transmission, Guangzhou, China, January 2014. Euro Surveill 19:20837

Yassine HM, Al-Natour MQ, Lee CW et al (2007) Interspecies and intraspecies transmission of triple reassortant H3N2 influenza A viruses. Virol J 4:129

Yen HL, Webster RG (2009) Pandemic influenza as a current threat. Curr Top Microbiol Immunol 333:3–24

Yoon S, Webby RJ, Webster RG (2014) Evolution and ecology of influenza A viruses. In: Compans RW, Oldstone MBA (eds) Current topics in microbiology and immunology – influenza pathogenesis and control. Springer, Basel

Yu H, Cowling BJ, Feng L et al (2013) Human infection with avian influenza A H7N9 virus: an assessment of clinical severity. Lancet 382:138–145

Zeller HG, Schuffenecker I (2004) West Nile virus: an overview of its spread in Europe and the Mediterranean Basin in contrast to its spread in the Americas. Eur J Clin Microbiol Infect Dis 23:147–156

Zhang Q, Shi J, Deng G et al (2013) H7N9 influenza viruses are transmissible in ferrets by respiratory droplet. Science 341:410–414

Chapter 5
Molecular Tools for Monitoring and Source-Tracking *Salmonella* in Wildlife and the Environment

Anita Wright, Amber Ginn, and Zhiyao Luo

Abstract *Salmonella* causes an estimated 1.2 million cases of gastroenteritis per year (Scallan et al. 2011) or 16.42 illnesses per 100,000 persons (Gilliss et al. 2013). Salmonellosis is also associated with the largest incidence of food-associated hospitalizations and deaths (Gilliss et al. 2013). Furthermore, no significant change in the incidence of *Salmonella* infections has occurred since the beginning of surveillance during 1996–1998 (Gould et al. 2013; Gilliss et al. 2013). Many foodborne diseases, including salmonellosis, were once classified as strictly zoonotic infections because they were mostly attributed to meats and products derived from domesticated animals (Chisholm et al. 1999; van Duijkeren et al. 2002). However, recent outbreaks of salmonellosis from produce grown on farms with minimal or no contact with domesticated livestock exemplify the contribution of environmental reservoirs of infections (Gould et al. 2011; Hanning et al. 2009; Danyluk et al. 2007). In fact, the number of illnesses per outbreak is often greater for produce than for any other food product (Franz and van Bruggen 2008). Therefore, wildlife as a source of preharvest contamination of produce with human pathogens is under consideration, including reptiles, amphibians, and birds that may harbor potentially virulent strains of *Salmonella* (Gorski et al. 2013; Reche et al. 2003; Pfleger et al. 2003).

Keywords Environmental health • Microarray • Molecular epidemiology • PCR • Pulsed-field gel electrophoresis • Salmonella • Sample preparation • Source tracking • Whole genome sequencing • Wildlife

---

A. Wright (✉) • A. Ginn • Z. Luo
Food Science and Human Nutrition Department, University of Florida, Gainesville, FL, USA
e-mail: acw@ufl.edu

M. Jay-Russell, M.P. Doyle (eds.), *Food Safety Risks from Wildlife*, Food Microbiology and Food Safety, DOI 10.1007/978-3-319-24442-6_5

Salmonella and Foodborne Disease

Salmonella causes an estimated 1.2 million cases of gastroenteritis per year (Scallan et al. 2011) or 16.42 illnesses per 100,000 persons (Gilliss et al. 2013). Salmonellosis is also associated with the largest incidence of food-associated hospitalizations and deaths (Gilliss et al. 2013). Furthermore, no significant change in the incidence of *Salmonella* infections has occurred since the beginning of surveillance during 1996–1998 (Gould et al. 2013; Gilliss et al. 2013). Many foodborne diseases, including salmonellosis, were once classified as strictly zoonotic infections because they were mostly attributed to meats and products derived from domesticated animals (Chisholm et al. 1999; van Duijkeren et al. 2002). However, recent outbreaks of salmonellosis associated with produce grown on farms with minimal or no contact with domesticated livestock exemplify the contribution of environmental sources of infections (Gould et al. 2011; Hanning et al. 2009; Danyluk et al. 2007). In fact, the number of illnesses per outbreak is often greater for produce than for any other food product (Franz and van Bruggen 2008). Therefore, wildlife as a source of preharvest contamination of produce with human pathogens is under consideration, including reptiles, amphibians, and birds that may harbor potentially virulent strains of *Salmonella* (Gorski et al. 2013; Reche et al. 2003; Pfleger et al. 2003).

Rivers and ponds provide natural habitats for a variety of wildlife and presumably become contaminated through the introduction of fecal material of infected animals (Plusquellec et al. 1994; Ijabadeniyi et al. 2011; Pachepsky et al. 2011). Once present, these pathogens can become established in these environments and persist for long periods of time (Winfield and Groisman 2003; Barak and Liang 2008). Therefore, tracking sources of disease outbreaks need to include not only examination of postharvest processing equipment, distribution protocol, and food products, but also the surveillance of preharvest environments that may harbor extremely diverse populations of pathogens from various sources, including wildlife.

The initially proposed Produce Safety Rules within Food Safety Modernization Act (FSMA) would require periodic testing for generic *Escherichia coli* in agricultural water intended for preharvest contact with the edible portion of fresh produce (FDA 2013). Unfortunately, fecal indicator bacteria may not provide reliable estimates of *Salmonella* contamination due to the greater resistance of this pathogen to the stressful environmental conditions relative to that of indicator organisms (Pianetti et al. 2004; Polo et al. 1998). Thus, direct monitoring of this pathogen may be needed to adequately assess potential disease risk associated with agricultural reservoirs. An impediment to evaluating these risks and tracing back sources of exposure is that detection methods are often unable to detect low levels of pathogens that are generally present in aquatic environments (Escartin et al. 2002; Madsen 1994). This chapter reviews currently available molecular tools to detect and source-track *Salmonella* in wildlife hosts and agricultural environments.

Molecular Tools for *Salmonella* Detection from Wildlife

Detection and Enumeration of Salmonella

Detection of *Salmonella* from environmental sources generally involves some type of enrichment to increase the efficiency of detection, recovery of stressed/injured bacteria, and/or provide larger sample size, as described by the US Food and Drug Administration (http://www.fda.gov/Food/FoodScienceResearch/LaboratoryMethods/ucm070149.htm).

Samples of water or sediment are measured, diluted, and used to inoculate broth cultures. *Salmonella* in wildlife can be also assessed directly but wildlife agencies and university IACUC should be notified to see if permits are required to trap and handle the animals. Animals are sampled using swabs to determine external or rectal contamination. Alternatively, small aquatic animal(s) can be sampled by immersion in sterile solutions and assaying the wash water. Swabs and solutions are then applied directly to enrichment broth or concentrated as described below. Typically, a dual enrichment strategy employs inoculation into a primary nonselective enrichment broth, such as buffered peptone or lactose broth, followed by overnight growth and transfers to a more selective broth medium. Presumptive positive colonies are isolated from broth cultures on various *Salmonella* selective media which have been reviewed elsewhere (Odumeru and León-Velarde 2012). Subsequent species confirmation can be achieved by numerous methods, including biochemical analysis, immunological identification, and/or species-specific PCR. The most probable number (MPN) method provides quantitative analysis based on a statistical table that calculates the MPN per gram or ml from the number of replicate tubes that are positive for each dilution. Alternatively, an MPN calculator is available for calculations (http://www.i2workout.com/mcuriale/mpn/).

Increased recovery of pathogens in water samples can be facilitated by concentration of samples. Concentration techniques include centrifugation (Basel et al. 1983), filtration (Farber and Sharpe 1984), lectin-based biosorbents (Payne et al. 1992), and immunomagnetic separation (IMS) technologies (Shaw et al. 1998). Filtration methods include nitrocellulose (usually Millipore, 0.2 μm) filter filtration (NFF), tangential flow filtration (TFF), and modified Moore swab filtration (MSF). NFF is the standard water testing method of the American Public Health Association (APHA 1992), and generally 100 ml of retentate on the filter is applied directly to an agar plate for the enumeration of colonies. TFF involves flowing water across a filter membrane, and the MSF method involves pumping water through rolled cotton gauze inserted into plastic pipes. The latter methods offer the advantage of filtering much larger volumes (up to 100 L) of water (Mull and Hill 2009; Gibson and Schwab 2011; Bisha et al. 2011). Samples concentrated by IMS use paramagnetic beads with attached *Salmonella*-specific antibody to remove bacteria from suspensions.

PCR and qPCR for Species Confirmation

Many biochemical assays can be used for species-specific identification of presumptive positive isolates of *Salmonella* (Odumeru and León-Velarde 2012). However, due to improved reliability, cost-effectiveness, and accessibility, PCR is becoming the "gold standard" for *Salmonella* species confirmation (McKillip and Drake 2004). One of the common species-specific targets for PCR of *Salmonella* is the *invA* gene, a component of the Type III secretion system and widely distributed among strains of *Salmonella* (Rajabi et al. 2011; Stone et al. 1994). Real-time or quantitative PCR (qPCR) uses fluorescent labeling of products for more sensitive detection of *Salmonella* (Espy et al. 2006; Li et al. 2014b). Bacteria can be enumerated by qPCR directly from filtered or unfiltered samples that are calibrated by a standard curve from known concentrations of target DNA. Alternatively, PCR can be used in combination with MPN. The BAX PCR system (DuPont-Qualicon Inc.) has been adopted by USDA-FSIS as a tool for screening and quantifying *Salmonella* in a variety of food and food ingredients. Other systems include ADIAFOOD Rapid Pathogen Detection System (AES Chemunex), the Assurance Genetic Detection System GDS (Biocontrol Inc.) utilizing a post-enrichment IMS step followed by real-time PCR, iQ- CheckTM *Salmonella* II (BioRad Laboratories, S.A.), and R.A.P.I.D. LT system (Idaho Technology Inc.). The detection limit for these assays is usually around 10^2–10^4 CFU/g (Jasson et al. 2010). Another commercially available technology (Roka Bioscience) targets RNA genes and takes advantage of their high copy number to improve detection sensitivity. It should be noted that PCR cannot distinguish between live and dead cells and has the potential to provide false-positive results, necessitating the application of enrichment for viability testing (Mandal et al. 2011).

McEgan et al. (2013) recently compared with qPCR several of the concentration methods described above and determined that nitrocellulose filters become clogged when approximately 500 ml of sample was processed (McEgan et al. 2013). However, TFF concentrated up to 10 L, and *Salmonella* was consistently detected from inocula of 1–760 CFU/L from enriched filtrate in lactose broth or from use of IMS beads (Dynabeads or Pathatrix). Furthermore, using the combination of TFF with IMS beads at *Salmonella* concentrations of <10 CFU/L, qPCR (Applied Biosystems MicroSEQ) had greater sensitivity compared to conventional PCR. Comparable results were obtained with MSF but more rigorous detection methods were required. These methods greatly facilitate the detection of pathogens through their capacity to screen larger volumes and more representative samples in a relatively short period of time (about 2 days).

DNA Microarrays for Species Confirmation

DNA microarrays provide recognition oligonucleotides in discrete locations on a solid matrix (Rasooly and Herold 2008). A primary feature of these assays is species-specific binding to multiple target molecules (McLoughlin 2011; Rasooly and Herold 2008). Gene targets include both species-specific genes and/or virulence-related genes from foodborne pathogens, which can be detected and genotyped from food samples using genomic DNA extracted from target cell or food samples. Target DNA is amplified by PCR, labeled with a fluorescent dye(s), and hybridized to the microarray. After array washing and scanning, the location, color, and intensity of fluoresce signal on the array data can reflect the characteristics of the target DNA (Rasooly and Herold 2008). Compared to other methods, DNA microarray has the advantages of rapid detection, high-throughput screening, multi-target analysis, and access to virulence information that goes beyond the species identification level (Rasooly and Herold 2008; Shin et al. 2014). For example, Shin et al. (2014) developed a DNA-based microarray with a *car*B gene to detect and differentiate three serotypes (Choleraesuis, Enteritidis, and Typhimurium) of *Salmonella enterica*, with a minimum range of sensitivity between 1.6 and 3.1 nM. Additionally, they reported the DNA microarray did not detect any nonspecific signals and did not have cross-reactivity with other common pathogenic bacteria or other serotypes of *Salmonella* when testing from a mixed culture. However, pathogen detection arrays have primarily been used in a research context, and disadvantages are mainly the cost and the need to periodically update with new strains or targets (McLoughlin 2011).

Alternatives to PCR

PCR can also be used to develop and evaluate methods that are less expensive and offer higher throughput. For example, we recently described an agar cross-streaking technique that employs sequential isolation on two selective agars for confirmation of *Salmonella* from enrichment broths (Luo et al. 2014). This protocol was combined with an MPN assay using lactose broth as the pre-enrichment medium and tetrathionate (TT) as secondary enrichment media to evaluate *Salmonella* isolation from irrigation water. Multiple downstream methods were evaluated for *Salmonella* species identification. The validity of presumptive isolation of typical *Salmonella* colonies from MPN enrichment on Xylose-Lysine-Tergitol4 (XLT4) agar, followed by subsequent confirmation by cross-streaking to CHROMagar™ *Salmonella* plus (CSP), was examined for assay sensitivity and specificity, as well a positive and negative predictive value (PPV/NPV), using *invA* PCR confirmation as the "gold standard." This method is described in Fig. 5.1 and had 99.95 % agreement with PCR confirmation, with only a single false-negative strain on XLT4 ($n=1640$ isolates). This cost-effective alternative to PCR offers increased throughput as the multiple steps required for DNA extraction, sample preparation, and PCR product detection are eliminated. Furthermore, the capacity of *Salmonella* evaluation is increased because less rigorous technical training and specialized equipment are required for equivalent results.

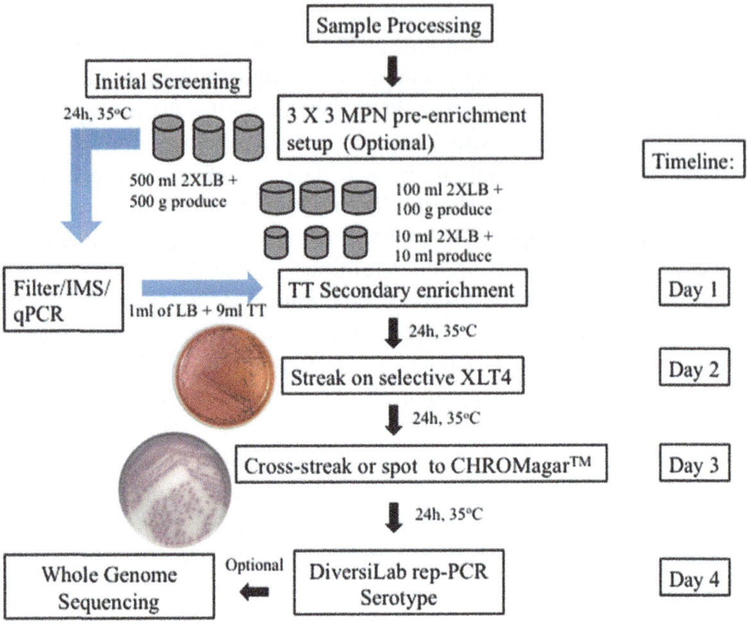

Fig. 5.1 Proposed 4-day method for detection, enumeration, and typing of *Salmonella*

Molecular Serotyping of *Salmonella*

Traditional serotyping is based on the Kauffmann-White-Le Minor scheme and has evolved over the past 80 years as the basis for the classification of *Salmonella* subspecies (Wattiau et al. 2011). Agglutination is used to indicate attachment of bacterial surface antigens to *Salmonella*-specific antibodies. Three primary antigens are used for *Salmonella* serotyping: O, H, and Vi antigens. The O antigen is the somatic lipopolysaccharides (LPS) on the external surface of the bacterial outer membrane. Sixty-four O antigens have been identified in *Salmonella,* and several O antigens can be expressed concurrently in a single cell (Wattiau et al. 2011). The H antigen is the flagellar protein, and 114 H antigens variants have been identified. Most*Salmonella* contain two copies of genes encoding the H antigens; therefore most isolates present diphasic antigens, i.e., H1 and H2. Although only one H antigen is expressed in one cell at a time, both phases can be detected in a colony as they can be expressed by different bacteria in the same culture/colony (Wattiau et al. 2011). Vi or capsular polysaccharide antigen is used to identify certain serotypes, as it only occurs with Typhi, Paratyphi C, and Dublin.

Serotyping is still commonly used for initial screening of *Salmonella* identification, but it is not considered to be specific or sensitive enough to correctly reflect the relationship among strains for outbreak investigations. Problems with *Salmonella* serology include the loss of O antigen expression, whereby the isolate becomes "rough" or "untypeable" (D'Aoust and John 2007). Also, false-positive reactions can occur due to auto-agglutination or cross-reactivity to non-*Salmonella* antigens (Schrader et al. 2008; Strawn et al. 2014). Genetic analysis has revealed that traditional serotyping methods did not correctly identify serotypes for all *Salmonella* subspecies, and approximately 10 % of isolates tested ($n=754$) are mistyped (Shi et al. 2013; Strawn et al. 2014). Strains within the same serotype can present with different genetic lineages; conversely, strains with very similar phylogeny can express different serotypes (Wise et al. 2009; Wattiau et al. 2011).

Molecular serotyping, also called "DNA-based serotyping," has been proposed as a rapid and high-throughput alternative to aid in the identification of serotype (Strawn et al. 2014). This method uses PCR to identify either LPS- and flagellar-specific structural genes or genomic markers that are common to strains of the same serotype (Wattiau et al. 2011). Porwollik et al. (2004) characterized 79 *Salmonella* isolates with different serovars by a whole genome profile DNA microarray and identified genomic regions that differentiated different serotypes (Porwollik et al. 2004). Based on their research, Kim et al. (2006) developed a multiplex PCR assay, choosing 12 genetic loci that identified the 28 most common clinical *Salmonella* serovars. They performed a blind screening and reported 98/111 of clinical isolates were correctly typed (Kim et al. 2006). Afterwards, the same research group improved this assay and developed a high-throughput multiplex PCR system by incorporating fluorescently labeled primers. This new system targeted 16 genomic regions and successfully identified 89.6 % (673/751) *Salmonella* isolates from the top 50 most common serovars associated with human infections (Leader et al. 2009). Additional molecular typing methods described in detail below have been adapted for *Salmonella* serovar prediction (Ranieri et al. 2013; Wise et al. 2009; Chenu et al. 2012; Kerouanton et al. 2007). The advantages of molecular over traditional serotyping include increased reproducibility and independence from antigen agglutination performance. However, due to the diversity and variability of genes at these loci, many strains, especially from environmental sources, including wildlife, may be either misidentified or untypeable (Wattiau et al. 2011).

Genotyping Methods for Source Tracking of *Salmonella*

Molecular typing tools play a very important role in epidemiologic investigations for tracing back to the primary sources of contamination, elucidating the pathways of transmission, identifying the virulence potential of isolates, and intervening in the distribution of pathogens (Foley et al. 2007, 2009; Weigel et al. 2004). Common molecular typing methods for microbial source tracking of Gram-negative bacterial foodborne pathogens are summarized in Table 5.1. These methods can be categorized into three groups based on their genomic targets: (1) restriction enzyme sites, (2) PCR amplification of repetitive DNA, and (3) DNA sequence of multiple loci or whole genomes. Some of these methods provide typing resolution at the serotype level, whereas others can differentiate the genomic variations within the same

Table 5.1 Brief descriptions of common molecular typing methods

Name	Abbreviation	Mechanism	Genetic marker	Reference
Plasmid analysis	–	Restriction-based	Restriction sites in plasmids	(Foley et al. 2009)
Ribotyping	–	Restriction-based	Restriction sites in ribosomal RNA (rRNA) genes	(Bouchet et al. 2008)
Insertion sequence-restriction fragment length polymorphism analysis	IS-RFLP	Restriction-based	Insertion sequences in bacterial genomes	(Foley et al. 2009)
Pulsed-field gel electrophoresis	PFGE	Restriction-based	Rare restriction sites in bacterial genome	(Weigel et al. 2004)
Amplified fragment length polymorphism	AFLP	Restriction and amplification-based	Restriction sites in bacterial genomes	(Foley et al. 2009)
Randomly amplified polymorphic DNA PCR	RAPD-PCR	Amplification-based	Random primers with 6–10 base pair length	
Repetitive element PCR	Rep-PCR	Amplification-based	Repeated DNA sequence elements distributed along the bacterial genomes	(Wise et al. 2009)
Variable number of tandem repeats analysis	VNTR	Amplification-based	Directly repeated DNA motifs	(Foley et al. 2009)
Multilocus sequence typing	MLST	Sequencing-based	House-keeping genes	(Maatallah et al. 2013)
Single nucleotide polymorphism analysis	SNP	Sequencing-based	Single nucleotide polymorphisms	(Foley et al. 2009)

serovars (Wise et al. 2009; Wattiau et al. 2011). Key features for evaluating appropriate typing methods include: (1) the discriminatory ability to distinguish between nonclonal isolates; (2) the typeability to generate interpretable results; and (3) the reproducibility of results among different personnel and laboratories. These parameters need to be balanced against one another when choosing the most appropriate method for typing (Foley et al. 2007). In this section, pulsed-field gel electrophoresis (PFGE), multilocus sequence typing (MLST), repetitive element PCR (Rep-PCR), and clustered regularly interspaced short palindromic repeats (CRISPR) and CRISPR-associated sequences (CRISPR-cas) are reviewed in depth because of their demonstrated discriminatory value for molecular typing different serovars of *Salmonella*.

Pulsed-Field Gel Electrophoresis (PFGE)

PFGE is still the most widely used method for subtyping foodborne pathogens and is the current gold standard for the nationwide Foodborne Diseases Active Surveillance Network (FoodNet) and PulseNet (Foley et al. 2009; Olive and Bean 1999), which has been adopted by the Centers for Disease Control and Prevention (CDC) for *Salmonella* source tracking and epidemiologic investigations (Foley et al. 2007). It employs restriction fragment length polymorphism (RFLP) of chromosomal fragments derived from restriction enzyme digestion of the whole genome. These enzymes recognize specific short DNA sequences and cut the DNA whenever these sequences occur. For example, *Xba*I, the most common restriction enzyme in *Salmonella* PFGE, recognizes the DNA sequence "TCTAGA" (Patchanee et al. 2010; Li et al. 2014a). Subsequently, fragments are separated by electrophoresis that changes the polarity of the current at regular intervals and has the capability of separating relatively large DNA fragments from 20 to 800 kb (van Belkum et al. 2007; Lukinmaa et al. 2004; Foley et al. 2009). Thus, PFGE patterns of DNA bands generate unique profiles that identify different "pulsotypes."

General guidelines to determine the genetic relatedness among isolates have been proposed and suggest "closely related" strains are isolates that differ in a single genetic event and present only two to three band differences: "possible related" strains were those that differ in two genetic events, whereas "unrelated" strains were those that have more than two genetic events (Tenover et al. 1995). These criteria, adopted by CDC for outbreak investigations, also consider information such as background history of the *Salmonella* species/subspecies/serovar (human clinical infections, food vehicles), uniqueness of the pulsotype pattern for that serovar, and limitations of PFGE reproducibility (Barrett et al. 2006). PFGE patterns are stored locally in state labs or submitted to the national database through the PulseNet system. These databases are regularly searched for pattern matches in order to rapidly identify clusters of related isolates. PFGE has been applied to many outbreak investigations and has been instrumental in successfully identifying the sources of contamination (Behravesh et al. 2011; Greene et al. 2008; Buchholz et al. 2011).

PFGE has been used extensively to delineate the genetic diversity, virulence potential, and transmission pathways of *Salmonella* strains (Gorski et al. 2011, 2013; Foley et al. 2009). Also, it is widely accepted as a valid tool to predict *Salmonella* serovars (Li et al. 2014a; Strawn et al. 2014; Kerouanton et al. 2007). For example, *Salmonella* ($n=51$) isolates from surface water in Southeastern United States had 17 different PFGE types using two restriction enzymes, *Xba*I and *Bln*I, and were identified as nine serovars (Li et al. 2014a). Strains identified as Enteritidis ($n=6$, 11.8 %), Javiana ($n=3$, 5.9 %), and Thompson ($n=2$, 3.9 %) had indistinguishable PulseNet patterns with isolates from prior outbreaks. *Salmonella* isolates from various environmental samples collected from agricultural regions in California had indistinguishable, PFGE types among both water and wild amphibian and reptile isolates, suggesting the potential contamination of surface water was from wildlife exposure (Gorski et al. 2011). Specifically, *Salmonella* Arizonae (IIIa) and Diarizonae (IIIb), subspecies previously associated with amphibian and reptile hosts, were predominant in the study.

Although PFGE is considered a higher-resolution genotyping technique that is more discriminatory than serotyping (Gorski et al. 2011), different serotypes may have similar PFGE patterns due to slight genetic changes that do not alter the macrorestriction profiles (Patchanee et al. 2010). Conversely, strains of the same serotype and identical PFGE patterns may actually be derived from distinct genomic and evolutionary backgrounds. Furthermore, major drawbacks to the use of PFGE include labor intensity and time constraints, as well as the persistent appearance of "untypeable" strains that are resistant to enzyme restriction under the standard conditions of this assay.

PCR-Based Typing

The inability of serotyping and PFGE to correctly group *Salmonella* spp. strains has led to increased reliance on alternative molecular tools for subtyping this species. PCR-based molecular typing methods generally target repetitive elements that are distributed throughout the bacteria genome. There are three major types of repetitive elements in Gram-negative bacteria and include repetitive intergenic consensus (ERIC), repetitive extragenic palindromic (REP), and BOX element. REP elements consist of a 38-bp consensus sequence that contains a 5-bp variable loop in the middle region located within untranslated regions of operons (Yang and Ames 1988). Initially considered as potential regulatory sequences due to their palindromic character and their ability to form stable stem-loop structures in transcribed RNA (Higgins et al. 1982), several proposed functions of REP elements in gene regulation include mRNA stability, transcription termination, and recognition sites for DNA replication proteins (Yang and Ames 1988; Versalovic et al. 1991). ERIC regions are larger 126-bp elements, which are not related to the REP elements, and are also located in the extragenic regions of a highly conserved inverted repeat in the center (Versalovic et al. 1991). BOX elements are also inverted repeat elements, consisting of three subunits (Foley et al. 2009; Martin et al. 1992). These noncoding repetitive DNA elements share a high degree of evolutionary conservation (Versalovic et al. 1991) and may play an important role on DNA interactions, such as binding the DNA replication proteins gyase and polymerase I (Yang and Ames 1988; Gilson et al. 1986). These repetitive sequences can maintain themselves as "selfish DNA" and evolve through mild internal changes by gene conversion (Magee et al. 1992; Higgins et al. 1988).

DiversiLab is a commercial, semi-automated system for rep-PCR analysis. It offers a standardized, internet-based analytical platform, using microcapillary electrophoresis to achieve higher reproducibility and resolution compared to traditional agarose gel-based assays (Healy et al. 2005). Rep-PCR has been widely used for genetic analysis of various bacteria (Rajabi et al. 2011; Rademaker et al. 2000; Goto and Yan 2011; Roussel et al. 2010). Rajabi et al. (2011) isolated *Salmonella* ($n = 110$) from the Suwannee River and compared those isolates to other environmental or clinical strains using the DiversiLab rep-PCR system (Rajabi et al. 2011). These isolates were distributed into 16 rep-PCR genogroups (>85 % similarity per genogroup). Most (74 %) were clustered into the environmental genogroups, whereas 12 % were clustered with primarily clinically associated genogroups (Rajabi et al. 2011). Previous research revealed that DiversiLab rep-PCR was capable of predicting serotypes of *Salmonella* (Wise et al. 2009; Weigel et al. 2004). The major advantage of this platform over the traditional PFGE assay is reduced labor intensity, better assessment of horizontal gene transfer (Wattiau et al. 2011), and more reliable discrimination among some serotypes (Weigel et al. 2004).

Multiple-locus variable-number tandem repeat analysis (MLVA) and spacer oligotyping, or "spoligotyping," have also been used to improve discrimination of closely related subtypes (Kamerbeek et al. 1997). MLVA examines size variation of PCR products targeting multiple, well-characterized loci. Another family of repeated DNA identified in many prokaryotes is termed "clustered regularly interspaced short palindromic repeats" or CRISPRs (Jansen et al. 2002). *Salmonella* genomes have conserved CRISPR loci, but with hypervariable polarized insertions and deletions of spacers that are responsible for acquisition of foreign DNA (Pourcel et al. 2005). CRISPRs in *Salmonella* have been used as a high-throughput assay that has practical use in public health laboratories. Effectiveness of CRISPRs for typing *Salmonella* was determined by examining 783 isolates belonging to 130 serotypes, which revealed >3800 uniquely identified spacers that strongly correlated to both serotype and MLST genotype (Fabre et al. 2012). Similar results were obtained using spacer regions of CRISPRs 1 and 2 (Jansen et al. 2002; Touchon and Rocha 2010). CRISPR loci and CRISPR-associated sequence (cas) genes comprise the CRISPR-Cas system and had greater discriminatory power in contrast to serotyping or lineages derived through PFGE, MLST, and MLVA. Possible "biogeographic" molecular markers were identified since that may provide evolutionary details of possible source environments (Pettengill et al. 2014).

Multilocus Sequence Typing (MLST)

MLST defines sequence types based on comparative identity of genetic alleles that generally encode some type of "housekeeping" function. Sequences are derived from internal PCR amplification of informative regions of multiple genes or from whole genome sequencing. Defined gene targets for *Salmonella* MLST include *thrA* (aspartokinase homoserine dehydrogenase), *purE* (phosphoribosylaminoimidazole carboxylase), *sucA* (alpha ketoglutarate dehydrogenase), *hisD* (histidinol dehydrogenase), *aroC* (chorismate synthase), *hemD* (uroporphyrinogen III cosynthase), and *dnaN* (DNA polymerase III beta subunit). MLST databases and primers are publically available (http://pubmlst.org/databases.shtml). The multiple sequences are concatenated, aligned, and added to a custom database. Clustal, MUSCLE (Edgar

2004), and MEGA are widely used multiple alignment programs for fasta sequences (Dereeper et al. 2008), and tree building programs such as MEGA 6 (Tamura et al. 2013) or E-Burst (Achtman et al. 2012) are used to construct phylogenies of parsed BLAST hits of MLST loci. Maximum Likelihood scores among multiple replicates of trees are used to validate the accuracy of MLST methods (Pettengill et al. 2014).

The results of MLST are highly reproducible and have been validated as a molecular typing technique in many *Salmonella* studies. For example, Koteitishvili et al. (2002) compared the performance of MLST with PFGE for 182 environmental *Salmonella* isolates from poultry farms and 61 clinical isolates and reported MLST based on three genes had better discriminatory ability than PFGE. Furthermore, some *Salmonella* Hadar isolates within the same PFGE pulsotype were further divided into several MLST types, but not vice versa. Conversely, Fakhr et al. (2005) obtained much better discriminatory power with PFGE for 85 *Salmonella* Typhimurium strains isolated from cattle that had 50 PFGE patterns for which there was 100 % MLST identity based on four genes (*spaM*, *pduF*, *glnA*, and *manB*). Such inconsistencies may be due to the origin of isolates, genetic variation within serotypes, and/or gene target selection; however, it is noted that the application of MLST needs to be validated with great caution, and the inclusion of virulence genes is sometimes required to increase discriminatory ability (Foley et al. 2007).

The advantage of using housekeeping genes for MLST is that they are present in all strains within a species, as they are necessary for the basic maintenance of cellular function. However, because their function is critical to bacterial survival, these genes tend to be more conserved and less subject to selective pressure. The lack of discriminatory ability sometimes disqualifies them as good genetic markers for an epidemiologic tool. In this situation, the introduction of virulence genes may be desired, as they can increase the sensitivity and distinguish isolates that are closely related (Foley et al. 2007). Although MLST frequently clusters strains of the same serotypes, it can also indicate genetic similarity of strains with different serotypes but with the same evolutionary descent. Conversely, unrelated strains of the same serotype can also be identified by this method (Achtman et al. 2012). Unfortunately, some serotypes appear monophylogenic by this method; hence finer discriminatory tools are needed to delineate their lineages (Allard et al. 2013). The overall utility of the MLST vs. repetitive element-based typing is likely to vary with the lineage of strains within a species, and the most accurate and relevant method will likely be determined by whole genome sequence (WGS) data analysis.

Whole Genome Sequence Analysis (WGS)

With the advent of next-generation sequencing platforms, high-throughput sequencing is not only feasible but is rapidly becoming cost-effective. It is anticipated that WGS data will soon become routine for traceback investigations. Recent response to outbreaks has highlighted the power of WGS analysis (Hawkey et al. 2013). Whereas current CDC protocol for PFGE requires approximately 2 weeks, strain characterization by WGS can be completed in less than 1 week and provides more informative data.

WGS data can be compared against existing or inputted databases in using fasta sequences in GenBank (http://blast.ncbi.nlm.nih.gov/Blast.cgi). Genome Workbench (http://www.ncbi.nlm.nih.gov/tools/gbench/) is a publically available application offered through NCBI, using MUSCLE to perform multiple alignments locally. A common tool to construct phylogenies of homologous sequences with multiple alignments is the Phylogeny.fr platform (http://phylogeny.lirmm.fr/phylo_cgi/simple_phylogeny.cgi) (Dereeper et al. 2008), which aims to shorten the computation time and uses multiple programs to construct phylogenies with a "One Click" approach to alignment and tree building. Various algorithms are used to detect single nucleotide polymorphisms (SNPs) as reviewed by Nielsen et al. (2011). Genotyping methods take into account the likelihood with which the variants will be located based on counting allelic frequencies at specific sites on a reference genome and assigning a cutoff threshold. More advanced methods employ statistical frameworks to compute conditional probabilities, i.e., Bayesian methods. Filtering SNP candidates is required, as false-positive results are frequently observed (Nielsen et al. 2011; Altmann et al. 2012). In conjunction with SNP filtering, the use of appropriate alignment algorithms is essential to allow an acceptable number of mismatches between the reference and sequenced genome (Altmann et al. 2012). This dynamic approach to choosing an appropriate number of allowable mismatches is species specific and requires alignments to be sorted with respect to chromosomal position. SNP-based typing of WGS data is relatively new and has yet to be validated as a universal traceback method; however, informative SNPs found in gene clusters hold an ability to provide additional markers for epidemiologic investigations (Cao et al. 2013).

Recent comparison of several molecular typing methods to whole genome sequencing has revealed superior discrimination power for WGS compared to other typing methods (Cao et al. 2013). PFGE was not able to discriminate highly clonal strains, and neither PFGE nor MLST correctly identified the relationship among lineages within different strains of *Salmonella* Newport that were apparent by WGS. Evaluation of WGS for outbreak detection revealed that SNP analysis outperformed PFGE, but noted that the validation of this approach requires additional evaluation of sequencing platforms, analytical procedures, and larger databases (Leekitcharoenphon et al. 2014). As more genomes become available and methods become standardized, WGS should provide a better understanding of the evolution and ecology of *Salmonella* subspecies and serotypes. Genomic sequencing will likely provide the basis for the evolution of PulseNet into a similar network based on WGS. A "GenomTrakr" WGS database is currently under construction through a collaboration of CDC, FDA, USDA, and various state health departments and academic institutions. This system aims to provide rapid and much more accurate analysis of outbreak events based on WGS comparison of a database consisting of thousands of strains from locations throughout the world. WGS has the power to give high-throughput resolution of genomic information and may become a routine tool as a substitution for traditional bacteria typing methods (Leekitcharoenphon et al. 2012).

Evaluation of Virulence Potential

Finally, potential risks posed by *Salmonella* in agricultural systems cannot be adequately addressed without considering the virulence potential of these organisms. Serological methods do not adequately discriminate the virulence potential of environmental isolates or clearly delineate their relationship to outbreak lineages. Genetic typing methods have major benefits over serotyping, but can also have limitations imposed by discriminatory power and reproducibility (Foley et al. 2009; Lim et al. 2005). *Salmonella* strains vary widely in their virulence potential based on the relatively small number of genetically related strains that are associated with most human disease relative to the expansive diversity of strains from environmental sources, including domestic and wild animals. Unfortunately, identification of virulence markers that predict virulence potential of nontyphoid *Salmonella* gastroenteritis is hampered by the lack of an appropriate animal model. For example, many isolates of *Salmonella enterica* serotype Typhimurium derived from human disease do not cause disease in a mouse model of infection and may lack an essential "virulence" plasmid (Heithoff et al. 2008). Although experimental data in bovine models have implicated genes in *Salmonella* pathogenicity island 1 (SPI1), deletion of virulence-associated genes on SP1 did not eliminate the ability to cause human disease (Hu et al. 2008). Furthermore, variability in the presence/absence of virulence genes does not reliably correspond to genotype or serotype.

Recent studies have exploited WGS to identify genes associated with virulence and determine genetic lineages and biomarkers that could be used in traceback investigations during outbreaks. For example, Allard et al. (2013) found variable genes within serotype Enteritidis strains that are associated with virulence pathways and could be exploited for the development of rapid surveillance and typing methods (Allard et al. 2013). Division of *Salmonella* subsp. into at least two genetic populations, clades A and B, was based on 93 randomly selected loci and phylogenetic composition of core SNPs in the pan genome (den Bakker et al. 2014). Serotypes Typhi and Paratyphi A shared a recent common ancestry, which was attributed to convergent evolution due to adaptation in the human host. Niche differentiation between these clades was supported by segregation of genes encoding fimbriae (clade A) vs. glucuronidases (clade B), involved in adhesion and vertebrate host carbon utilization, respectively. Interestingly, a metalloprotease specific for clade B was associated with bacteria that use insects as alternative hosts. Furthermore, serotypes primarily involved in human gastroenteritis (Enteritidis, Typhimurium, Newport, and Javiana) were also in clade A, whereas the less common serotypes (Infantis, Montevideo, Schwarzengrund, Miami, and Muenster) were in clade B. Although sample size was limited in this study, congruent tree topologies implicated gene families that may be related to the subspecies adaptation to warm-blooded hosts.

Conclusions

Many options are available for detecting *Salmonella* in wildlife and other environmental sources. Technological advances have reduced confirmation time, but generally multiple enrichments and subculturing are still required due to the low levels of pathogens in environmental sources. Hence, the need for improved detection technology remains. We have proposed a 4-day method that uses common isolation approaches combined with a novel cross-plating technology for higher throughput and more cost-effective confirmation (Fig. 5.1). This method can be integrated with MPN for quantitative analysis and use of DiversiLab rep-PCR for strain typing. Although PFGE is still the "gold standard" for strain discrimination in traceback studies, recent collaborative efforts are building an international whole genome sequence database that will eventually permit public access to literally thousands of strains from different geographic locations, food products, agricultural, veterinary, and clinical sources. The prospect of a WGS approach not only holds promise from more rapid and accurate source tracking during outbreaks, but will potentially help to define virulence potential of the wide repertoire of diverse *Salmonella* populations for more rapid source tracking and better informed, science-based policy decisions and management strategies.

References

Achtman M, Wain J, Weill FX et al (2012) Multilocus sequence typing as a replacement for serotyping in *Salmonella enterica*. PLoS Pathog 8(6):e1002776. doi:10.1371/journal.ppat.1002776

Allard MW, Luo Y, Strain E et al (2013) On the evolutionary history, population genetics and diversity among Isolates of *Salmonella* Enteritidis PFGE Pattern JEGX01.0004. PLoS One 8(1):e55254. doi:10.1371/journal.pone.0055254

Altmann A, Weber P, Bader D et al (2012) A beginners guide to SNP calling from high-throughput DNA-sequencing data. Hum Genet 131(10):1541–1554. doi:10.1007/s00439-012-1213-z

APHA (1992) Standard methods for the examination of water and wastewater. American Public Health Association, Washington, DC

Barak JD, Liang AS (2008) Role of soil, crop debris, and a plant pathogen in *Salmonella* contamination of tomato plants. PLoS One 3(2):e1657

Barrett TJ, Gerner-Smidt P, Swaminathan B (2006) Interpretation of pulsed-field gel electrophoresis patterns in foodborne disease investigations and surveillance. Foodborne Pathog Dis 3(1):20–31. doi:10.1089/fpd.2006.3.20

Basel RM, Richter ER, Banwart GJ (1983) Monitoring microbial numbers in food by density centrifugation. Appl Environ Microbiol 45(3):1156–1159

Behravesh CB, Mody RK, Jungk J et al (2011) 2008 outbreak of *Salmonella* Saintpaul infections associated with raw produce. N Engl J Med 364(10):918–927. doi:10.1056/NEJMoa1005741

Bisha B, Perez-Mendez A, Danyluk MD et al (2011) Evaluation of modified moore swabs and continuous flow centrifugation for concentration of *Salmonella* and Escherichia coli O157: H7 from large volumes of water. J Food Prot 74(11):1934–1937. doi:10.4315/0362-028x.jfp-11-133

Bouchet V, Huot H, Goldstein R (2008) Molecular genetic basis of ribotyping. Clin Microbiol Rev 21(2):262. doi:10.1128/cmr.00026-07

Buchholz U, Bernard H, Werber D et al (2011) German outbreak of Escherichia coli O104:H4 associated with sprouts. N Engl J Med 365(19):1763–1770. doi:10.1056/NEJMoa1106482

Cao G, Meng J, Strain E et al (2013) Phylogenetics and differentiation of *Salmonella* Newport lineages by whole genome sequencing. PLoS One 8(2):e55687. doi:10.1371/journal.pone.0055687

Chenu JW, Cox JM, Pavic A (2012) Classification of *Salmonella enterica* serotypes from Australian poultry using repetitive sequence-based PCR. J Appl Microbiol 112(1):185–196. doi:10.1111/j.1365-2672.2011.05172.x

Chisholm SA, Crichton PB, Knight HI et al (1999) Molecular typing of *Salmonella* serotype Thompson strains isolated from human and animal sources. Epidemiol Infect 122(1):33–39. doi:10.1017/s0950268898001836

D'Aoust J-Y, John M (2007) *Salmonella* species. In: Doyle M-P, Beuchat L-R (eds) Food microbiology—fundamentals and frontiers, 3rd edn. ASM Press, Washington, DC, pp 187–236

Danyluk MD, Jones TM, Abd SJ et al (2007) Prevalence and amounts of *Salmonella* found on raw California almonds. J Food Prot 70(4):820–827

den Bakker HC, Allard MW, Bopp D et al (2014) Rapid whole-genome sequencing for surveillance of *Salmonella* enterica serovar enteritidis. Emerg Infect Dis 20(8):1306–1314. doi:10.3201/eid2008.131399

Dereeper A, Guignon V, Blanc G et al (2008) Phylogeny.fr: robust phylogenetic analysis for the non-specialist. Nucleic Acids Res 36(Web Server issue):W465–W469. doi:10.1093/nar/gkn180

Edgar RC (2004) MUSCLE: a multiple sequence alignment method with reduced time and space complexity. BMC Bioinformatics 5:113. doi:10.1186/1471-2105-5-113

Escartin EF, Lozano JS, Garcia OR et al (2002) Potential *Salmonella* transmission from ornamental fountains. J Environ Health 65(4):9–12

Espy MJ, Uhl JR, Sloan LM et al (2006) Real-time PCR in clinical microbiology: applications for routine laboratory testing. Clin Microbiol Rev 19(1):165–256. doi:10.1128/cmr.19.1.165-256.2006

Fabre L, Zhang J, Guigon G et al (2012) CRISPR typing and subtyping for improved laboratory surveillance of *Salmonella* infections. PLoS One 7(5):e36995. doi:10.1371/journal.pone.0036995

Farber JM, Sharpe AN (1984) Improved bacterial recovery by membrane filters in the presence of food debris. Appl Environ Microbiol 48(2):441–443

FDA (2013) Food Safety Modernization Act (FSMA). http://www.fda.gov/Food/GuidanceRegulation/FSMA/ucm304045.htm

Fakhr MK, Nolan LK, Logue CM (2005) Multilocus sequence typing lacks the discriminatory ability of pulsed-field gel electrophoresis for typing Salmonella enterica serovar Typhimurium. J Clin Microbiol 43(5):2215–2219

Foley SL, Lynne AM, Nayak R (2009) Molecular typing methodologies for microbial source tracking and epidemiological investigations of Gram-negative bacterial foodborne pathogens. Infect Genet Evol 9(4):430–440. doi:10.1016/j.meegid.2009.03.004

Foley SL, Zhao S, Walker RD (2007) Comparison of molecular typing methods for the differentiation of *Salmonella* foodborne pathogens. Foodborne Pathog Dis 4(3):253–276. doi:10.1089/fpd.2007.0085

Franz E, van Bruggen AHC (2008) Ecology of E. coli O157:H7 and *Salmonella* enterica in the primary vegetable production chain. Crit Rev Microbiol 34(3–4):143–161. doi:10.1080/10408410802357432

Gibson KE, Schwab KJ (2011) Tangential-flow ultrafiltration with integrated inhibition detection for recovery of surrogates and human pathogens from large-volume source water and finished drinking water. Appl Environ Microbiol 77(1):385–391. doi:10.1128/aem.01164-10

Gilliss D, Cronquist AB, Cartter M et al (2013) Incidence and trends of infection with pathogens transmitted commonly through food - foodborne diseases active surveillance network, 10 US sites, 1996-2012. MMWR Morb Mortal Wkly Rep 62(15):283–287

Gilson E, Perrin D, Clement JM et al (1986) Palindromic units from E. coli as binding sites for a chromoid-associated protein. FEBS Lett 206(2):323–328

Gorski L, Jay-Russell MT, Liang AS et al (2013) Diversity of pulsed-field gel electrophoresis pulsotypes, serovars, and antibiotic resistance among *Salmonella* isolates from wild amphibians and reptiles in the California central coast. Foodborne Pathog Dis 10(6):540–548. doi:10.1089/fpd.2012.1372

Gorski L, Parker CT, Liang A et al (2011) Prevalence, distribution, and diversity of *Salmonella enterica* in a major produce region of California. Appl Environ Microbiol 77(8):2734–2748. doi:10.1128/aem.02321-10

Goto DK, Yan T (2011) Genotypic diversity of Escherichia coli in the water and soil of tropical watersheds in Hawaii. Appl Environ Microbiol 77(12):3988–3997. doi:10.1128/aem.02140-10

Gould LH, Nisler AL, Herman KM et al (2011) Surveillance for foodborne disease outbreaks-United States, 2008 (Reprinted from MMWR, vol 60, pg 1197-1202, 2011). JAMA 306(20):2212–2214

Gould LH, Walsh KA, Vieira AR et al (2013) Surveillance for foodborne disease outbreaks - United States, 1998-2008. MMWR Surveill Summ 62(2):1–34

Greene SK, Daly ER, Talbot EA et al (2008) Recurrent multistate outbreak of *Salmonella* Newport associated with tomatoes from contaminated fields, 2005. Epidemiol Infect 136(2):157–165. doi:10.1017/s095026880700859x

Hanning IB, Nutt JD, Ricke SC (2009) Salmonellosis outbreaks in the United States due to fresh produce: sources and potential intervention measures. Foodborne Pathog Dis 6(6):635–648. doi:10.1089/fpd.2008.0232

Hawkey J, Edwards DJ, Dimovski K et al (2013) Evidence of microevolution of *Salmonella* Typhimurium during a series of egg-associated outbreaks linked to a single chicken farm. BMC Genomics 14(1):800, info:pmid/24245509

Healy M, Huong J, Bittner T et al (2005) Microbial DNA typing by automated repetitive-sequence-based PCR. J Clin Microbiol 43(1):199–207. doi:10.1128/jcm.43.1.199-207.2005

Heithoff DM, Shimp WR, Lau PW et al (2008) Human *Salmonella* clinical isolates distinct from those of animal origin. Appl Environ Microbiol 74(6):1757–1766. doi:10.1128/aem.02740-07

Higgins CF, Ames GF-L, Barnes WM et al (1982) A novel intercistronic regulatory element of prokaryotic operons. Nature 298(5876):760–762. doi:10.1038/298760a0

Higgins CF, McLaren RS, Newbury SF (1988) Repetitive extragenic palindromic sequences, mRNA stability and gene expression: evolution by gene conversion? A review. Gene 72(1–2):3–14

Hu Q, Coburn B, Deng W et al (2008) *Salmonella* enterica serovar Senftenberg human clinical isolates lacking SPI-1. J Clin Microbiol 46(4):1330–1336. doi:10.1128/jcm.01255-07

Ijabadeniyi OA, Debusho LK, Vanderlinde M et al (2011) Irrigation water as a potential preharvest source of bacterial contamination of vegetables. J Food Saf 31(4):452–461. doi:10.1111/j.1745-4565.2011.00321.x

Jansen R, Embden JD, Gaastra W et al (2002) Identification of genes that are associated with DNA repeats in prokaryotes. Mol Microbiol 43(6):1565–1575

Jasson V, Jacxsens L, Luning P et al (2010) Alternative microbial methods: an overview and selection criteria. Food Microbiol 27(6):710–730. doi:10.1016/j.fm.2010.04.008

Kamerbeek J, Schouls L, Kolk A et al (1997) Simultaneous detection and strain differentiation of Mycobacterium tuberculosis for diagnosis and epidemiology. J Clin Microbiol 35(4):907–914

Kerouanton A, Marault M, Lailler R et al (2007) Pulsed-field gel electrophoresis subtyping database for foodborne *Salmonella* enterica serotype discrimination. Foodborne Pathog Dis 4(3):293–303. doi:10.1089/fpd.2007.0090

Kim S, Frye JG, Hu J et al (2006) Multiplex PCR-based method for identification of common clinical serotypes of *Salmonella* enterica subsp. enterica. J Clin Microbiol 44(10):3608–3615. doi:10.1128/jcm.00701-06

Kotetishvili M, Stine OC, Kreger A, Morris JG Jr, Sulakvelidze A (2002) Multilocus sequence typing for characterization of clinical and environmental Salmonella strains. J Clin Microbiol 40:1626–1635

Leader B, Frye J, Hu J et al (2009) High-throughput molecular determination of *Salmonella enterica* serovars by use of multiplex PCR and capillary electrophoresis analysis. J Clin Microbiol 47(5):1290–1299. doi:10.1128/JCM.02095-08

Leekitcharoenphon P, Lukjancenko O, Friis C et al (2012) Genomic variation in *Salmonella enterica* core genes for epidemiological typing. BMC Genomics 13:88. doi:10.1186/1471-2164-13-88

Leekitcharoenphon P, Nielsen EM, Kaas RS et al (2014) Evaluation of whole genome sequencing for outbreak detection of *Salmonella enterica*. PLoS One 9(2):e87991. doi:10.1371/journal.pone.0087991

Li B, Vellidis G, Liu H et al (2014a) Diversity and antimicrobial resistance of Salmonella enterica isolated from surface water in southeastern U.S. Appl Environ Microbiol 80:6355–6365

Li B, Vellidis G, Liu H et al (2014b) Diversity and antimicrobial resistance of Salmonella enterica isolated from surface water in Southeastern U.S. Appl Environ Microbiol. doi:10.1128/aem.02063-14

Lim H, Lee KH, Hong CH et al (2005) Comparison of four molecular typing methods for the differentiation of *Salmonella* spp. Int J Food Microbiol 105(3):411–418. doi:10.1016/j.ijfoodmicro.2005.03.019

Lukinmaa S, Nakari UM, Eklund M et al (2004) Application of molecular genetic methods in diagnostics and epidemiology of food-borne bacterial pathogens. APMIS 112(11–12):908–929. doi:10.1111/j.1600-0463.2004.apm11211-1213.x

Luo Z, Gu G, Giurcanu MC et al (2014) Development of a novel cross-streaking method for isolation, confirmation, and enumeration of *Salmonella* from irrigation ponds. J Microbiol Methods 101:86–92. doi:10.1016/j.mimet.2014.03.012

Maatallah M, Bakhrouf A, Habeeb MA et al (2013) Four genotyping schemes for phylogenetic analysis of pseudomonas aeruginosa: comparison of their congruence with multi-locus sequence typing. PLOS One 8(12):e82069. doi:10.1371/journal.pone.0082069

Madsen M (1994) Enumeration of *Salmonella* in crocodile pond water by direct plate counts and by the MPN technique. Water Res 28(9):2035–2037. doi:10.1016/0043-1354(94)90180-5

Magee PT, Bowdin L, Staudinger J (1992) Comparison of molecular typing methods for Candida albicans. J Clin Microbiol 30(10):2674–2679

Mandal PK, Biswas AK, Choi K et al (2011) Methods for rapid detection of foodborne pathogens: an overview. Am J Food Technol 6(2):87–102. doi:10.3923/ajft.2011.87.102

Martin B, Humbert O, Camara M et al (1992) A highly conserved repeated DNA element located in the chromosome of Streptococcus pneumoniae. Nucleic Acids Res 20(13):3479–3483

McEgan R, Rodrigues CAP, Sbodio A et al (2013) Detection of *Salmonella* spp. from large volumes of water by modified Moore swabs and tangential flow filtration. Lett Appl Microbiol 56(2):88–94. doi:10.1111/lam.12016

McKillip JL, Drake M (2004) Real-time nucleic acid-based detection methods for pathogenic bacteria in food. J Food Prot 67(4):823–832

McLoughlin KS (2011) Microarrays for pathogen detection and analysis. Brief Funct Genomics 10(6):342–353. doi:10.1093/bfgp/elr027

Mull B, Hill VR (2009) Recovery and detection of Escherichia coli O157:H7 in surface water, using ultrafiltration and real-time PCR. Appl Environ Microbiol 75(11):3593–3597. doi:10.1128/aem.02750-08

Nielsen R, Paul JS, Albrechtsen A et al (2011) Genotype and SNP calling from next-generation sequencing data. Nat Rev Genet 12(6):443–451. doi:10.1038/nrg2986

Odumeru JA, León-Velarde CG (2012) Salmonella detection methods for food and food ingredients. In: Mahmoud BSM (ed) Salmonella—a dangerous foodborne pathogen. InTech, Rijeka, Croatia. doi:10.5772/29526

Olive DM, Bean P (1999) Principles and applications of methods for DNA-based typing of microbial organisms. J Clin Microbiol 37(6):1661–1669

Pachepsky Y, Shelton DR, McLain JET et al (2011) Irrigation waters as a source of pathogenic microorganisms in produce: a review. Adv Agron 113(113):73–138. doi:10.1016/b978-0-12-386473-4.00007-5

Patchanee P, Molla B, White N et al (2010) Tracking *Salmonella* contamination in various watersheds and phenotypic and genotypic diversity. Foodborne Pathog Dis 7(9):1113–1120. doi:10.1089/fpd.2010.0572

Payne MJ, Campbell S, Patchett RA et al (1992) The use of immobilized lectins in the separation of Staphylococcus aureus, Escherichia coli Listeria and Salmonella spp from pure cultures and foods. J Appl Bacteriol 73(1):41–52

Pettengill JB, Timme RE, Barrangou R et al (2014) The evolutionary history and diagnostic utility of the CRISPR-Cas system within *Salmonella enterica* ssp. enterica. PeerJ 2:e340. doi:10.7717/peerj.340

Pfleger S, Benyr G, Sommer R et al (2003) Pattern of *Salmonella* excretion in amphibians and reptiles in a vivarium. Int J Hyg Environ Health 206(1):53–59. doi:10.1078/1438-4639-00184

Pianetti A, Sabatini L, Bruscolini F et al (2004) Faecal contamination indicators, *Salmonella*, Vibrio and Aeromonas in water used for the irrigation of agricultural products. Epidemiol Infect 132(2):231–238. doi:10.1017/s095026880300181x

Plusquellec A, Beucher M, Lelay C et al (1994) Uptake and retention of *Salmonella* by bivalve shellfish. J Shellfish Res 13(1):221–227

Polo F, Figueras MJ, Inza I et al (1998) Relationship between presence of *Salmonella* and indicators of faecal pollution in aquatic habitats. FEMS Microbiol Lett 160(2):253–256. doi:10.1111/j.1574-6968.1998.tb12919.x

Porwollik S, Boyd EF, Choy C et al (2004) Characterization of *Salmonella* enterica subspecies I genovars by use of microarrays. J Bacteriol 186(17):5883–5898. doi:10.1128/jb.186.17.5883-5898.2004

Pourcel C, Salvignol G, Vergnaud G (2005) CRISPR elements in Yersinia pestis acquire new repeats by preferential uptake of bacteriophage DNA, and provide additional tools for evolutionary studies. Microbiology 151(Pt 3):653–663. doi:10.1099/mic.0.27437-0

Rademaker JLW, Hoste B, Louws FJ et al (2000) Comparison of AFLP and rep-PCR genomic fingerprinting with DNA-DNA homology studies: Xanthomonas as a model system. Int J Syst Evol Microbiol 50:665–677

Rajabi M, Jones M, Hubbard M et al (2011) Distribution and genetic diversity of Salmonella enterica in the upper Suwannee river. Int J Microbiol. doi:10.1155/2011/461321

Ranieri ML, Shi CL, Switt AIM et al (2013) Comparison of typing methods with a new procedure based on sequence characterization for *Salmonella* serovar prediction. J Clin Microbiol 51(6):1786–1797. doi:10.1128/jcm.03201-12

Rasooly A, Herold KE (2008) Food microbial pathogen detection and analysis using DNA microarray technologies. Foodborne Pathog Dis 5(4):531–550. doi:10.1089/fpd.2008.0119

Reche MP, Echeita MA, de los Rios JEG et al (2003) Comparison of phenotypic and genotypic markers for characterization of an outbreak of Salmonella serotype Havana in captive raptors. J Appl Microbiol 94(1):65–72. doi:10.1046/j.1365-2672.2003.01791.x

Roussel S, Felix B, Colaneri C et al (2010) Semi-automated repetitive-sequence-based polymerase chain reaction compared to pulsed-field gel electrophoresis for Listeria monocytogenes subtyping. Foodborne Pathog Dis 7(9):1005–1012. doi:10.1089/fpd.2009.0450

Scallan E, Hoekstra RM, Angulo FJ et al (2011) Foodborne illness acquired in the United States-major pathogens. Emerg Infect Dis 17(1):7–15. doi:10.3201/eid1701.P11101

Schrader KN, Fernandez-Castro A, Cheung WKW et al (2008) Evaluation of commercial antisera for *Salmonella* serotyping. J Clin Microbiol 46(2):685–688. doi:10.1128/jcm.01808-07

Shaw SJ, Blais BW, Nundy DC (1998) Performance of the dynabeads anti-*Salmonella* system in the detection of *Salmonella* species in foods, animal feeds, and environmental samples. J Food Prot 61(11):1507–1510

Shi C, Singh P, Ranieri ML et al (2013) Molecular methods for serovar determination of *Salmonella*. Crit Rev Microbiol. doi:10.3109/1040841x.2013.837862

Shin HH, Hwang BH, Seo JH et al (2014) Specific discrimination of three pathogenic *Salmonella enterica* subsp. enterica serotypes by carB-based oligonucleotide microarray. Appl Environ Microbiol 80(1):366–373. doi:10.1128/aem.02978-13

Stone GG, Oberst RD, Hays MP et al (1994) Detection of *Salmonella* serovars from clinical-samples by enrichment broth cultivation PCR procedure. J Clin Microbiol 32(7):1742–1749

Strawn LK, Danyluk MD, Worobo RW et al (2014) Distributions of *Salmonella* subtypes differ between two US produce-growing regions. Appl Environ Microbiol 80(13):3982–3991. doi:10.1128/aem.00348-14

Tamura K, Stecher G, Peterson D et al (2013) MEGA6: molecular evolutionary genetics analysis version 6.0. Mol Biol Evol 30(12):2725–2729. doi:10.1093/molbev/mst197

Tenover FC, Arbeit RD, Goering RV et al (1995) Interpreting chromosomal DNA restriction patterns produced by pulsed-filed gel electrophoresis-criteria for bacterial strain typing. J Clin Microbiol 33(9):2233–2239

Touchon M, Rocha EP (2010) The small, slow and specialized CRISPR and anti-CRISPR of Escherichia and *Salmonella*. PLoS One 5(6):e11126. doi:10.1371/journal.pone.0011126

van Belkum A, Tassios PT, Dijkshoorn L et al (2007) Guidelines for the validation and application of typing methods for use in bacterial epidemiology. Clin Microbiol Infect 13:1–46. doi:10.1111/j.1469-0691.2007.01786.x

van Duijkeren E, Wannet WJB, Houwers DJ et al (2002) Serotype and phage type distribution of *Salmonella* strains isolated from humans, cattle, pigs, and chickens in the Netherlands from 1984 to 2001. J Clin Microbiol 40(11):3980–3985. doi:10.1128/jcm.40.11.3980-3985.2002

Versalovic J, Koeuth T, Lupski JR (1991) Distribution of repetitive DNA-sequences in eubacteria and application to fingerprinting of bacterial genomes. Nucleic Acids Res 19(24):6823–6831. doi:10.1093/nar/19.24.6823

Wattiau P, Boland C, Bertrand S (2011) Methodologies for *Salmonella enterica* subsp. enterica subtyping: gold standards and alternatives. Appl Environ Microbiol 77(22):7877–7885. doi:10.1128/aem.05527-11

Weigel RM, Qiao BZ, Teferedegne B et al (2004) Comparison of pulsed field gel electrophoresis and repetitive sequence polymerase chain reaction as genotyping methods for detection of genetic diversity and inferring transmission of *Salmonella*. Vet Microbiol 100(3–4):205–217. doi:10.1016/j.vetmic.2004.02.009

Winfield MD, Groisman EA (2003) Role of nonhost environments in the lifestyles of *Salmonella* and *Escherichia coli*. Appl Environ Microbiol 69(7):3687–3694. doi:10.1128/Aem.69.7.3687-3694.2003

Wise MG, Siragusa GR, Plumblee J et al (2009) Predicting *Salmonella enterica* serotypes by repetitive sequence-based PCR. J Microbiol Methods 76(1):18–24. doi:10.1016/j.mimet.2008.09.006

Yang Y, Ames GF (1988) DNA gyrase binds to the family of prokaryotic repetitive extragenic palindromic sequences. Proc Natl Acad Sci U S A 85(23):8850–8854

Chapter 6
Reducing the Risk of Foodborne Transmission of Nipah Virus

Stephen P. Luby, Nazmun Nahar, and Emily S. Gurley

Abstract Nipah virus is a paramyxovirus whose wildlife host is large fruit bats in the genus *Pteropus*. Antibodies against Nipah virus and closely related Henipaviruses are common among old world fruit bats that live in Australia, across South and Southeast Asia and sub-Saharan Africa, but human infections with Nipah virus are uncommon. When humans are infected with Nipah virus 40 – 70 % die. People who are infected with Nipah virus can transmit the infection to other people. In the first recognized and largest Nipah outbreak, Nipah virus was transmitted from bats to pigs in Malaysia. A widespread outbreak among pigs led to infections among people who had close contact with infected pigs. The outbreak was arrested by culling over 900,000 pigs. Human Nipah virus infections have been identified in Bangladesh nearly every year from 2001 through 2014. The most common pathway of human Nipah infection is from drinking raw date palm sap. *Pteropus* bats frequently visit trees at night where sap is being collected and lick the sap stream as it flows into the collection pot. Drinking fresh date palm sap is a widely enjoyed seasonal delicacy in Bangladesh. Focused intensive interventions in limited areas discouraging people from drinking raw date palm sap or encouraging sap harvesters to use skirts to prevent bats access to the sap stream have reduced but not eliminated high risk practices.

Keywords Behavior change • Environmental health • Epidemiology • PCR • Source tracking • Wildlife

---

S.P. Luby (✉)
Center for Innovation in Global Health, Stanford University, Stanford, CA, USA
e-mail: sluby@stanford.edu

N. Nahar
Swiss Tropical Public Health Institute, Basel, Switzerland

E.S. Gurley
Centre for Communicable Diseases, ICDDR, B, Dhaka, Bangladesh

M. Jay-Russell, M.P. Doyle (eds.), *Food Safety Risks from Wildlife*,
Food Microbiology and Food Safety, DOI 10.1007/978-3-319-24442-6_6

Introduction

Nipah virus is a member of the recently discovered Henipavirus genus of the family *Paramyxoviridae*. It is closely related to Hendra virus, a Henipavirus that has caused numerous outbreaks in horses and occasional human infections in Australia (Hess et al. 2011). The wildlife reservoir for both Nipah and Hendra virus is large fruit bats in the genus *Pteropus*. Various species of *Pteropus* bats live across South and Southeast Asia, Australia, and in Madagascar (Nowak 1994).

Nipah virus infection does not lead to clinical illness in the bats (Middleton et al. 2007). Indeed, the virus likely co-evolved with these bats over millions of years (Field et al. 2007). Human infection with Nipah virus commonly causes severe disease, including encephalitis and respiratory insufficiency (Goh et al. 2000; Hossain et al. 2008). Between 40 and 70 % of people infected with Nipah virus die (Chua et al. 2000; Luby et al. 2009a). Moreover, when a person becomes infected with Nipah virus, they can pass the infection onto other people. In a review of the 122 Nipah virus cases identified in Bangladesh from 2001 through 2007, 62 (51 %) developed illness 5–15 days after close contact with another Nipah virus patient (Luby et al. 2009a).

Nipah virus is an example of a stage III zoonotic disease (Wolfe et al. 2007), which is a zoonotic infection that, when it spills over into humans, can be transmitted from person-to-person. The average number of people infected by each new case of Nipah virus in humans is <1, so person-to-person outbreaks of Nipah virus are characterized by stuttering chains of transmission that eventually burn themselves out (Lloyd-Smith et al. 2009). During human infection, this RNA virus is in an environment that would select for mutations that would adapt the virus for more efficient human infection and person-to-person transmission. Thus, each human infection with Nipah virus presents a risk for virus evolution that could produce a strain of Nipah virus adapted to humans with the potential to generate a devastating global pandemic. Hence, Nipah virus not only is a local problem among people who encounter this bat virus in Asia but also represents a broader global risk. This chapter borrows background information from earlier reviews written by the authors (Luby and Gurley 2012; Luby 2013, 2014), but focuses on issues of foodborne transmission from wildlife and efforts at preventing this transmission.

Human Nipah Virus Outbreaks

Malaysia/Singapore

Human Nipah virus infection was first recognized in a large outbreak among domestic pig (*Sus scrofa*) farmers in peninsular Malaysia from September 1998 through May 1999 (Paton et al. 1999; Chua et al. 2000; Chua 2003). Compared with people who lived on the same farms but were not infected, Nipah virus-infected patients were more likely to have direct contact with pigs that appeared sick and to have close contact with pigs through feeding pigs, processing baby pigs, assisting in breeding of pigs, assisting in the birth of pigs, injecting or medicating pigs, and handling dead pigs.

Between March 10 and 19, 1999, 11 workers in one of Singapore's abattoirs became infected with Nipah virus and developed encephalitis or pneumonia (Paton et al. 1999). One worker died. Compared to controls who were also abattoir workers, cases were more likely to be exposed to pig urine or feces from pigs imported from Malaysia during the Malaysian Nipah virus outbreak. Nipah virus RNA recovered from autopsy specimens from the one deceased worker had a nucleotide sequence that was identical to isolates from humans and from pigs in Malaysia (Paton et al. 1999).

Most pigs infected with Nipah virus had mild illness, with ≤5 % of infected adult pigs dying of the disease (Mohd Nor et al. 2000). Severely affected pigs had extensive involvement of their lungs with a giant cell pneumonia. The multinucleated syncytial cells of the lungs and the epithelial cells lining the upper airways contained Nipah virus antigen (Chua et al. 2000). Nipah virus was recovered from respiratory secretions of infected pigs, and Nipah virus antigen was also detected in renal tubular epithelial cells (Chua et al. 2000; Middleton et al. 2002).

The isolation of Nipah virus from pigs' lungs and respiratory secretions combined with the observation that human cases of Nipah virus infection had more contact with pigs' secretions and excretions than controls suggests that Nipah virus was transmitted from infected pigs to humans through contaminated saliva and possibly urine. The human outbreak of Nipah virus infection ceased after widespread deployment of personal protective equipment to people contacting sick pigs, restrictions on livestock movements, and culling over 900,000 pigs (Uppal 2000). Since the outbreak ended in 1999, through 2014 no human or porcine Nipah virus infections have been reported from Malaysia. Pork consumption was not associated with Nipah virus infection in either Malaysia or Singapore. Rather, exposure to infected pigs was the primary pathway of transmission.

India/Bangladesh

The epidemiology of human Nipah virus infection in Bangladesh/India has been quite different than in Malaysia. Since 2001, human infections with Nipah virus have been recognized in South Asia most years (Fig. 6.1). The cases in Bangladesh have largely clustered in western/northwestern Bangladesh (Fig. 6.2). The two recognized Indian outbreaks occurred in West Bengal, close to where cases have been repeatedly identified in Bangladesh (Fig. 6.2).

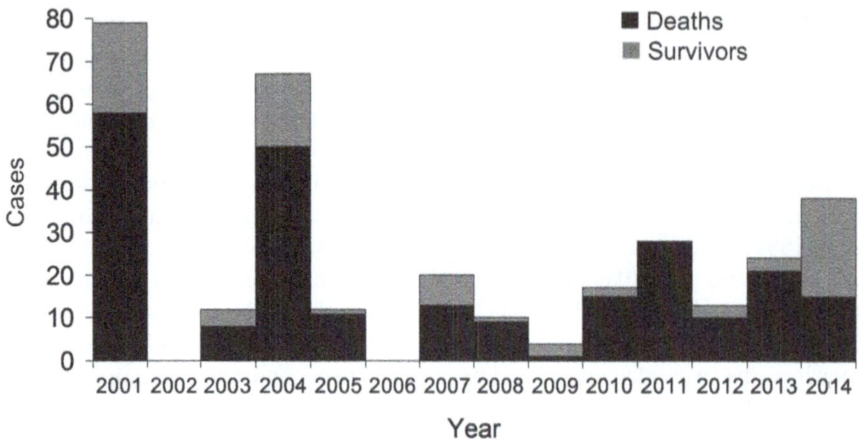

Fig. 6.1 Human infections with Nipah virus infection in Bangladesh and India by year

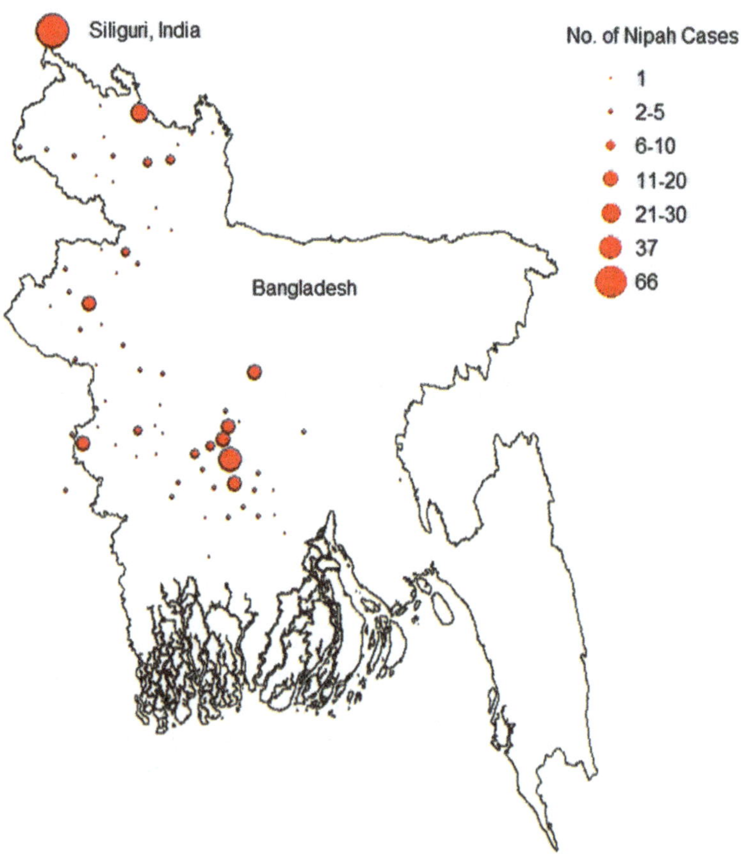

Fig. 6.2 Location of human Nipah virus infections in Bangladesh and India 2001–2014

Nipah Virus Foodborne Transmission Through Date Palm Sap

Outbreak investigations in Bangladesh have identified drinking raw date palm sap as the most common pathway of Nipah virus transmission from *Pteropus* bats to people. The 2005 outbreak investigation in Tangail, Bangladesh, found that Nipah virus cases were 7.9 times more likely to report drinking raw date palm sap in the 10 days before they developed illness than neighborhood matched controls (Luby et al. 2006). Similarly in the 2008 outbreak in Manikgonj and Rajbari districts in Bangladesh, cases were 25 times more likely than controls to report drinking raw date palm sap (Rahman et al. 2012). In outbreaks in Faridpur, Bangladesh, in 2010, and in Lalmonirhat in 2011, cases were again significantly more likely than controls to report drinking raw date palm sap in the 2 weeks prior to the onset of illness (Chakraborty 2011; Sazzad et al. 2013). The outbreaks of human Nipah virus infection in Bangladesh and India coincided with the date palm sap harvesting season (Luby et al. 2009a).

In Bangladesh, date palm sap harvesters most commonly begin collecting sap following the first cold night in November or December and continue collecting most regularly through January and early February, though some harvesters continue to collect in at least a few trees through March, and some harvesters collect sap for fermentation year-round. At the beginning of the season, the bark is shaved off of one side of the tree (*Phoenix sylvestris*) near the top in a V-shape, and a small hollow bamboo tap is placed at the base of the V (Nahar et al. 2010). In the late afternoon, the date palm sap harvester climbs the tree, scrapes the area where the bark is denuded so the sap can flow freely, and ties a 2–4-l clay pot underneath the tap. During the night as the sap rises to the top of the tree, some sap oozes out from where the bark is denuded, flows through the tap, and drips into the clay pot. Palm sap harvesters climb the trees at daybreak to gather the clay pots.

Most date palm sap in Bangladesh is cooked and made into molasses that is a popular sweetener for cakes and other desserts (Halim et al. 2008; Nahar et al. 2010). Some date palm sap is sold raw for immediate consumption. Harvesters will often share raw sap as a treat with family members or neighbors, or walk house to house near where the sap was collected and offer it for sale to neighbors. Sometimes people come to the harvester's home to purchase sap, or the harvester sells his raw sap in the local market. A few hours after sunrise, presumably because of progressive fermentation, the date palm sap is less sweet and so sap sellers lower the price.

Sap harvesters and villagers report that bats and other animals sometimes visit the trees during sap collection. Sap harvesters commonly find bat feces outside of the clay pot or floating in the sap and occasionally find drowned dead bats floating in the pots (Luby et al. 2006; Nahar et al. 2010). Infrared wildlife photography confirms that *Pteropus* bats, the presumed reservoir of Nipah virus in Bangladesh, commonly visit date palm trees during collection and lick the sap stream (Khan et al. 2010) (Fig. 6.3). Infrared cameras placed in the seven trees that were the source of

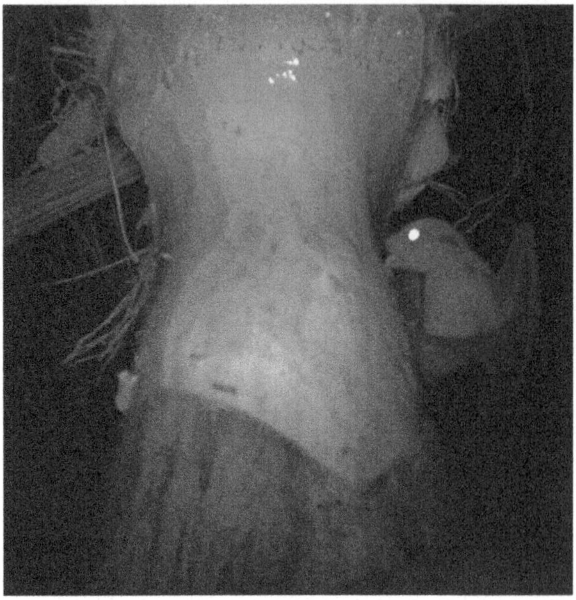

Fig. 6.3 Night-time infrared photograph of a fruit bat licking a stream of date palm sap

raw date palm sap drunk by the human Nipah virus cases in the 2008 Manikgonj/Rajbari outbreak identified an average of four *Pteropus* bat visits per tree where the bat licked the sap stream, per night of observation (Rahman et al. 2012).

Date palm sap is a plausible vehicle for transmission of Nipah virus from *Pteropus* bats to people. *Pteropus* bats occasionally shed Nipah virus in their saliva (Wacharapluesadee et al. 2005; Anthony et al. 2013). The infrared camera studies confirm that *Pteropus* bats directly lick raw date palm sap and occasionally urinate in the sap collection pot (Khan et al. 2010; Rahman et al. 2012). Nipah virus inoculated in fruit juice was recoverable at high concentrations 3 days later (Fogarty et al. 2008). Nipah virus that was inoculated into a solution of 14 % sucrose and 0.21 % bovine serum albumin to mimic date palm sap survived for 8 days at 22 °C with no reduction in titer (de Wit et al. 2014); two of eight hamsters that drank this artificial date palm sap developed Nipah virus infection and died (de Wit et al. 2014). Nipah virus has not been isolated directly from date palm sap. This is not surprising because *Pteropus* shedding of Nipah virus is intermittent (Wacharapluesadee et al. 2010), and with the median 10-day incubation period from exposure to date palm sap to illness (Rahman et al. 2012), and the time required to recognize an outbreak of Nipah virus, outbreak investigation teams have only been able to collect sap samples from trees weeks after the likely transmission event.

Some date palm sap in Bangladesh is fermented into palm wine (*tari*). One Nipah virus case in India (Arankalle et al. 2011) and an outbreak in Bangladesh (Islam 2012) have been tied to drinking this fermented date palm sap. Apparently, at least in some cases, the alcohol content of the fermented sap is insufficient to inactivate the virus.

Nipah Virus Person-to-Person Transmission

Person-to-person transmission of Nipah virus has been repeatedly identified in Bangladesh/India. The first Nipah virus outbreak recognized in the Indian subcontinent was a large outbreak affecting 66 people in Siliguri, India, in 2001. The outbreak apparently originated from an unidentified patient admitted to Siliguri District Hospital who transmitted infection to 11 additional patients, all of whom were transferred to other facilities. In two of the facilities, subsequent transmission infected 25 staff and 8 visitors (Chadha et al. 2006). The longest sustained chain of person-to-person transmission of Nipah virus so far identified in Bangladesh occurred in an outbreak in Faridpur District in 2004. Friends and family members who provided direct care to Nipah virus-infected patients, or helped to carry them or transport them to health facilities when they were near death, sustained a chain of transmission through five generations (Gurley et al. 2007).

Nipah virus RNA has been frequently identified in the saliva of Nipah virus-infected patients (Chua et al. 2001; Harcourt et al. 2005). Outbreak investigations in Bangladesh suggest that exposure to respiratory secretions is the primary vehicle of person-to-person transmission of Nipah virus (Gurley et al. 2007). Across all recognized outbreaks in Bangladesh from 2001 through 2007, Nipah virus patients who had difficulty breathing during their illness were more likely to transmit Nipah virus than Nipah virus patients who did not have difficulty breathing (12 % vs. 0 %, $p=0.03$) (Luby et al. 2009a).

Other Plausible Pathways of Nipah Virus Transmission

There are a number of plausible pathways of Nipah virus transmission from *Pteropus* bats to people that have been explored in outbreak investigations in Bangladesh, but have not been implicated as pathways of transmission. One such plausible pathway is living underneath a bat roost. *Pteropus* bats intermittently shed Nipah virus in their urine (Wacharapluesadee et al. 2010). Although many villages have *Pteropus* bat roosts within or just outside the village boundary, outbreak investigators infrequently identify homes underneath or immediately adjacent to a bat roost, and such close proximity to bat roosts is no more common among Nipah cases than among controls (Luby et al. 2009b).

Another plausible pathway of transmission is consumption of bat bitten fruit. Both birds and fruit bats often drop fruit after taking a single bite. In Bangladesh, where child malnutrition is widespread (NIPORT 2013), ripe tasty dropped fruit is commonly picked up and eaten by rural residents. In each of the outbreak investigations in Bangladesh, consumption of dropped fruit has been evaluated as a potential exposure, but in none of the outbreaks has cases been reported to have eaten dropped fruit significantly more commonly than controls (Hegde et al. 2013). Similarly, eating molasses, a product cooked at high temperature from date palm sap, has not been implicated in Nipah virus transmission.

Discouraging Consumption of Raw Date Palm Sap

Investigations of the first three recognized human Nipah outbreaks in Bangladesh did not implicate drinking raw date palm sap as a risk factor for infection (Hsu et al. 2004; Gurley et al. 2007). When outbreak investigations in 2005 and 2008 implicated raw date palm sap as a pathway of transmission, government health workers disseminated messages in the affected communities to avoid drinking raw date palm sap. These messages were not disseminated nationally. Ministry of Health personnel noted that drinking date palm sap was popular in rural Bangladesh and the absolute risk of a single exposure was low. They expressed concern for the livelihood of sap harvesters. After successive high mortality outbreaks with outbreak investigations repeatedly implicating raw date palm sap as the pathway of transmission of Nipah virus from bats to people (Chakraborty 2011; Sazzad et al. 2013), in 2012 the Bangladesh Ministry of Health and Family Welfare began recommending that people not drink raw date palm sap.

With support from USAID, a research team collaborated with the Government of Bangladesh to evaluate the impact of a professionally developed behavior change intervention in reducing raw date palm sap consumption. This "no raw sap" approach was implemented in 342 villages in Rajbari District during the 2012/2013 date palm sap collection season (Nahar et al. 2014b). A local nongovernmental organization convened meetings with 281 opinion leaders to seek support for the campaign to discourage raw date palm sap consumption. They noted that consuming molasses that was made from cooking date palm sap was safe and that this message would be included to preserve the livelihood of sap harvesters. The local nongovernmental organization conducted 304 community meetings discouraging drinking raw date palm sap. The intervention included a professionally developed public service announcement that presented a short docudrama, providing an engaging story that explained the risks of raw date palm sap and discouraged consumption. The public service announcement was broadcast on local cable networks that served these communities. Intervention implementers also distributed 3000 posters (Fig. 6.4) in the villages and 1500 calendars reinforcing the message distributed to opinion leaders.

The effect of the intervention was assessed by comparing knowledge and behavior in "no raw sap" intervention villages with knowledge and behavior in nonintervention villages selected from a different district where Nipah virus outbreaks had also occurred (Nahar et al. 2014b). After the intervention, 60 % of community respondents in the intervention villages reported knowing about a disease that could be transmitted from bats to humans, compared with 20 % in control villages. The "no raw sap" intervention was implemented at the beginning of the date palm sap collection season. At the end of the season, 18 % of residents of intervention villages reported consuming raw date palm sap during the season, compared with 40 % of residents of nonintervention villages. This difference in reported behavior was also noted in observational assessments. In control villages, observers noted that 53 % of sap harvesters sold raw date palm sap for immediate consumption.

Fig. 6.4 English translation of poster from "No Raw Sap" campaign

By contrast, in intervention villages observers noted that 22 % of sap harvesters sold raw sap. These results suggest that communicating the risk of consumption of raw date palm sap reduced the population's exposure to raw sap and so likely to Nipah virus.

Improving the Safety of Raw Date Palm Sap

Some residents of rural Bangladesh continued to drink raw date palm sap even after being informed of the risk. Such responses to warnings are commonly observed following public health interventions. For example, even after explaining health risks, some people continue to smoke cigarettes, some people continue to inject drugs, and some people habitually consume excess calories. Thus, we began to explore strategies to reduce the risk of Nipah virus contamination of raw date palm sap for those who continued to consume it.

During focused anthropological studies, many date palm sap harvesters indicated that fruit bats were a nuisance (Nahar et al. 2010). They noted that the bats drank some of the sap, sometimes fouled the sap with their secretions and defecation, and sometimes even drowned within the pot of sap. Sap harvesters occasionally used different strategies to protect the sap, including applying lime around the shaved part of the tree or attaching a bamboo skirt, which was a variation on commonly made fishing nets to prevent bat access to the sap stream and collection pot (Fig. 6.5) (Nahar et al. 2010).

Fig. 6.5 Bamboo skirt covering the sap stream and entrance to a date palm sap collection pot

To assess the acceptability of making skirts to protect the sap stream, field researchers described the process of skirt making and encouraged ten sap harvesters to make them. They used old dried bamboo, which was available locally, and tied it together with twine. The total materials cost less than $.10 per skirt, but they required an average of 102 min to assemble and 70 s to apply the skirt to a date palm tree (Nahar et al. 2013). The intervention team recommended applying skirts to all of the trees, but sap harvesters used skirts selectively. Harvesters targeted specific trees that produced the sweetest sap for raw consumption and so for use of skirts. For the majority of trees that generated sap for molasses, the harvesters did not bother using skirts. In follow-up qualitative assessments, the sap harvesters expressed favorable impressions of the skirt. One explained, "*This is the first time the owner of the date palm trees took sap from me for raw consumption, as the sap was better in quality*." A second harvester opined, "*When people will come to know that I use a bamboo skirt, they will be more interested to buy molasses which I make.*" We concluded that sap harvesters could make the skirts and there was a potential added value to them for investing the time.

We evaluated whether applying lime or skirts reduced bat contact with date palm sap. Field researchers applied lime around the shaved part of a tree. In four nights of infrared camera observation, 60 bats visited the shaved part of the tree. We concluded that the lime was not an effective deterrent. Next, we evaluated the efficacy of the skirts (Khan et al. 2012). In our first effort we enrolled a single villager to make skirts. Depending on the particular tree and the placement of the pot, 35 % of the time the short skirts that he produced did not completely cover the sap stream. Field researchers placed infrared cameras with motion detectors in the trees to record the activities of bats approaching the trees. On nights that trees were protected at least partially by a skirt, bats visited the tree and directly contacted the sap stream a mean of two times per night compared with a mean of 32 contacts per night when the sap stream was not covered with a skirt (Khan et al. 2012).

In the subsequent season we attempted to improve the effectiveness of the skirts in preventing bat access to date palm sap (Khan et al. 2012). We ensured that the skirts were long enough to completely cover the pot and the shaved part of the tree. In addition, in piloting efforts to encourage harvesters to make their own skirts, some harvesters, instead of using bamboo, used stalks from different locally available plants or sheets of polyethylene (Nahar et al. 2014a). Thus, for the second evaluation of the effectiveness of skirts, we evaluated four different types of materials to make skirts—stalks from three locally available plants—bamboo, jute or *doincha* or polyethylene sheets. Field researchers identified 277 tall date palm sap-producing trees in a single village, and randomly selected 60 trees. They identified 60 matched trees of similar height, and shaving pattern. They randomly assigned one of the matched pair to be the control tree and assigned the other tree to receive one of the intervention skirts. Field researchers again set up infrared cameras with motion detectors to record the activities of bats approaching the trees. The skirts were remarkably effective in preventing bat access to the sap stream. For the control trees, 3556 episodes of bat contact with the sap stream was observed on the infrared cameras, but there were zero contacts with the sap streams protected by a bamboo, doincha, or polyethylene skirt. One jute skirt had sufficient space between two of the jute stalks that a few persistent bats were able to contact sap on 11 occasions. We concluded that skirts made from a variety of materials, if properly applied, would prevent bat access to date palm sap.

While the efficacy study of the skirts was ongoing, we also worked with sap harvesters to evaluate their willingness to use skirts when they collected sap. The first skirt uptake trial was conducted between December 2009 and February 2010. The intervention team targeted 1303 tree owners and 168 sap harvesters in 15 villages in Faridpur District. During intervention meetings, field researchers explained the risk of Nipah virus, demonstrated how to make skirts, and encouraged the sap harvesters and tree owners to use them. In the baseline evaluation, 3 % of sap harvesters and 1 % of tree owners reported using a skirt in the previous season. In the season following the intervention, 30 % of sap harvesters and 10 % of tree owners reported using skirts at least once, but in the subsequent sap collection season 1 year later, only 9 % of sap harvesters and 4 % tree owners reported using any skirts on their trees (Sultana et al. 2013). We concluded that some sap harvesters could be motivated to use skirts to protect those trees used for selling raw date palm sap, but without stronger incentives and ongoing promotion, uptake would be limited and leave most consumers of raw date palm sap at risk for exposure to Nipah and other bat-borne viruses.

To assess the feasibility, acceptability, and effectiveness of a larger scale approach to reduce risk, researchers expanded their collaboration with the Government of

Fig. 6.6 English translation of poster from the "Only Safe Sap" campaign

Bangladesh and developed and implemented a risk reduction intervention in Faridpur District in parallel to the "no raw sap" intervention described above. The core messages of the "only safe sap" intervention were that drinking raw date palm sap exposed consumers to the risk of a deadly virus and that the safest practice was to avoid drinking raw date palm sap, but if people were intent on drinking date palm sap, they should insist that the sap be collected from a skirt-protected tree.

The elements of the "only safe sap" intervention were similar to the "no raw sap" intervention. The intervention team produced a longer version of the docudrama public service announcement that included the message of only drinking sap if it was protected with a skirt. A nongovernment organization convened 220 community meetings in rural villages and 381 meetings with opinion leaders. They placed 6000 posters (Fig. 6.6) in the intervention area and distributed 5000 calendars to community leaders that contained the key messages. The intervention organization reached out separately to sap harvesters and tree owners. During three rounds of visits to 1160 sap harvesters, they transmitted the central messages and lead sessions where they taught harvesters how to make skirts out of local materials and encouraged them to use them regularly.

After the intervention, 71 % of community respondents in the "only safe sap" communities reported knowing about a disease that could be transmitted from bats to humans compared with 20 % in control villages. At baseline a similar proportion of respondents in "only safe sap" intervention villages (44 %) and nonintervention villages (40 %) reported consuming raw date palm sap during the preceding season. At endline, respondents in the "only safe sap" intervention villages were much more likely to report drinking raw date palm sap from a skirt-protected tree (43 %) compared with baseline (3 %) (Nahar et al. 2014b).

We concluded that a focused effort to encourage using skirts increased the proportion of raw date palm sap that was protected from bat exposure, though the impact on total consumption of raw date palm sap was less clear.

Nipah Transmission Through Horsemeat in the Philippines

During 2014 on the island of Mindanao in the Philippines, health authorities investigated a cluster of 17 ill persons with symptoms of encephalitis or severe respiratory illness that occurred following ten horse deaths in two villages (Ching et al. 2015). Nine (53 %) of the human cases died. All of the horses were found dead; nine of the ten showed neurological signs (head tilting, circling, ataxia) before death. Seven of the human cases slaughtered the horses or ate the horsemeat and three ate horsemeat, but did not participate in slaughtering. Five human cases had no exposure to horsemeat, but cared for people who became ill. Four cats and one dog that ate meat from the same horses also died within 5 days of consumption. Available serum from three patients had neutralizing antibodies against Nipah virus. This outbreak demonstrates another pathway for Nipah virus foodborne transmission.

Broader Risks

There is little direct foodborne risk of Nipah virus to the global human population. Consumption of meat from animals that die of illness is a local practice in many impoverished communities (Chakraborty et al. 2012; Sultana et al. 2012), but is less likely to contaminate the global food supply. Palm sap is a locally produced, locally consumed product that perishes rapidly and so is not available in distant markets. Various communities in Africa and Asia collect palm sap, ferment it, and subsequently sell and consume it as palm wine (Okereke 1982; Bennett et al. 1998; Lebbie and Guries 2002). Outbreaks in both India (Arankalle et al. 2011) and Bangladesh (Islam 2012) have been linked to drinking this fermented date palm sap. Apparently, within the sugary sap matrix of at least some of the fermented sap, the alcohol concentration is insufficient to inactivate Nipah virus. Henipaviruses or strands of RNA closely related to known strains of Henipaviruses have been identified from *Pteropus* bats and the closely related *Eidolon* family of fruit bats across sub-Saharan Africa, South and Southeast Asia (Drexler et al. 2009, 2012; Wacharapluesadee et al. 2010; Field et al. 2011; Anthony et al. 2013). Hence, Nipah virus or another bat virus that causes human disease could be passed from bats to people through palm wine.

Recently, scientists have identified serum that neutralized Nipah virus among people in Cameroon who butcher bats (Pernet et al. 2014). Although no disease has yet been linked to these apparent Henipavirus infections, they demonstrate a potential pathway of exposure to Henipavirus that may include foodborne transmission.

More generally, foodborne transmission of Nipah virus to people in South Asia means that this RNA virus is exposed to an environment that selects for mutations that favor adaptation to human infection and human transmission. Although the probability of this type of transformation from a virus that causes only stuttering chains of person-to-person transmission to a virus that is adapted for sustained person-to-person transmission is unknown, this transition was apparently accomplished by measles virus, another paramyxovirus, that apparently evolved between the eleventh and twelfth century from the progenitor of rinderpest virus, a pathogen of cattle and other hoofed animals (Furuse et al. 2010).

An individual's decision to drink date palm sap not only places his/her own health at risk, but because of the risk of person-to-person transmission, places his or her community at risk. Thus, society has an interest in reducing exposure to Nipah virus. The communication materials developed in the intervention trials to discourage consumption of raw date palm sap or only to drink protected sap apparently contributed to changing behavior, so they could be disseminated to communities across Bangladesh where date palm sap is harvested. If further investigation of transmission patterns of bat-borne viruses implicates palm wine in others settings, then deploying skirts to protect sap streams from bat contamination or other strategies to reduce the risk of transmission could be recommended more broadly.

An alternative strategy to reduce the risk of foodborne transmission of Nipah virus would be to vaccinate people who are at risk of exposure. Animal studies have demonstrated that vaccines can protect against Henipavirus infection (Bossart et al. 2011; Pallister et al. 2011). Australia has deployed a vaccine against Hendra virus to protect horses and the people who come in contact with them (Middleton et al. 2014). Hendra virus is closely related to Nipah virus and the vaccine is likely cross protective (Bossart et al. 2012). Although a human vaccine would not be a cost-effective strategy to prevent 10 or 20 deaths per year in Bangladesh, if the risk of pandemic transformation of the virus is high enough, such vaccination may be a sound investment to reduce the risk of a pandemic strain of Nipah virus gaining a foothold in South Asia and spreading globally. Because of the risk of person-to-person transmission and the unknown risk of pandemic transformation, efforts to reduce the risk of foodborne transmission of Nipah virus are a global concern.

References

Anthony SJ, Epstein JH, Murray KA et al (2013) A strategy to estimate unknown viral diversity in mammals. mBio 4:e00598–e00513. doi:10.1128/mBio.00598-13

Arankalle VA, Bandyopadhyay BT, Ramdasi AY et al (2011) Genomic characterization of Nipah virus, West Bengal, India. Emerg Infect Dis 17:907–909. doi:10.3201/eid1705.100968

Bennett LA, Campillo C, Chandrashekar CR et al (1998) Alcoholic beverage consumption in India, Mexico, and Nigeria: a cross-cultural comparison. Alcohol Health Res World 22:243–252

Bossart KN, Geisbert TW, Feldmann H et al (2011) A neutralizing human monoclonal antibody protects African green monkeys from Hendra virus challenge. Sci Transl Med 3:105ra103. doi:10.1126/scitranslmed.3002901

Bossart KN, Rockx B, Feldmann F et al (2012) A Hendra virus G glycoprotein subunit vaccine protects African green monkeys from Nipah virus challenge. Sci Transl Med 4:146ra107. doi:10.1126/scitranslmed.3004241

Chadha MS, Comer JA, Lowe L et al (2006) Nipah virus-associated encephalitis outbreak, Siliguri, India. Emerg Infect Dis 12:235–240

Chakraborty A (2011) Nipah outbreak in Lalmonirhat district, 2011. Health Sci Bull 9:13–19

Chakraborty A, Khan SU, Hasnat MA et al (2012) Anthrax outbreaks in Bangladesh, 2009-2010. Am J Trop Med Hyg 86:703–710. doi:10.4269/ajtmh.2012.11-0234

Ching PKG, de los Reyes CV, Sucaldito MN et al (2015) Outbreak of henipavirus infection, Philippines, 2014. Emerg Infect Dis 21:328–331

Chua KB (2003) Nipah virus outbreak in Malaysia. J Clin Virol 26:265–275

Chua KB, Bellini WJ, Rota PA et al (2000) Nipah virus: a recently emergent deadly paramyxovirus. Science 288:1432–1435

Chua KB, Lam SK, Goh KJ et al (2001) The presence of Nipah virus in respiratory secretions and urine of patients during an outbreak of Nipah virus encephalitis in Malaysia. J Infect 42:40–43

de Wit E, Prescott J, Falzarano D et al (2014) Foodborne transmission of Nipah virus in Syrian hamsters. PLoS Pathog 10:e1004001. doi:10.1371/journal.ppat.1004001

Drexler JF, Corman VM, Gloza-Rausch F et al (2009) Henipavirus RNA in African bats. PLoS One 4:e6367. doi:10.1371/journal.pone.0006367

Drexler JF, Corman VM, Muller MA et al (2012) Bats host major mammalian paramyxoviruses. Nat Commun 3:796. doi:10.1038/ncomms1796

Field H, de Jong C, Melville D et al (2011) Hendra virus infection dynamics in Australian fruit bats. PLoS One 6:e28678. doi:10.1371/journal.pone.0028678, PONE-D-11-11577 [pii]

Field HE, Mackenzie JS, Daszak P (2007) Henipaviruses: emerging paramyxoviruses associated with fruit bats. Curr Top Microbiol Immunol 315:133–159

Fogarty R, Halpin K, Hyatt AD et al (2008) Henipavirus susceptibility to environmental variables. Virus Res 132:140–144

Furuse Y, Suzuki A, Oshitani H (2010) Origin of measles virus: divergence from rinderpest virus between the 11th and 12th centuries. Virol J 7:52. doi:10.1186/1743-422X-7-52

Goh KJ, Tan CT, Chew NK et al (2000) Clinical features of Nipah virus encephalitis among pig farmers in Malaysia. N Engl J Med 342:1229–1235

Gurley ES, Montgomery JM, Hossain MJ et al (2007) Person-to-person transmission of Nipah virus in a Bangladeshi community. Emerg Infect Dis 13:1031–1037

Halim MA, Chowdhury MSH, Muhamed N et al (2008) Sap production from khejur palm (phoenix sylvestris roxb) husbandry: a substantial means of seasonal livelihood in rural Bangladesh. For Trees Livelihoods 18:305–318

Harcourt BH, Lowe L, Tamin A et al (2005) Genetic characterization of Nipah virus, Bangladesh, 2004. Emerg Infect Dis 11:1594–1597

Hegde ST, Sazzad HM, Hossain MJ et al (2013) Rick factor analysis for Nipah infection in Bangladesh 2004 to 2012. Am J Trop Med Hyg 89:1

Hess IM, Massey PD, Walker B et al (2011) Hendra virus: what do we know? N S W Public Health Bull 22:118–122. doi:10.1071/NB10077, NB10077 [pii]

Hossain MJ, Gurley ES, Montgomery JM et al (2008) Clinical presentation of nipah virus infection in Bangladesh. Clin Infect Dis 46:977–984

Hsu VP, Hossain MJ, Parashar UD et al (2004) Nipah virus encephalitis reemergence, Bangladesh. Emerg Infect Dis 10:2082–2087

Islam MS (2012) Nipah transmission from bats to humans associated with drinking traditional liquor (tari) in northern Bangladesh, 2011. Health Sci Bull 10:16–20

Khan MS, Hossain J, Gurley ES et al (2010) Use of infrared camera to understand bats' access to date palm sap: implications for preventing Nipah virus transmission. Ecohealth 7:517–525. doi:10.1007/s10393-010-0366-2

Khan SU, Gurley ES, Hossain MJ et al (2012) A randomized controlled trial of interventions to impede date palm sap contamination by bats to prevent nipah virus transmission in Bangladesh. PLoS One 7:e42689. doi:10.1371/journal.pone.0042689

Lebbie AR, Guries RP (2002) The palm wine trade in Freetown, Sierra Leone: production, income, and social construction. Econ Bot 56:246–254. doi:10.1663/0013-0001(2002)056[0246:TPWTIF]2.0.CO;2

Lloyd-Smith JO, George D, Pepin KM et al (2009) Epidemic dynamics at the human-animal interface. Science 326:1362–1367. doi:10.1126/science.1177345

Luby S (2014) Nipah virus. In: Motarjemi Y (ed) Encyclopedia of food safety. Elsevier, Waltham, MA, pp 214–217

Luby S, Hossain J, Gurley E et al (2009a) Recurrent zoonotic transmission of Nipah virus into humans, Bangladesh, 2001–2007. Emerg Infect Dis 15:1229–1235

Luby SP (2013) The pandemic potential of Nipah virus. Antiviral Res 100:38–43. doi:10.1016/j.antiviral.2013.07.011

Luby SP, Gurley ES (2012) Epidemiology of henipavirus disease in humans. Curr Top Microbiol Immunol 359:25–40. doi:10.1007/82_2012_207

Luby SP, Gurley ES, Hossain MJ (2009b) Transmission of human infection with Nipah virus. Clin Infect Dis 49:1743–1748. doi:10.1086/647951

Luby SP, Rahman M, Hossain MJ et al (2006) Foodborne transmission of Nipah virus, Bangladesh. Emerg Infect Dis 12:1888–1894

Middleton D, Pallister J, Klein R et al (2014) Hendra virus vaccine, a one health approach to protecting horse, human, and environmental health. Emerg Infect Dis 20:372–379. doi:10.3201/eid2003.131159

Middleton DJ, Morrissy CJ, van der Heide BM et al (2007) Experimental Nipah virus infection in pteropid bats (Pteropus poliocephalus). J Comp Pathol 136:266–272

Middleton DJ, Westbury HA, Morrissy CJ et al (2002) Experimental Nipah virus infection in pigs and cats. J Comp Pathol 126:124–136

Mohd Nor MN, Gan CH, Ong BL (2000) Nipah virus infection of pigs in peninsular Malaysia. Rev Sci Tech 19:160–165

Nahar N, Mondal UK, Hossain MJ et al (2014a) Piloting the promotion of bamboo skirt barriers to prevent Nipah virus transmission through date palm sap in Bangladesh. Glob Health Promot. doi:10.1177/1757975914528249

Nahar N, Mondal UK, Sultana R et al (2013) Piloting the use of indigenous methods to prevent Nipah virus infection by interrupting bats' access to date palm sap in Bangladesh. Health Promot Int 28:378–386. doi:10.1093/heapro/das020

Nahar N, Sultana R, Gurley ES et al (2010) Date palm sap collection: exploring opportunities to prevent Nipah transmission. Ecohealth 7:196–203. doi:10.1007/s10393-010-0320-3

Nahar N, R Sultana, R Paul et al (2014) A community intervention trial utilizing behavior change to reduce the risk of Nipah spillover through date palm sap in Bangladesh. Project report. icddr,b, Dhaka

NIPORT (2013) Bangladesh demographic and health survey 2011. NIPORT, Dhaka, Bangladesh/Calverton, MD

Nowak R (1994) Walker's bats of the world. Johns Hopkins University Press, Baltimore

Okereke O (1982) The traditional system of oil palm wine production in Igbo Eze local government area of Anambra state of Nigeria. Agric Syst 9:239–253, doi: http://dx.doi.org/10.1016/0308-521X(82)90079-8

Pallister J, Middleton D, Wang LF et al (2011) A recombinant Hendra virus G glycoprotein-based subunit vaccine protects ferrets from lethal Hendra virus challenge. Vaccine 29:5623–5630. doi:10.1016/j.vaccine.2011.06.015, S0264-410X(11)00879-6 [pii]

Paton NI, Leo YS, Zaki SR et al (1999) Outbreak of Nipah-virus infection among abattoir workers in Singapore. Lancet 354:1253–1256

Pernet O, Schneider BS, Beaty SM et al (2014) Evidence for henipavirus spillover into human populations in Africa. Nat Comm 5:5342. doi:10.1038/ncomms6342

Rahman MA, Hossain MJ, Sultana S et al (2012) Date palm sap linked to Nipah virus outbreak in Bangladesh, 2008. Vector Borne Zoonotic Dis 12:65–72. doi:10.1089/vbz.2011.0656

Sazzad HM, Hossain MJ, Gurley ES et al (2013) Nipah virus infection outbreak with nosocomial and corpse-to-human transmission, Bangladesh. Emerg Infect Dis 19:210–217. doi:10.3201/eid1902.120971

Sultana R, Mondal UK, Abedin J et al (2013) Evaluating long-term behavior change resulting from an intervention to prevent Nipah virus transmission from bats to humans in Bangladesh International Meeting on Emerging Diseases and Surveillance. International Society for Infectious Diseases, Vienna, Austria, p 41

Sultana R, Rimi NA, Azad S et al (2012) Bangladeshi backyard poultry raisers' perceptions and practices related to zoonotic transmission of avian influenza. J Infect Dev Ctries 6:156–165

Uppal PK (2000) Emergence of Nipah virus in Malaysia. Ann N Y Acad Sci 916:354–357

Wacharapluesadee S, Boongird K, Wanghongsa S et al (2010) A longitudinal study of the prevalence of Nipah virus in Pteropus lylei bats in Thailand: evidence for seasonal preference in disease transmission. Vector Borne Zoonotic Dis 10:183–190. doi:10.1089/vbz.2008.0105

Wacharapluesadee S, Lumlertdacha B, Boongird K et al (2005) Bat Nipah virus, Thailand. Emerg Infect Dis 11:1949–1951

Wolfe ND, Dunavan CP, Diamond J (2007) Origins of major human infectious diseases. Nature 447:279–283. doi:10.1038/nature05775

Chapter 7
A Survey of How Growers, Shippers, and Handlers Address Food Safety Risks from Wildlife in Leafy Greens

Henry Giclas and Diane Wetherington

Abstract The September 2006 *E. coli* spinach outbreak eroded consumer confidence, costing the leafy green industry millions of dollars. In response in 2007, commodity-specific food safety guidelines for lettuce and leafy greens were adopted by the California Leafy Green Products Handler Marketing Agreement (CALGMA). The guidelines, known as the *Commodity Specific Food Safety Guidelines for the Production and Harvest of Lettuce and Leafy Greens* (CSGLLG) (Western Growers, http://www.wga.com/issues/food-safety, 2013), address food safety concerns associated with animals and animal events, and provide guidance for reducing potential crop contamination associated with wildlife risks. Today, CALGMA estimates that 99 % of California leafy green production volume, and roughly 75 % of leafy green production in the USA, are grown using these guidelines (CALGMA, http://www.caleafygreens.ca.gov/about-us/annual-reports, 2010).

Keywords California • Conservation • Good agriculture practices • Food industry • Leafy Greens Marketing Agreement • Produce food safety • Spinach • Survey • Wildlife

Introduction

The September 2006 *E. coli*spinach outbreak eroded consumer confidence, costing the leafy green industry millions of dollars. In response in 2007, commodity-specific food safety guidelines for lettuce and leafy greens were adopted by the California Leafy Green Products Handler Marketing Agreement (CALGMA). The guidelines,

H. Giclas (✉)
Western Growers, Irvine, CA, USA
e-mail: hgiclas@wga.com

D. Wetherington
iDecisionSciences, Seattle, WA, USA

M. Jay-Russell, M.P. Doyle (eds.), *Food Safety Risks from Wildlife*,
Food Microbiology and Food Safety, DOI 10.1007/978-3-319-24442-6_7

known as the *Commodity Specific Food Safety Guidelines for the Production and Harvest of Lettuce and Leafy Greens* (CSGLLG) (Western Growers 2013), address food safety concerns associated with animals and animal events, and provide guidance for reducing potential crop contamination associated with wildlife risks. Today, CALGMA estimates that 99 % of California leafy green production volume, and roughly 75 % of leafy green production in the USA, are grown using these guidelines (CALGMA 2010).

Guidance compliance is monitored by the California Department of Food and Agriculture (CDFA) auditors using a checklist based on the CSGLLG. Forty-five out of more than 200 questions in the checklist relate to wildlife. Of concern to environmental organizations and others (Wild Farm Alliance 2008; RCD Monterey County 2007) is that some growers' practices for addressing potential contamination have adverse impacts on wildlife and the environment. In particular, environmental organizations have concerns about the impact of these practices on wildlife habitat and water quality. One concern was that some species might have migration routes affected by fencing, or endangered species might be indiscriminately killed by trapping, baiting, or other methods. Another concern was that buffer strips (removal of vegetation) could have an adverse impact on stream or wetland quality.

Hence, Western Growers conducted an extensive research project that included a review of relevant peer-reviewed scientific literature, and a survey of California leafy green growers regarding their wildlife and conservation practices. This research was conducted to determine whether the audit questions might promote negative impacts on habitat and animal populations, and if so to obtain information for modification of best practices and audit questions to reduce or remove potential conservation concerns while protecting the microbiological safety of produce. This research was conducted through a survey of the industry to determine common practices used in leafy greens operations, and identify those practices that actually posed wildlife and environmental concerns. Areas of the CSGLLG that were unclear or that raised concerns were identified and solutions were proposed based on current science. Ultimately, as a result of this research, California vegetable growers benefited by having more effective and efficient practices identified, while unnecessary practices were eliminated from food safety guidelines, thereby saving growers both time and money, as well as reducing pressure on the surrounding habitat and animal populations. From these research findings, the Commodity Specific Food Safety Guidelines for the Production and Harvest of Lettuce and Leafy Greens were modified and an expert panel was convened to review and refine recommended changes.

The recommendations were submitted to the California Leafy Greens Marketing Agreement Technical Subcommittee and subsequently adopted in an effort to strengthen the relationship between co-management and food safety in the CALGMA metrics. Many of the recommendations are reflective of what the industry was already doing in accordance with the survey results. While strengthening food safety, the changes should reduce pressure on wildlife and wildlife habitats in leafy green production areas throughout California. In essence, these changes shifted the focus from concerns about discrete lists of animals and animal intrusion to an emphasis on fecal matter in the field. The modifications are science-based, auditable changes to metrics that have the support of industry leaders, wildlife and environmental experts, and food safety scientists. Additionally, all changes were vetted with both the United States Department of Agriculture (USDA) and the Food and Drug Administration (FDA).

Materials and Methods

The project included a grower survey to (1) identify current production practices and their perceived impact on wildlife and the environment, (2) a review of the scientific literature addressing food safety concerns in relation to wildlife and environmental risks, and (3) assembly of a food safety and environmental expert panel to review and recommend changes to the CSGLLG. The recommended changes were reviewed, commented on, further clarified, and then incorporated in the CSLLLG by a process managed by Western Growers.

Survey

The survey was designed to identify current California leafy green food safety guidelines and co-management practices having potential adverse effects on wildlife and/or the environment. In 2007, the Resource Conservation District (RCD) of Monterey conducted an environmental practices survey for row crops (RCMD 2007) and this leafy green survey was structured based on the approach taken by the RCD. The in-depth survey consisted of 84 questions and employed skip logic. Prior to initiating the survey, it was reviewed with industry members and conservation experts.

Survey Distribution

Between July 2010 and January 2011, iDecisionSciences (IDS) designed and conducted the industry study using an Internet survey approach. Leafy green industry members were informed about the questionnaire through e-mails and newsletters from various industry groups, including the California Leafy Greens Products Handler Marketing Agreement, the Grower-Shipper Association of Central California, the Grower-Shipper Associations of Santa Barbara and San Luis Obispo, the Imperial Valley Vegetable Growers Association, and Western Growers. Efforts were made to reach as many leafy green producers growing under food safety programs, such as the CALGMA, as possible.

Data Collection and Analysis

Responses from 53 questionnaires were used for the analysis. Based on a sample population of 197 growers, the number of returned questionnaires represented 26.9 % of leafy green growers in California. Once the database was validated, a statistical analysis of the survey results was performed using Microsoft Excel.

The majority of respondents (77.4 %) planted more than 500 total crop acres in 2009. Acreage planted includes all crops and not just leafy greens. Respondents who planted fewer than 500 crop acres were equally divided by acreage: 0–50 acres (7.4 %), 51–200 acres (7.4 %), and 201–500 acres (7.4 %). Applying the SBA's definition of a small business (growers earning less than $750,000 per year, roughly estimated to be equivalent to 500 production acres), it appears as if three-quarters of the questionnaire respondents were large growers (77 %) and the remainder of the questionnaire population consisted of small- to medium-sized growing operations (23 %). Most of the respondents indicated that their crops were conventionally grown (67.3 %), although some grew organic and conventional crops (28.8 %). A small percentage of the respondents grew only organic crops (3.8 %).

Review of Scientific Literature

Over 120 articles, websites, and studies relating to co-management issues associated with leafy green food safety and conservation practices were reviewed for relevance. Preference was given to peer-reviewed and other scientific journal articles. In addition to the research summaries, government agency guidelines and agency recommendations for conservation practices were searched, and several government agencies were contacted to obtain further information on these topics. The research was conducted to better understand co-management issues facing growers and to provide background information for the expert panel.

Expert Panel Review

Eight expert panel members were selected to represent small, medium, and large growers, wildlife non-governmental organizations (NGOs), wildlife academics, produce shippers, produce processors, and food safety academics. Government representatives from the USDA and the FDA participated as observers and offered insights where appropriate. Expert panel members met 12 times between August 2011 and March 2012.

After a review of the survey responses and the scientific literature, expert panel members used professional judgment to develop recommendations for refining the CSGLLG food safety and wildlife guidelines.

Survey Results

Food Safety Programs

Growers were asked about the food safety programs they use. All of the growers responding to the questionnaire have a food safety program in place, and many are following multiple programs. The LGMA and the PrimusLabs.com GAP programs are named more than any other program (Table 7.1). Growers used buyer-specific food safety programs and other third-party programs such as the Safe Quality Food Institute's SQF program. Only 15.1 % of the respondents were using the USDA GAP/GHP verification program; 20.8 % were using GLOBALG.A.P.

Most growers as opposed to shippers (44.2 %) or buyers (34.9 %) receive specific details about individual food safety programs from auditors (69.8 %). If requirements are conflicting, growers managed the conflicts by applying the most stringent requirements to all operations as opposed to applying the individual requirements to specific acreage.

Growers described the process they use to identify areas where their leafy green acreage may be at risk from wildlife concerns. While most growers look for signs of animal presence, the process and frequency of monitoring for animal activity varied. Some growers cited the need for a pre-season assessment followed by routine monitoring that may occur weekly or daily. There was not an agreed-upon approach to how the assessments were done, when they were done, and how frequently routine monitoring occurred.

Table 7.1 Food safety programs currently in place for leafy greens grown in California, July 2010 to January 2011($n = 53$)

Program[a]	Number of respondents	Percent
CALGMA	47	88.7
PrimusLabs.com GAP Program	33	62.3
GLOBALG.A.P.	11	20.8
Buyer-specific program	10	18.9
AZ LGMA	9	17.0
USDA-AMS GAP/GHP Audit Verification Program	8	15.1
NSF Davis Fresh	5	9.4
Other	3	5.7
SQF	1	1.9
None	0	0.0

CALGMA California Leafy Green Products Handler Marketing Agreement, *GAP* Good Agricultural Practices, *GLOBALG.A.P.* certification system for international Good Agricultural Practices, *AZ LGMA* Arizona Leafy Green Products Shipper Marketing Agreement, *USDA-AMS GAP/GHP* USDA Agricultural Marketing Service Good Agricultural Practices/Good Handling Practices, *NSF Davis Fresh* is a division of NSF International now called NSF Agriculture, *SQF* Safe Quality Food is a certification system from the Safe Quality Food Institute

When asked how they jointly with their buyers identify areas where leafy green acreage may be at risk from wildlife concerns, the grower responses varied greatly. The answers ranged from using common sense to pre-planting, pre-season assessment followed by regular monitoring.

Regardless of how the risk was identified, nearly 64 % of growers who observed wildlife did not plant land because of wildlife concerns. Reasons for not planting included the proximity to grazing domestic animals, riparian areas, and Concentrated Animal Feeding Operations (CAFOs) and buyer requirements.

Conservation Practices

To determine what impact the CALGMA food safety program had on conservation measures, growers were asked to not only name the conversation practices they were currently following, but also note how the CALGMA had impacted those practices. More respondents (82.2 %) had implemented conservation practices in their leafy green-growing environments than in their overall growing environment (78.7 %). The most frequently implemented conservation and food safety practices included cover crops, irrigation water management, and nutrient management. The adoption of CALGMA did not result in the reduction or elimination of conservation measures for 82.6 % of respondents who have implemented conservation practices in leafy green crops. In fact, some growers (23.4 %) implemented conservation practices as a result of the CALGMA guidelines. For these growers, the CALGMA led to the introduction of cover crops, critical planting areas, and hedgerows.

For those respondents who eliminated or decreased conservation practices because of the CALGMA (17.4 %), they described the changes made as follows: mowing grasses in filter strips, removing grass filter strips in some areas due to frog presence or other reasons, constructing bare roads along waterways (reducing cropped acreage and beneficial habitat), removing vegetation around fields to reduce habitat for rodents, removing trees to reduce the presence of birds and their droppings, removing some water catchment basins, and not reusing recovered tailwater because of possible contamination.

Several respondents participated in government-sponsored or -supported conservation programs from the USDA Natural Resources Conservation Service, the University of California Cooperative Extension, and the USDA Farm Services.

Discussion

Using the survey results, analysis of relevant scientific literature, and input from a peer review, several changes were recommended to the CSGLLG including the following:

Animals of Significant Risk

The first recommended change was to remove the "animals of significant risk" list from the document. The animal list, consisting of cattle, sheep, goats, pigs (domestic and wild), and deer, was developed for the 2007 CSGLLG based on Centers for Disease Control and Prevention (CDC) publications identifying those animals as reservoirs for *E. coli* 0157:H7 and hence posing the greatest risk. This list was established and written into the original document because the primary focus during the development of guidance was on *E. coli* O157:H7 as the human pathogen of most significant concern. Since that time, numerous studies have revealed the need to include other potential human pathogens such as *Salmonella*, *Campylobacter*, and *Listeria monocytogenes*. As the list of human pathogens has expanded, so has the number of animals identified as potential pathogen vectors (Fenlon 1985; Ferens and Hovde 2011; Gorski et al. 2011; Jay et al. 2007; Keene et al. 1997; LeJeune et al. 2008; Perz and Le Blancq 2001).

Based on the scientific literature and findings of this survey, the expert panel concluded that updating or adding to the existing list of animals of significant risk would be counterproductive. Research findings since 2006 reveal that the current list is inadequate from a food safety perspective, and the panel felt a new list would be too long. Therefore, the recommendation was made to remove the list of animals.

As additional rationale, it was apparent from the survey responses that growers perceived animals other than ones on the "animals of significant risk" list as threats to produce safety and were acting accordingly. When asked about the types of animals growers observe and the frequency of observations, it is not surprising that birds were seen more frequently than any other animal on a daily basis (Table 7.2). Frogs, rodents, rabbits, and dogs were also sighted daily and monthly according to questionnaire responses. Deer and wild pig sightings occurred once or several times a month, and there were no sightings of cows in leafy green fields.

Table 7.2 Animal presence observed in leafy green fields in California ($n=46$)

Answer options	Number respondents	Daily throughout the year	Daily during mating season	Daily during migration	Several times a month	Maybe once a month	Not at all
Birds	45	31	2	4	6	1	1
Cows	41	0	0	0	0	0	41
Deer	42	0	0	0	6	10	26
Dogs	42	2	0	0	6	20	14
Frogs	37	1	1	0	6	12	17
Rodents	45	9	3	0	20	11	2
Rabbits	44	9	2	0	14	13	6
Wild pigs	41	0	0	0	1	10	30
Other	18	4	0	0	5	2	7

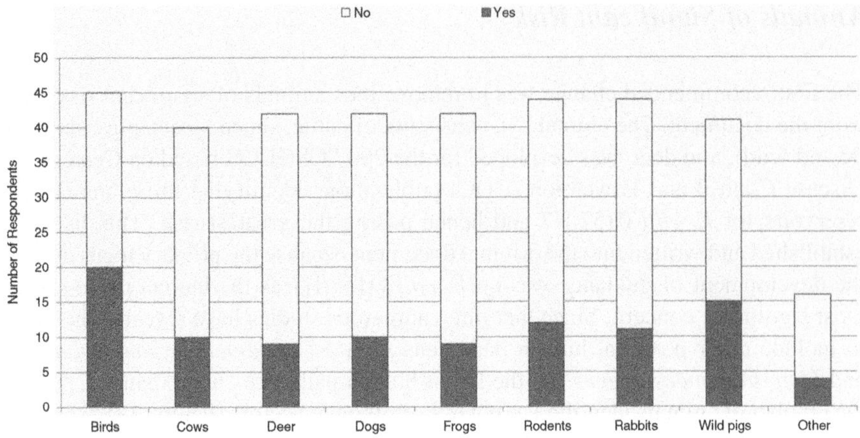

Fig. 7.1 Are animals observed on your land a threat to the safety of your leafy green crops ($n=45$)

Table 7.3 Auditors specifying wildlife as a food safety concern ($n=27$)

	Which animals were suggested or specified as being a hazard/concern?						
Answer options	Birds	Deer	Domestic animals	Feral pigs	Field rodents	Response percent	Response count
CDFA auditor for LGMA	6	3	8	1	5	50.0	9
Primus auditor	5	4	7	2	7	66.7	12
Davis Fresh (NSF Int'l) auditor	4	2	3	1	2	27.8	5
GlobalGAP auditor	1	0	0	0	0	5.6	1
SQF auditor	0	0	0	0	0	0.0	0
Retailer	0	2	2	2	1	11.1	2
Handler	6	2	4	2	6	38.9	7
Food service operator	1	1	2	1	2	11.1	2
Other	0	1	1	0	1	5.6	1

Among all reported animals, birds were most frequently observed and were perceived as the greatest wildlife risk to produce safety (44 % of the growers observing birds in their leafy green fields) (Fig. 7.1). The second most frequently cited animal thought to be a food safety concern was wild pigs (36.6 %) followed by rodents (27.3 %). In many cases, growers did not plant crops because of animal concerns.

Similarly, auditors indicated they had animal-related food safety concerns (Table 7.3), including birds and domestic animals cited by several audit companies. Based on grower feedback and the scientific literature, removing the list of "animals of significant risk" strengthens food safety risk assessment and risk management by addressing other potential animal and pathogen concerns. While environmentalists may express concerns that this modification has the potential to cause an overemphasis on food safety practices related to animals, additional recommendations were developed to ensure that this would not be the case (see below).

Animal Intrusion

The second recommended change was to replace "intrusion by animal of significant risk that might impact produce safety" with "any fecal contamination that may present a risk to the production block or crop." This recommendation by the expert panel was made based on the current scientific literature, co-management concerns, and existing grower and USDA/FDA practices. The concern from a food safety perspective, based on the literature, is that a list of potential animal vectors would include much of the animal kingdom—yet the real issue is not with animal intrusion but with feces and the potential for fecal contamination. From a wildlife perspective, any list is perceived as targeting species on the list and potentially endangering the animals and their habitats. By removing "animal intrusion" language and focusing on feces or indicators of fecal presence (e.g., crop damage), the panel felt that specific animals would no longer be targeted, de-emphasizing the significance of sighting any one animal species. Even if animals are present, there may or may not be fecal matter or evidence of potential microbial hazards due to feeding. At the same time, the proposed language revisions were in keeping with the current approach used by USDA and FDA auditors; namely, regardless of the origin of feces, if any fecal contamination is found—crops will need to be destroyed.

Adjacent Land

The third recommended change was to delete the wording, "Locate production blocks to minimize potential access by animals of significant risk and maximize distances to possible sources of microbial contamination. For example, consider the proximity to water (i.e., riparian areas), animal of significant risk harborage, open range lands, non-contiguous blocks, urban centers, etc. Periodically monitor these factors and assess during preseason and pre-harvest assessments." The new wording suggested was: "The designated food safety professional or other trained personnel should evaluate the potential for microbial contamination from adjacent areas. A risk assessment shall be performed to determine the risk level as well as to evaluate potential strategies to control or reduce the introduction of human pathogens. Periodically monitor these factors and assess during the preseason and pre-harvest assessments ...." The change acknowledges differences between ranches in what can and cannot be done as well as a consideration of local fish and game and water quality initiatives. Instead of prescribing solutions that may not be suitable for the majority of users, the modification recognizes the role of the designated food safety professional and points to supporting resources the food safety professional can use.

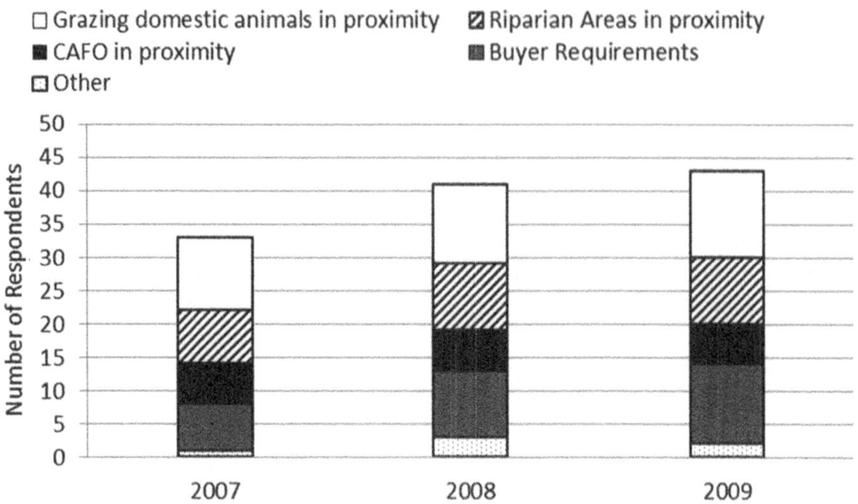

Fig. 7.2 Reasons for not planting relating to wildlife concerns ($n = 12$)

Adjacent land concerns do result in growers not planting in certain locations (Fig. 7.2). From a food safety perspective, risk assessments are conducted to assess adjacent land hazards prior to planting, and, if deemed necessary, risk mitigation efforts can provide data for making more informed planting decisions when faced with adjacent land concerns.

Equipment and the Potential for Contamination

The fourth recommended change was to address equipment that may come into contact with animals or areas of a field that may have been contaminated by animals. The recommendation was to change "Such equipment should not be used in proximity to or in areas where it may contact edible portions of lettuce and or leafy greens" by adding "without proper sanitation" at the end of the sentence.

Crop Damage

Since the recommended changes shifted the focus from the presence of animals to actual "crop damage," the fifth recommended change was to include a definition of crop damage as "any damage to the crop that renders the crop adulterated and thus unfit for harvest and/or consumption by humans." Adulteration can include but is not limited to:

1. Animal-induced damage through eating, trampling, or any other noticeable physical damage to the crop.
2. Contamination from animal feces, urine, body fluids, or animal parts and/or matter due to acts such as molting or shedding.

Implementation of Expert Panel Recommendations

In April and June 2012, presentations to the CALGMA Technical Committee were made based on the recommended changes. Results of the meeting were mixed. While the recommended removal of the animals of significant risk list was agreed upon by the Technical Committee, it was overturned by the CALGMA board. Other changes were tabled until the next Technical Committee meeting in order to address concerns raised by board and committee members.

Western Growers engaged in a series of discussions with growers as well as public and private experts in the fields of food safety and conservation. It was clear that growers were fundamentally concerned with how the new proposed changes would be interpreted and how they could effectively implement the shift in emphasis from "presence" to "damage."

After months of discourse and debate, a flowchart and corresponding descriptive statements that graphically depict the process of monitoring and evaluation that was being recommended in the proposed language were developed (Fig. 7.3). This "flowchart" helped in understanding the concepts being proposed. While there was some additional work to perfect specific language in the CALGMA guidance and to ensure that the text of the guidance was synchronous with the flowchart, it was this tool that ultimately led to the expert panel recommendations being adopted by the CALGMA board, as well as the Arizona Leafy Green Marketing Association (AZLGMA) board.

Observations

In order to effect change and promote the adoption and implementation of new food safety practices on the farm, it is imperative that affected stakeholders have a direct role in the development and vetting of concepts that they will be asked to employ. These individuals provide practical input in how to ensure that programs and or requirements can be crafted so that they both accomplish food safety objectives and can be applied in the field or facility. An honest and open exchange of ideas between food safety professionals, academic experts, and growers and handlers is necessary to achieve multiple objectives that together promote and advance science-based food safety practices.

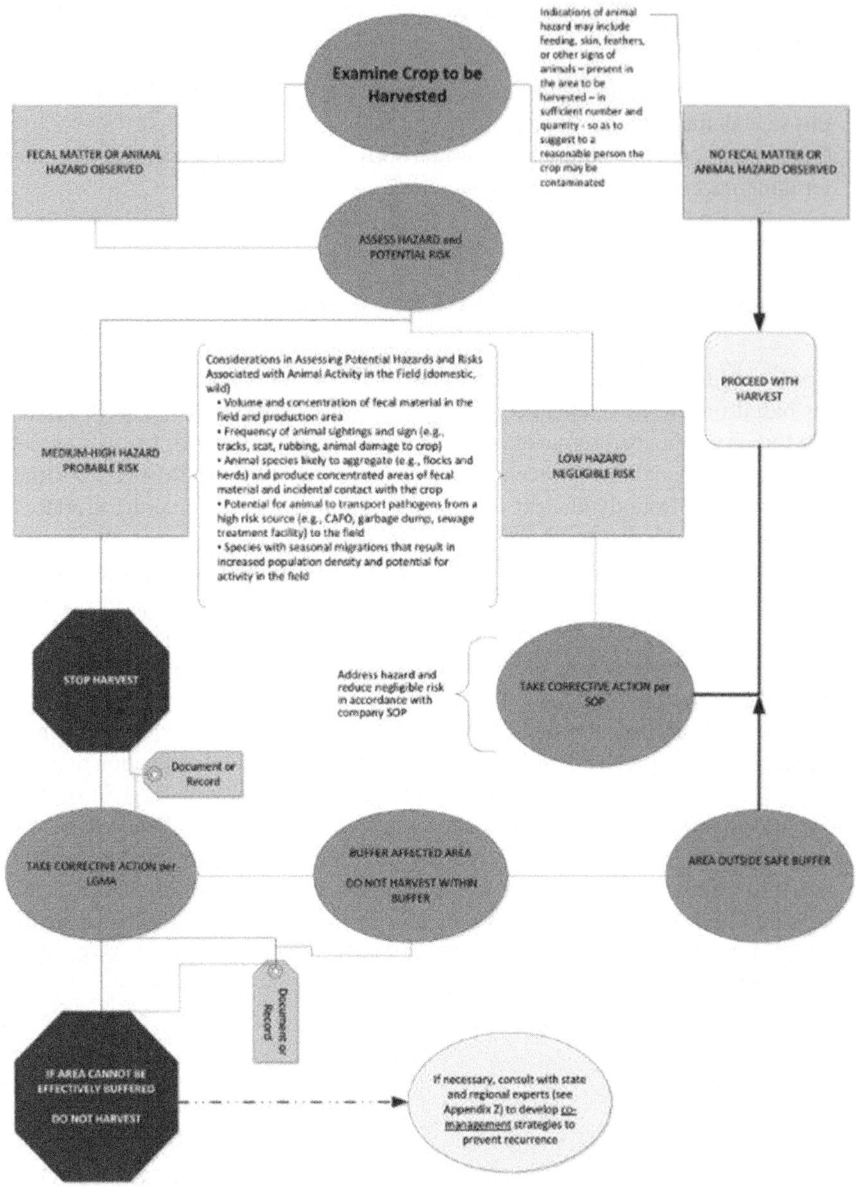

Fig. 7.3 Preharvest and harvest assessment—animal hazard/fecal matter decision tree (Western Growers 2013)

Flowcharts, decision trees, hierarchies, and other graphical methods of presenting information along with explanatory notes and text are useful in obtaining an understanding of complex ideas. This was a lynchpin tool for this project that enabled a series of focused discussions with all parties involved and helped to avoid the possibility of getting caught up in "language" of what it may or may not mean.

Acknowledgements We thank the members of the expert panel—Mary Bianchi, Chris Fischer, Sharan Lanini, Afreen Malik, Brendan Miele, Laura Mills, Michele Jay-Russell, and Ken Stearns. We are also grateful to Cecilia Crowley and Michael Mahovic from the FDA, Gregory McNair, and the auditors who reviewed the proposed changes from the USDA, Karen Lowell, and Vicki Scott from the AZLGMA. This study was supported by funding from the California Department of Food and Agriculture, grant #SCB09003, as part of the USDA's Specialty Crop Block Grant program.

References

California Leafy Green Products Handler Marketing Agreement (2010) 2009/2010 Annual report. http://www.caleafygreens.ca.gov/about-us/annual-reports. Accessed 3 May 2011

Fenlon DR (1985) Wild birds and silage as reservoirs of Listeria in the agricultural environment. J Appl Bacteriol 59:537–543

Ferens WA, Hovde CJ (2011) Escherichia coli O157:H7: human reservoir and sources of human infection. Foodborne Pathog Dis 8(4)

Gorski L, Parker CT, Liang A et al (2011) Prevalence, distribution, and diversity of Salmonella enterica in a major produce region of California. Appl Environ Microbiol 77(8):2734–2748

Jay MT, Cooley M, Carychao D et al (2007) Escherichia coli O157:H7 in feral swine near spinach fields and cattle, central California coast. Emerg Infect 13(12):1908–1911

Keene WE, Sazie E, Kok J et al (1997) An outbreak of Escherichia coli O157:H7 infections traced to jerky made from deer meat. J Am Med Assoc 277(15):1229–1231

LeJeune J, Homan J, Pearl DL (2008) Role of the European starling in the transmission of E. coli O157 on dairy farms. In: Proceedings of the vertebrate pest conference, San Diego, 2008

Perz JF, Le Blancq SM (2001) Cryptosporidium parvum infection involving novel genotypes in wildlife from lower New York State. Appl Environ Microbiol 1154–1162

Resource Conservation District of Monterey County (2007) A grower survey reconciling food safety and environmental protection

Western Growers (2013) Commodity specific food safety guidelines for the production and harvest of lettuce and leafy greens. http://www.wga.com/issues/food-safety. Accessed 10 Sep 2013

Wild Farm Alliance (2008) Unintended consequences make food less safe. http://www.wildfarmalliance.org/Press%20Room/press_room_bulletin2.htm. Accessed 29 Aug 2012

Chapter 8
Keeping Wildlife Out of Your Food: Mitigation and Control Strategies to Reduce the Transmission Risk of Food-Borne Pathogens

Alan B. Franklin and Kurt C. VerCauteren

Abstract In this chapter, we provide a general framework for developing strategies to mitigate the contamination of agricultural operations with pathogens carried by wildlife. As part of this framework, we present adaptive management as a viable approach to developing these strategies to reduce the uncertainty over time as to whether management methods are being effective. We provide the general steps to developing an adaptive management strategies as well as generic mitigation methods that can be applied to agricultural operations as part of an adaptive management strategy.

Keywords Adaptive management • Agriculture • Food safety • Habitat modification • Human-wildlife conflict • Mitigation • Population control • Risk assessment • Wildlife • Wildlife damage management

Introduction

In the past few decades, wildlife has been increasingly recognized as a threat to food safety because of their ability to transmit pathogens to agricultural crops and livestock (Langholz and Jay-Russell 2013; Miller et al. 2013). Although the risk and extent of this problem still need to be clarified, increased regulation of agricultural producers has been predicated on the assumption that wildlife has a high probability of contaminating produce fields and livestock, primarily with their feces (U.S. Department of Health and Human Services 2013), which may or may not contain pathogens posing a risk to humans consuming agricultural products.

A.B. Franklin (✉) • K.C. VerCauteren
USDA-APHIS Wildlife Services National Wildlife Research Center, Fort Collins, CO, USA
e-mail: alan.b.franklin@aphis.usda.gov

M. Jay-Russell, M.P. Doyle (eds.), *Food Safety Risks from Wildlife*,
Food Microbiology and Food Safety, DOI 10.1007/978-3-319-24442-6_8

Our intent in this chapter is not to provide a litany of methods that can be used to keep wildlife from contaminating agricultural operations, but to provide an overview that agricultural producers can use as a starting point in developing strategic programs to deal with the issue of wildlife contamination of agricultural operations with food-borne pathogens. While we provide some broad categories of tools that can be used, it is not an exhaustive list. An important caveat in the use of some of these tools is that most were developed to prevent or mitigate physical wildlife damage. Thus, the effectiveness of many wildlife damage management methods in preventing or mitigating contamination of agricultural operations with food-borne pathogens has not been evaluated, primarily because this problem has only become a focus in recent years (Langholz and Jay-Russell 2013).

General Strategies

We advocate strategies that are proactive, including a number of what we consider to be essential components and allow for adaptive management (Fig. 8.1). Adaptive management is a programmatic approach, which was originally developed in natural resource management to deal with problems where uncertainty was present in a system (Walters 1986; Walters and Holling 1990; Nichols et al. 1995). Our general strategy (Fig. 8.1) includes the following key components, each of which we will cover in more detail further on:

1. Identifying the problem—Are wildlife a problem in contaminating agricultural operations with food-borne pathogens?

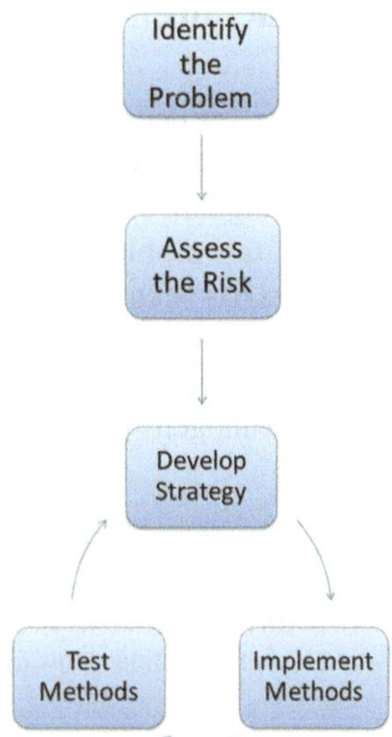

Fig. 8.1 General flowchart for developing and applying methods to mitigate contamination of agricultural operations by food-borne pathogens carried by wildlife

2. Assessing the risk—If wildlife are a problem, what is the level of risk and consequences (i.e., what is the magnitude of the problem)?
3. Developing a strategy—If wildlife pose a risk, how will the problem be dealt with?
4. Implementing mitigation methods—In conjunction with developing a strategy, what are the specific options available for mitigating contamination by wildlife?
5. Evaluation of management effort (testing methods)—Once the general strategy and mitigation methods are implemented, are they working as expected in mitigating or eliminating the problem?

Adaptive Management

The feedback loop in the bottom of Fig. 8.1 represents part of the adaptive management component of the process. Although the use of adaptive management has been proposed for use in wildlife damage management (Reidinger and Miller 2013), it has rarely been applied to management of wildlife-borne pathogens (Miller et al. 2013). One exception that closely resembles adaptive management is an ongoing program to reduce transmission of bovine tuberculosis from wildlife to cattle in Michigan (Box 8.1).

Box 8.1: Example of a Strategic Process Resembling Adaptive Management to Minimize Transmission of Bovine Tuberculosis from Wildlife to Cattle in Michigan

In Michigan, state and federal agencies and universities have been challenged with assisting producers in modifying their practices to reduce potential for exposure to *Mycobacterium bovis* from wildlife to cattle. First, they identified the problem and monitored wildlife and cattle herds to determine its pathways and magnitude (Bruning-Fann et al. 2001; Kaneene et al. 2002; Palmer et al. 2004a, b; Walter et al. 2014). Second, they conducted research to learn about the ecology of the pathogen (Palmer and Whipple 2006; Fine et al. 2011) and wildlife species involved (Atwood et al. 2009; Walter et al. 2013). Finally, they developed methods for addressing the issues (VerCauteren et al. 2012b; Phillips et al. 2012; Vercauteren et al. 2010) and then implemented a cooperative adaptive management program that was tailored for each specific producer. This program has been ongoing and monitoring and adjustment is under way (Walter et al. 2012).

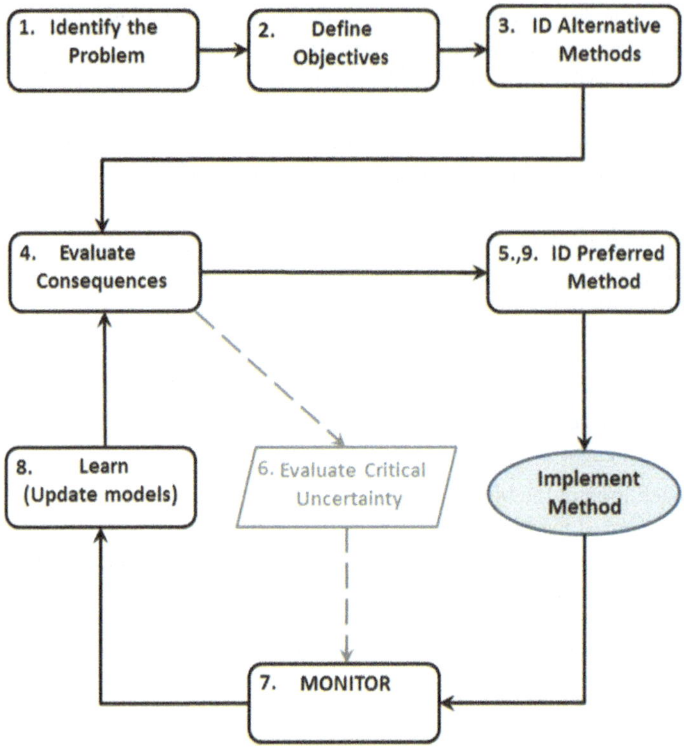

Fig. 8.2 Conceptual framework of adaptive management for managing wildlife contamination of agricultural operations (modified from Runge 2011)

Adaptive management is a formal, learning-based approach for dealing with wildlife management problems (Knutson et al. 2010). It is a formal framework in the sense that it incorporates a structured process of iterative decision making, which is often mathematical in nature (Runge 2011). This is in contrast to management by trial and error where management options are attempted, and if unsuccessful then some other management option is implemented, with no systematic mechanism of "learning by doing" to guide alternative options (Williams and Brown 2012). The adaptive management process includes the steps we outlined previously but puts certain aspects into a more formal framework (Fig. 8.2), which we will discuss further.

Some may argue that the adaptive management process is too time consuming, complicated, costly, and slow (i.e., we need to act now). However, this argument needs to be balanced against the effects of product recalls, restrictive policies for agricultural producers, and other economic costs accrued by not adequately addressing and solving the problem. Thus, we argue that an adaptive management framework is ideal for solving problems of pathogen contamination of agricultural operations by wildlife.

Strategic Processes

Identifying the Problem

The first step in any management issue is to address the following:

1. Is there a problem?
2. If there is a problem, what is the degree and magnitude of the problem?

Wildlife have recently become a concern for spread of food-borne pathogens to agricultural operations, such as produce fields (Langholz and Jay-Russell 2013), concentrated animal feeding operations (Carlson et al. 2011b), and dairy operations (LeJeune et al. 2008). While outbreaks of human illness have been attributed to wildlife contaminating produce fields with food-borne pathogens (Erickson and Doyle 2012), few studies have adequately documented the magnitude of wildlife contamination. Thus, the first step for any agricultural operation is to identify whether wildlife are a potential risk for contaminating their product. This includes identifying which wildlife species are involved, what is the magnitude of their visitation rates to the operation, and what pathogens they are carrying that might affect human food safety.

Most wildlife populations around agricultural operations are synanthropic (peridomestic) species, which are those species that easily coexist with humans. Examples of native synanthropic species (those species indigenous to a particular area) include white-tailed and mule deer, raccoons, skunks, coyotes, cottontail rabbits, and foxes (Clark 2014; Rice 2014). Thus, the first identification of wildlife problems will probably focus initially on these types of wildlife species.

Assessing the Risk

Risk of contamination of agricultural operations from wildlife is a function of:

1. The species of wildlife visiting the facility.
2. The pathogens these wildlife species are infected with.
3. The prevalence of pathogens of concern in these wildlife species.
4. The amount of pathogens they can shed (either orally or through feces) when visiting agricultural facilities (pathogen loads).
5. How often they visit (visitation rates).
6. How many animals visit.
7. What time of year they visit.
8. The contact rates (direct or indirect) between wildlife and agricultural products.
9. The vulnerability of the products to microbial contamination based on type of processing (raw, minimally processed, treated with a kill step) and the production/harvest methods (hand vs. mechanical).
10. Whether there is substantial long-term variation in characteristics 1–8 above.

Understanding the characteristics outlined above requires understanding both the ecology and transmission mechanisms of the wildlife species that may impact a particular agricultural operation. For example, European starlings have been associated with *Salmonella* contamination of livestock feedlots (concentrated animal feeding operations, or CAFOs) (Carlson et al. 2011b) and *Escherichia coli* O157 on dairy farms (Cernicchiaro et al. 2012). In particular, the extent of contamination on CAFOs and dairy farms has been associated with numbers of starlings visiting facilities (Carlson et al. 2011b; Cernicchiaro et al. 2012), which diminished once starling numbers were controlled on a facility (Carlson et al. 2011a). However, control of starlings on single facilities may not always be a cost-effective approach; starlings occupy roost areas away from facilities and often visit multiple facilities (Cernicchiaro et al. 2012; Homan et al. 2013; Gaukler et al. 2008). Thus, understanding the ecology of starlings beyond their impacts on individual operations is important because effective control will depend on the degree of their site fidelity to agricultural operations, their use of other agricultural operations, and roosting behavior (Homan et al. 2013).

In addition, each of the characteristics described above cannot be considered in isolation. For example, prevalence of *Escherichia coli* O157 is relatively low (3 %) in European starlings (LeJeune et al. 2008). However, the number of starlings visiting facilities can be very high, up to ~50,000 daily (Carlson et al. 2011b), which translates to a potential of 1500 starlings infected with *Escherichia coli* O157 visiting such a facility every day at certain times of the year.

Developing Strategies

Developing a strategy to deal with contamination of agricultural operations with pathogens can range from simple guidelines, such as those published by the Colorado State University Extension (2012), to more complex, adaptive strategies. Although adaptive management strategies have not been used specifically for addressing issues of wildlife contaminating agricultural facilities with food-borne pathogens, adaptive management has been attempted in other wildlife damage issues (Parkes et al. 2006; Bryce et al. 2011).

One drawback of adaptive management is that it requires a level of technical expertise to develop the framework of the strategy and the required monitoring effort (Doherty and McLean 2011; Parma 1998). In its truest form, adaptive management is couched in a formal statistical and sampling framework that requires statistical expertise to establish and implement (see Williams et al. 2002). However, the gains in knowledge in dealing with the problem far outweigh the requirement of statistical and scientific rigor required in designing and implementing the strategy. For example, Parkes et al. (2006) argued that adaptive management decreased uncertainty in complex problems or decreased the risk of failure by making uncertainty explicit when dealing with invasive species management.

Fig. 8.3 Portion of the Salinas Valley, California, showing different farm and ranch ownerships (*green boundaries*) in proximity to large expanses of public lands (*blue boundaries*)

Of critical importance in developing a strategy for dealing with pathogen contamination by wildlife is the scale of the plan. Few agricultural producers will likely be able to effectively develop an adaptive management plan for their single facilities. However, scales to be considered for effective management can range from local (e.g., county) scales to regional (e.g., state or combination of states) to national scales. The scale to be considered is dependent on the nature of the problem and the uniqueness of the situation. For example, the Salinas Valley in California is the top producer of leafy greens in the USA (Cooley et al. 2007), has a number of independent producers, and is relatively isolated from other similar growing regions (Fig. 8.3). Rather than having separate strategies for each agricultural operation, a common strategy encompassing the entire valley across all producers would probably be most effective, both economically and strategically.

Part of the strategy may include understanding the ultimate source (i.e., a resource that is responsible for contaminating wildlife) of contamination if pathogen contamination by wildlife is suspected as only a proximate source (i.e., is immediately responsible for the contamination). For example, deer were considered the ultimate source of contamination of strawberry fields with *Escherichia coli* O157 (Laidler et al. 2013), which subsequently infected humans consuming the strawberries. However, other ultimate sources, such as water, were not reported as potential causative factors that could have contaminated both the fields and the deer using those fields. In contrast, contamination of spinach fields with *Escherichia coli* O157 was more thorough but less clear with several ultimate sources implicated, including feral swine as a proximate source (Jay et al. 2007).

Implementing Mitigation Measures

Mitigation measures used in wildlife damage management have direct implications for managing pathogen transmission from wildlife and may often dovetail with issues where wildlife are involved in both damage and pathogen contamination. For example, feral swine cause considerable crop damage as well as pose a risk for pathogen transmission to agriculture (Bevins et al. 2014; Jay and Wiscomb 2008), suggesting that mitigation strategies could simultaneously deal with these two problems.

Mitigation measures can be classified into two primary categories, population control where wildlife populations are reduced or eliminated, and exclusionary measures where wildlife are excluded from agricultural operations (e.g., farm fields, dairies, and livestock facilities).

Population Control

The primary goal in population control is to reduce wildlife populations that represent a contamination threat around agricultural operations. There are three broad categories of population control: lethal control, reproductive control, and habitat modification.

Lethal Control

Lethal control is always an option in wildlife damage management but it has become increasingly difficult to justify with some native wildlife species in terms of ecological effects and has become much less politically and socially palatable (Bergstrom et al. 2014). In addition, we currently lack the ability to alleviate many wildlife damage problems in effective and economical ways using only nonlethal techniques (Conover 2001). For invasive species, such as European starlings and feral swine, the use of lethal control is considered more justifiable because it simultaneously resolves ecological and damage issues beyond just agricultural contamination by wildlife-borne pathogens and is, thus, more politically palatable.

For example, feral swine are effectual reservoirs of an array of diseases (Williams and Barker 2001) that could be transmitted to crop fields and domestic swine herds through interactions that have been documented to occur between wild and domestic populations (Wyckoff et al. 2012). Feral swine also wallow in and around water sources, thereby increasing potential for pathogen contamination (Atwill et al. 1997; Jay et al. 2007). For these reasons and other wildlife damage issues, a national program to eradicate feral swine throughout most of the USA has been recently implemented (Bevins and Franklin 2014). However, to be effective in the long term, we argue that the use of lethal control to remove some invasive species is ultimately a regional and national problem (e.g., feral swine, European starlings) with reduced effectiveness when control is solely at local levels.

Two examples of lethal control methods with relevance to wildlife in agricultural operations are regulated hunting with ungulates and Integrated Pest Management (IPM) with rodents. Regulated, managed hunting in rural settings is the most practical and effective method of managing overabundant ungulate populations and controlling damage. It is also the most ecologically, socially, and fiscally responsible method. Some states have special depredation permits that can be issued to landowners to remove deer in areas where they are causing damage or threatening to transmit pathogens to agricultural crops or livestock outside the normal hunting season, if sufficient control cannot be achieved during the hunting season. An IPM approach (Witmer 2007) is recommended for control of rodents and other small mammals. The IPM concept favors timely and strategic incorporation of a combination of cost-effective control techniques (lethal and nonlethal) to reduce the impact of species on valuable resources (Newman et al. 2012).

Reproductive Control

Reproductive control is where reproduction is inhibited in free-ranging wildlife populations through sterilization, contraceptives, or immune-contraceptive vaccines. There is a large body of literature on reproductive control and wildlife. However, except for a few species, it has largely been untested as a definitive management tool and is currently not being used effectively in managing wildlife species relative to agricultural production. Considerable effort has been expended to develop fertility control agents (contraceptives) and methods of delivery for primarily wild ungulates, geese, and feral pigeons (Fagerstone et al. 2002; Rhyan et al. 2013). Contraceptives for wildlife have the potential to be a complementary tool for population management in scenarios where current nonlethal management techniques are ineffective or unacceptable. In addition, Killian et al. (2007) argue that reproductive control should be used rather than lethal control to prevent pathogen transmission from wildlife because animals removed through lethal control may be replaced by others infected with pathogens. There are several contraceptive strategies, including chemosterilants, immunocontraceptives, intrauterine devices, and surgical procedures, that can all effectively result in decreased reproduction by individuals (Fagerstone et al. 2002, 2010). Orally delivered contraceptives as well as live vector (bacterial or viral) delivery are being explored further (Fagerstone et al. 2002; Conner et al. 2007). However, it is unlikely that fertility control will become a viable stand-alone management strategy (Dolbeer 1998; DeNicola et al. 2000) until better and more consistent delivery systems are developed, and research and registration of compounds to use with species other than deer, geese, and pigeons have been completed.

Habitat Modification

All animals are dependent on food and shelter. Therefore, elimination of one or both of these requirements may force wildlife to move from the immediate area. Habitat modification as a mitigation tool has been extensively criticized for its effects on wildlife conservation (Gennet et al. 2013). Using agricultural practices in the Salinas Valley as an example, Gennet et al. (2013) argue that habitat modification, especially in riparian systems, was based on reactive strategies resulting from sporadic outbreaks of food-borne pathogens in produce associated with wildlife (Jay et al. 2007). In addition, the proximity of large blocks of wildlife habitat (Fig. 8.3) precludes the effectiveness of localized habitat modification at smaller scales for wide-ranging wildlife species, such as wild ungulates and feral swine.

Given the above caveat, habitat modification can be useful when used judiciously and at small scales. For example, habitat modification can be implemented in many situations to make roosting, loafing, or feeding sites less attractive to birds, such as European starlings. Although the initial investment of time and money may be high, these modifications often provide long-lasting relief. Thinning or pruning vegetation can cause roosting birds such as blackbirds and starlings to move, often increasing the commercial or ecological aspects at the same time (Leitch et al. 1997). However, there is considerable uncertainty in ecological consequences from large-scale habitat modifications around agricultural facilities. For example, reduction of habitats supporting insectivorous birds and bats could result in increased pest insect populations with subsequent increases in crop damage.

Exclusionary Methods

Here, we view nonlethal, exclusionary methods as including physical barriers, scare devices, and repellants. VerCauteren et al. (2012a) and Reidinger and Miller (2013) provide an extensive review of exclusionary methods that can be used to keep wildlife away from agricultural operations. While many of these methods have been developed to mitigate wildlife damage, they also have direct applications toward mitigating contamination from pathogens carried by wildlife. Methods that prevent wildlife from entering agricultural facilities and crops, such as those evaluated by Johnson et al. (2014), are the most relevant to mitigating contamination with pathogens from wildlife.

Limited effectiveness and high cost of some nonlethal strategies frequently make them economically impractical, even when used in conjunction with lethal strategies. Frequently, the efficacy of nonlethal techniques is directly correlated to the level of motivation of the targeted individuals. For example, a simple frightening device employing sound and lights or a single strand of electric fence may be a sufficient deterrent to minimize deer use of a minimally desired resource. However, when stressed for food, deer can breech a 2.1-m-high woven-wire mesh fence to feed on and potentially contaminate stored crops, imposing risk for pathogen transmission to livestock (VerCauteren et al. 2003). Thus, the management technique chosen for a scenario under one level of motivation may have a different degree of success in dissimilar scenarios, so the level of motivation of the targeted wildlife must be considered prior to implementation of any nonlethal technique.

Frequently, fencing is the only long-term, nonlethal method to effectively minimize exposure of agricultural facilities and crops to wildlife. Many fence designs are available, although an effective yet low-cost design that keeps out multiple wildlife species has yet to be perfected. Fencing provides protection as a physical barrier, as a psychological barrier, or as a combination of the two. The standard deer fence, a 2.4-m-high woven-wire fence, is a physical barrier and greatly reduces the possibility of an animal passing through, over, or under. Conversely, a single- or double-strand electric poly-tape fence acts as a psychological barrier through aversive conditioning. Conditioning occurs when an animal attempts to breach the fence and receives a powerful electric shock. This training can be expedited with the use of bait such as peanut butter applied directly to the fence (Porter 1983). Plastic netting has been used as a cost-effective method to exclude birds from individual fruit trees or high-value crops such as blueberries or grapes (Fuller-Perrine and Tobin 1993), but is probably infeasible for large expanses of crops or feed bunks at large livestock facilities.

Scare devices, such as propane cannons, flashing lights, shell crackers, and other sonic devices, used near an agricultural facility can provide temporary relief from wildlife intrusions (Gilsdorf et al. 2002). Blackbird roosts containing up to several million birds can be moved by using a combination of devices, particularly recorded distress calls, shell crackers, rockets, and propane cannons (Mott 1980). Strobe lights placed in the roost are also helpful. However, some species, such as wild ungulates, adjust or habituate to frightening devices quickly, and these devices are generally not effective for an entire crop-growing season. Recent research has evaluated the efficacy of animal-activated frightening devices, revealing mixed results (Gilsdorf et al. 2004a, b; Belant et al. 1998; Beringer et al. 2003). Often these devices are most effective when used in combination with other methods rather than as a sole exclusionary method (Gilsdorf et al. 2002).

While repellants may minimize or prevent wildlife from damaging crops, they will not necessarily prevent potential contamination from pathogens in feces unless there is a strong negative habituation from repellants in the use of areas by wildlife. As with other nonlethal techniques, factors such as ungulate population density, availability of alternate foods, target plant species, weather, repellent concentration, and duration of the problem can influence the effectiveness of repellents.

One underutilized, but potentially effective, exclusionary method to eliminate or reduce wildlife intrusion into agricultural crop fields and facilities is the use of guard dogs. Guard dogs have been effectively used to minimize contact between wildlife and field crops (VerCauteren et al. 2005) and wildlife and livestock (VerCauteren et al. 2008, 2012b). Despite the initial cost and effort of training, guard dogs may be a long-term and cost-effective method for keeping wildlife, and hence pathogen transmission, out of livestock facilities and agricultural fields.

Testing Methods Through Monitoring

The last, but most important, step in implementing any strategy is monitoring to test whether the strategy is working and, if not, where it is failing. This is also an integral component of adaptive management and provides the "learning-by-doing" component (Knutson et al. 2010). Nichols and Williams (2006) distinguish between surveillance monitoring and targeted monitoring, where targeted monitoring has the advantage of being designed and includes rigorous monitoring that produces scientifically credible results. Monitoring alone does not make a strategy fit with adaptive management; adaptive management also involves the implementation and integration of multiple components in both assessment and adaptation (Fig. 8.2) (Williams and Brown 2012).

Monitoring is a critical step in the adaptive management process; the failure of most adaptive management programs is because the monitoring component has not been adequately supported (Knutson et al. 2010; Nichols and Williams 2006). Under adaptive management, the monitored attributes must be directly related to management objectives or else it will be difficult to ascertain whether the management objectives were met (Knutson et al. 2010).

Monitoring wildlife populations and their impacts is often problematic because wildlife are not completely detectable. This issue of incomplete detectability has generated considerable effort to develop population estimators that account for lack of complete detectability through estimation of detection probabilities (Thompson et al. 1998). The statistical and sampling issues surrounding detection of pathogens in wildlife in a monitoring program are further described conceptually by Doherty and McLean (2011) and analytically by McClintock et al. (2010).

Conclusions

Throughout this chapter, we have argued that an adaptive management approach is an appropriate, objective, scientifically based approach for mitigating or eliminating pathogen contamination of agricultural operations by wildlife. In addition, the flexibility of adaptive management allows for multiple objectives and also allows for balancing competing objectives (Knutson et al. 2010; Parma 1998; Williams and Brown 2012). For example, mitigating pathogen contamination and maintaining wildlife habitat are two seemingly competing objectives that can be evaluated and potentially balanced using an adaptive management approach (Gennet et al. 2013). In developing an adaptive management strategy, we suggest that agricultural producers:

1. Form localized coalitions among independent producers and groups to efficiently share resources.
2. Partner with university, state, and federal scientists familiar with adaptive management to develop effective strategic approaches.
3. Consider multiple methods for mitigating wildlife intrusion into agricultural facilities, which may include a combination of population control and exclusionary measures.

All of these points should be considered in terms of the scope and scale of the problem. For example, developing strategies for leafy green crops in the Salinas Valley may not be completely relevant to other leafy green production areas because of differences in landscapes, wildlife species, and pathogens of concern. However, the general framework of the strategy may be very similar, with only the specifics needing modification.

In considering population control as an option, we argue that lethal control should generally be used only when dealing with invasive species because it resolves both ecological and agricultural problems and, thus, is more palatable to the general public. Habitat modification is also difficult to justify without more scientific evidence in terms of its effectiveness (Gennet et al. 2013).

In summary, we argue that adaptive management strategies coupled with existing methods for preventing and mitigating wildlife damage have the greatest promise for achieving cost-effective and long-term practices that balance the needs of wildlife conservation while preventing their intrusion and subsequent contamination of agricultural facilities and crops.

References

Atwill ER, Sweitzer RA, Pereira MG et al (1997) Prevalence of and associated risk factors for shedding *Cryptosporidium parvum* oocysts and *Giardia* cysts within feral pig populations in California. Appl Environ Microbiol 63:3946–3949

Atwood TC, Deliberto TJ, Smith HJ et al (2009) Spatial ecology of raccoons related to cattle and bovine tuberculosis in northeastern Michigan. J Wildl Manage 73:647–654

Belant JL, Seamans TW, Tyson LA (1998) Evaluation of electronic frightening devices as white-tailed deer deterrents. In: Baker RO, Crabb AC (eds) Proceedings of the eighteenth vertebrate pest conference. University of California, Davis, CA, pp 107–110

Bergstrom BJ, Arias LC, Davidson AD et al (2014) License to kill: reforming federal wildlife control to restore biodiversity and ecosystem function. Cons Lett 7:131–142

Beringer J, VerCauteren KC, Millspaugh JJ (2003) Evaluation of an animal-activated scarecrow and a monofilament fence for reducing deer use of soybean fields. Wildl Soc Bull 31:492–498

Bevins SN, Franklin AB (2014) One Health in action: reducing feral swine damage and disease. One Health Newsl 7(2):4–6

Bevins SN, Pedersen K, Lutman MW et al (2014) Consequences associated with the recent range expansion of nonnative feral swine. Bioscience 64:291–299

Bruning-Fann CS, Schmitt SM, Fitzgerald SD et al (2001) Bovine tuberculosis in free-ranging carnivores from Michigan. J Wildl Dis 37:58–64

Bryce R, Oliver MK, Davies L et al (2011) Turning back the tide of American mink invasion at an unprecedented scale through community participation and adaptive management. Biol Conserv 144:575–583

Carlson JC, Engeman RM, Hyatt DR et al (2011a) Efficacy of European starling control to reduce *Salmonella enterica* contamination in a concentrated animal feeding operation in the Texas panhandle. BMC Vet Res 7:9

Carlson JC, Franklin AB, Hyatt DR et al (2011b) The role of starlings in the spread of Salmonella within concentrated animal feeding operations. J Appl Ecol 48:479–486

Cernicchiaro N, Pearl DL, McEwen SA et al (2012) Association of wild bird density and farm management factors with the prevalence of *E. coli* O157 in dairy herds in Ohio (2007–2009). Zoonoses Public Health 59:320–329

Clark L (2014) Disease risks posed by wild birds associated with agricultural landscapes. In: Matthews KR, Sapers GM, Gerba CP (eds) The produce contamination problem: causes and solutions, 2nd edn. Academic Press, San Diego, CA, pp 139–165

Colorado State University Extension (2012) Fundamentals of creating a Colorado farm food safety plan: food safety plan guide. Colorado State University, Fort Collins, CO

Conner MM, Baker DL, Wild MA et al (2007) Fertility control in free-ranging elk using gonadotropin-releasing hormone agonist leuprolide: effects on reproduction, behavior, and body condition. J Wildl Manage 71:2346–2356

Conover MR (2001) Resolving human-wildlife conflicts: the science of wildlife damage management. CRC Press, Boca Raton, FL

Cooley M, Carychao D, Crawford-Miksza L et al (2007) Incidence and tracking of *Escherichia coli* O157:H7 in a major produce production region in California. PLoS One 2(11), e1159

DeNicola AJ, VerCauteren KC, Curtis PD et al (2000) Managing white-tailed deer in suburban environments: a technical guide. Cornell Cooperative Extension, Ithaca, NY

Doherty PF, McLean RG (2011) Conceptual and practical issues in monitoring disease in wild animal populations: A review of avian influenza programs. In: Majumdar SK, Brenner FJ, Huffman JE et al (eds) Pandemic influenza viruses: science, surveillance and public health. The Pennsylvania Academy of Science, Easton, PA, pp 130–142

Dolbeer RA (1998) Population dynamics: The foundation of wildlife damage management for the 21st century. In: Baker RO, Crabb AC (eds) Proceedings of the 18th vertebrate pest conference. University of California, Davis, Davis, CA, pp 2–11

Erickson MC, Doyle MP (2012) Plant food safety issues: production agriculture with One Health. In: Institute of Medicine (ed) Improving food safety through a One Health approach. The National Academies Press, Washington, DC, pp 140–175

Fagerstone KA, Coffey MA, Curtis PD et al (2002) Wildlife fertility control. Wildlife society technical review. The Wildlife Society, Bethesda, MD

Fagerstone KA, Miller LA, Killian G et al (2010) Review of issues concerning the use of reproductive inhibitors, with particular emphasis on resolving human-wildlife conflicts in North America. Integr Zool 5:15–30

Fine AE, Bolin CA, Gardiner JC et al (2011) A study of the persistence of *Mycobacterium bovis* in the environment under natural weather conditions in Michigan, USA. Vet Med Int 2011:12

Fuller-Perrine LD, Tobin ME (1993) A method for applying and removing bird-exclusion netting in commercial vineyards. Wildl Soc Bull 21:47–51

Gaukler SM, Homan HJ, Dyer NW et al (2008) Pathogenic diseases and movements of wintering European starlings using feedlots in central Kansas. In: Timm RM, Madon MB (eds) 23rd vertebrate pest conference, University of California, San Diego, CA, pp 280–282

Gennet S, Howard J, Langholz J et al (2013) Farm practices for food safety: an emerging threat to floodplain and riparian ecosystems. Front Ecol Environ 11:236–242

Gilsdorf JM, Hygnstrom SE, VerCauteren KC (2002) Use of frightening devices in wildlife damage management. Integr Pest Manage Rev 7:29–45

Gilsdorf JM, Hygnstrom SE, VerCauteren KC et al (2004a) Propane exploders and electronic guards were ineffective at reducing deer damage in cornfields. Wildl Soc Bull 32:524–531

Gilsdorf J, Hygnstrom SE, VerCauteren KC et al (2004b) Evaluation of a deer-activated bioacoustic frightening device for reducing deer damage in cornfields. Wildl Soc Bull 32:515–523

Homan HJ, LeJeune JT, Pearl DL et al (2013) Use of dairies by postreproductive flocks of European starlings. J Dairy Sci 96:4487–4493

Jay MT, Wiscomb GW (2008) Food safety risks and mitigation strategies for feral swine (*Sus scrofa*) near agriculture fields. In: Timm RM, Madon MB (eds) Proceedings of the 23rd vertebrate pest conference. University of California, Davis, CA, pp 21–25

Jay MT, Cooley M, Carychao D et al (2007) *Escherichia coli* O157:H7 in feral swine near spinach fields and cattle, Central California Coast. Emerg Infect Dis 13:1908–1911

Johnson HE, Fischer JW, Hammond M et al (2014) Evaluation of techniques to reduce deer and elk damage to agricultural crops. Wildl Soc Bull 38:358–365

Kaneene JB, Bruning-Fann CS, Granger LM et al (2002) Environmental and farm management factors associated with tuberculosis on cattle farms in northeastern Michigan. J Am Vet Med Assoc 221:837–842

Killian G, Fagerstone K, Kreeger T et al (2007) Management strategies for addressing wildlife disease transmission: The case for fertility control. In: Nolte DL, Arjo WM, Stalman D (eds) Proceedings of the 12th wildlife damage management conference. National Wildlife Research Center, Fort Collins, CO, pp 265–271

Knutson M, Laskowski H, Moore C et al (2010) Defensible decision making: harnessing the power of adaptive resource management. Wildl Prof 4:58–62

Laidler MR, Tourdjman M, Buser GL et al (2013) *Escherichia coli* O157:H7 infections associated with consumption of locally grown strawberries contaminated by deer. Clin Infect Dis 57:1129–1134

Langholz JA, Jay-Russell MT (2013) Potential role of wildlife in pathogenic contamination of fresh produce. Hum Wildl Interact 7:140–157

Leitch JA, Linz GM, Baltezore JF (1997) Economics of cattail (*Typha* spp.) control to reduce blackbird damage to sunflower. Agric Ecosyst Environ 65:141–149

LeJeune J, Homan J, Linz G et al (2008) Role of the European starling in the transmission of *E. coli* O157 on dairy farms. In: Timm RM, Madon MB (eds) Proceedings of the 23rd vertebrate pest conference, University of California, San Diego, California. University of California Press, San Diego, CA, pp 31–34

McClintock BT, Nichols JD, Bailey LL et al (2010) Seeking a second opinion: uncertainty in disease ecology. Ecol Lett 13:659–674

Miller RS, Farnsworth ML, Malmberg JL (2013) Diseases at the livestock–wildlife interface: status, challenges, and opportunities in the United States. Prev Vet Med 110:119–132

Mott DF (1980) Dispersing blackbirds and starlings from objectionable roost sites. In: Clark JP, Marsh RE (eds) Proceedings of the 9th vertebrate pest conference. University of California, Davis, CA

Newman P, Salmon TP, Gorenzel WP (2012) Food safety and rodent control in leafy green crops. In: Timm RM, Madon MB (eds) Proceedings of the 25th vertebrate pest conference. University of California-Davis, Davis, CA, pp 107–112

Nichols JD, Williams BK (2006) Monitoring for conservation. Trends Ecol Evol 21:668–673

Nichols JD, Johnson FA, Williams BK (1995) Managing North American waterfowl in the face of uncertainty. Annu Rev Ecol Syst 26:177–199

Palmer MV, Whipple DL (2006) Survival of *Mycobacterium bovis* on feedstuffs commonly used as supplemental feed for white-tailed deer (*Odocoileus virginianus*). J Wildl Dis 42:853–858

Palmer MV, Waters WR, Whipple DL (2004a) Investigation of the transmission of *Mycobacterium bovis* from deer to cattle through indirect contact. Am J Vet Res 65:1483–1489

Palmer MV, Waters WR, Whipple DL (2004b) Shared feed as a means of deer-to-deer transmission of *Mycobacterium bovis*. J Wildl Dis 40:87–91

Parkes JP, Robley A, Forsyth DM et al (2006) Adaptive management experiments in vertebrate pest control in New Zealand and Australia. Wildl Soc Bull 34:229–236

Parma AM (1998) What can adaptive management do for our fish, forests, food, and biodiversity? Integr Biol 1:16–26

Phillips GE, Lavelle MJ, Fischer JW et al (2012) A novel bipolar electric fence for excluding white-tailed deer from stored livestock feed. J Anim Sci 90:4090–4097

Porter WF (1983) A baited electric fence for controlling deer damage to orchard seedlings. Wildl Soc Bull 11:325–327

Reidinger RF, Miller JE (2013) Wildlife damage management: prevention, problem solving and conflict resolution. Johns Hopkins University Press, Baltimore, MD

Rhyan JC, Miller LA, Fagerstone KA (2013) The use of contraception as a disease management tool in wildlife. J Zoo Wildl Med 44:S135–S137

Rice DH (2014) Produce contamination by other wildlife. In: Matthews KR, Sapers GM, Gerba CP (eds) The produce contamination problem: causes and solutions, 2nd edn. Academic Press, San Diego, CA, pp 139–165

Runge MC (2011) An introduction to adaptive management for threatened and endangered species. J Fish Wildl Manage 2:220–233

Thompson WL, White GC, Gowan C (1998) Monitoring vertebrate populations. Academic Press, San Diego, CA

U. S. Department of Health and Human Services (2013) Proposed rule for standards for the growing, harvesting, packing, and holding of produce for human consumption. Fed Reg 78:3504–3646

VerCauteren K, Pipas M, Peterson P et al (2003) Stored-crop loss due to deer consumption. Wildl Soc Bull 31:578–582

VerCauteren KC, Seward NW, Hirchert DL et al (2005) Dogs for reducing wildlife damage to organic crops: a case study. In: Nolte DL, Fagerstone KA (eds) Proceedings of the 11th wildlife damage management conference, Fort Collins, CO, pp 286–293

VerCauteren KC, Lavelle MJ, Phillips GE (2008) Livestock protection dogs for deterring deer from cattle and feed. J Wildl Manage 72:1443–1448

Vercauteren KC, Vandeelen TR, Lavelle MJ et al (2010) Assessment of abilities of white-tailed deer to jump fences. J Wildl Manage 74:1378–1381

VerCauteren KC, Dolbeer R, Gese E (2012a) Identification and management of wildlife damage. In: Silvy NJ (ed) The wildlife techniques manual, 7th edn. John Hopkins University Press, Baltimore, MD, pp 232–269

VerCauteren KC, Lavelle MJ, Gehring TM et al (2012b) Cow dogs: use of livestock protection dogs for reducing predation and transmission of pathogens from wildlife to cattle. Appl Anim Behav Sci 140:128–136

Walter WD, Anderson CW, Smith R et al (2012) On-farm mitigation of transmission of tuberculosis from white-tailed deer to cattle: Literature review and recommendations. Vet Med Intern 2012:15.

Walter WD, Smith R, Vanderklok M et al (2014) Linking bovine tuberculosis on cattle farms to white-tailed deer and environmental variables using Bayesian hierarchical analysis. PLoS One 9, e90925

Walters C (1986) Adaptive management of renewable resources. Macmillan Publishing Company, New York, NY

Walters CJ, Holling CS (1990) Large-scale management experiments and learning by doing. Ecology 71:2060–2068

Williams ES, Barker IK (2001) Infectious diseases of wild mammals. Iowa State University Press, Ames, IA

Williams BK, Brown ED (eds) (2012) Adaptive management: The U.S. Department of the interior applications guide. Adaptive Management Working Group, Washington, DC

Williams BK, Nichols JD, Conroy MJ (2002) Analysis and management of animal populations. Academic, San Diego, CA

Witmer GW (2007) The ecology of vertebrate pests and integrated pest management (IPM). In: Kogan M, Jepson P (eds) Perspectives in ecological theory and integrated pest management. Cambridge University Press, Cambridge, pp 393–410

Wyckoff AC, Henke SE, Campbell TA et al (2012) Movement and habitat use of feral swine near domestic swine facilities. Wildl Soc Bull 36:130–138

Chapter 9
Co-management: Balancing Food Safety, the Environment, and the Bottom Line

Mary Bianchi and Karen Lowell

Abstract Growers and distributors of fresh produce have long realized that reliably safe products and responsible use of resources inspire brand trust and consumer loyalty. Balancing food safety and resource conservation goals has become a vital element of produce industry management throughout the supply chain. Co-management is a process that seeks to balance food safety and sustainability goals in the context of maintaining a sound bottom line. The resources to develop effective co-management strategies lie within diverse communities of practice, including agricultural producers, food safety and wildlife professionals, conservation professionals, and academics with primary focus on any the above areas.The economic loss incurred from fields that must be abandoned before harvest due to fecal contamination can be significant. In addition, compliance with both food safety and conservation goals may generate additional operational costs. Key research questions remain, many defined during the critical conversations surrounding on-farm decisions regarding co-management. Additionally, responsibility lies with the research community for creating an open and integrated approach to interpretation, extension, and implementation of research results surrounding contamination, transport and survival of pathogens in the production environment.

Keywords Co-management • Conflict resolution • Conservation • Economic analysis • Food industry • Food safety • Natural resources • Sustainable agriculture • Wildlife

---

M. Bianchi (✉)
University of California Cooperative Extension, San Luis Obispo, CA, USA
e-mail: mlbianchi@ucanr.edu

K. Lowell
United States Department of Agriculture Natural Resources Conservation Service, Salinas, CA, USA

M. Jay-Russell, M.P. Doyle (eds.), *Food Safety Risks from Wildlife*, Food Microbiology and Food Safety, DOI 10.1007/978-3-319-24442-6_9

Introduction

Protecting fresh produce from pathogen contamination relies upon a robust risk analysis and preventative controls to mitigate the risks identified. Because many human pathogens are present in fecal matter, including wildlife feces, risks associated with wildlife in the production environment are an important element of risk analysis for a produce farm. Movement of domesticated animals can generally be controlled with fencing and other management techniques. Wildlife movement is much more difficult to control and presents unique challenges for those who grow crops that are consumed raw, and are thus more likely to lead to foodborne illness due to pathogen contamination. This chapter addresses how the produce industry has responded to an increasing focus on food safety risks at the field level, with particular emphasis on how the industry has learned to co-manage food safety and conservation practices with regard to wildlife. While the term conservation may cover many types of management practices, within this chapter it is used to refer to natural resource conservation.

Background

Growers and distributors of fresh produce have long realized that reliably safe products and responsible use of resources inspire brand trust and consumer loyalty. Balancing food safety and resource conservation goals has become a vital element of produce industry management throughout the supply chain (Bianchi and Lowell 2012). On their farms, growers are active stewards of the land, supporting wildlife populations by preserving their habitat in non-cropped areas. Growers also protect soil and water quality with a variety of management strategies, which may include features that support wildlife (e.g., vegetation and water bodies). At the same time, growers must ensure that their crops are protected from contamination by fecal matter, which may introduce pathogens that can cause foodborne illnesses. Balancing these unique management objectives, while maintaining a sound bottom line, is a central challenge for produce farmers.

Food safety concerns are not new in the produce industry; however, focus on the production environment increased markedly following an outbreak of foodborne illness linked to *E. coli* O157:H7-contaminated spinach in 2006 (RCD 2007, 2009). Produce growers faced challenges presented by an increased emphasis on adjacent land use or management practices that might increase the risk of introduced pathogens. Growers found themselves explaining farming and conservation practices to food safety professionals who often had little experience in fresh produce production at the field level (Johnston et al. 2014). Other challenges arose when individuals had experience in food safety, but little knowledge of important conservation goals such as water quality and the importance of protecting and enhancing wildlife habitat (Crohn and Bianchi 2008).

As growers responded to produce buyers' concerns, management practices shifted. In a survey completed in 2007 (RCD 2007), growers reported taking steps to eliminate wildlife, vegetation, and water bodies near crops in response to pressures from auditors, inspectors, and other food safety professionals. Eighty-nine percent of all growers who responded to the survey (RCD 2007) indicated that they had adopted at least one measure to actively discourage or eliminate wildlife from cropped areas in response to expressed food safety concerns. Of the growers who reported taking active steps to discourage or eliminate wildlife, 42 % reported the use of poison baits, 37 % reported removal of non-crop vegetation, and 21 % reported removing or abandoning conservation practices specifically installed for water quality (RCD 2007). Many conservation practices rely upon vegetation to protect soil, slow and filter surface water runoff, and provide diverse benefits to soil and water quality as well as wildlife. Similarly, water bodies provide powerful tools to protect water quality. For example, sediment retention basins capture runoff and reduce sediment loading in surface water, and treatment wetlands capture and filter a range of contaminants and may also provide habitat for wildlife. These contaminants may include nutrients, pesticides, sediment, and pathogens (Díaz and Dahlgren 2012; Lowell and Bianchi 2011). If open-source irrigation water (e.g., streams, ponds, irrigation reservoirs) is not protected from pathogen contamination, then contamination of produce may occur when crops are irrigated. Thus, removal of vegetation and water bodies that protect surface water has the potential to have adverse impacts on water quality and soil health as well as on broader conservation and food safety objectives.

Management challenges frequently emerge when farms are located in proximity to rivers and streams that may support wildlife populations, as the unique qualities of riparian zones may provide habitat for a particularly diverse population of wildlife (Kocher and Harris 2007; Naiman et al. 1993; Hilty and Merenlender 2004)). Management decisions on produce farms must consider laws at the federal, state, and county levels that relate to protection of water quality; water/wetland management; stream bank protection measures; and protection of birds/fish/animals/plants designated as endangered, threatened, or otherwise protected.

Produce industry leaders recognize that management decisions related to produce fields in response to food safety concerns may undermine conservation objectives, and they have worked to restore a balanced management approach (Western Growers 2012). Working collaboratively with research scientists, extension specialists, conservation and food safety professionals, and farmers, a process of co-managing for diverse management objectives has emerged. Co-management has been differently defined by several groups (Lowell et al. 2010; National Sustainable Agriculture Coalition 2014; CA LGMA 2013). The California Leafy Greens Handlers Marketing Agreement (CA LGMA) accepted a broadly vetted definition which reads: "Co-management minimizes the risk of fecal contamination and the resulting microbiological hazards associated with food production while simultaneously conserving soil, water, air, wildlife and other natural resources." (CA LGMA 2013). The importance of supporting co-management was included in the original language of the 2011 Food Safety Modernization Act.

A key development in co-management is an emphasis on evidence of contamination risk, specifically fecal matter or animal intrusion in the crop field (e.g., tracks, evidence of feeding). Initially food safety professionals focused on specific animal species (CA LGMA 2008). For example early versions of the LGMA referenced specific animals, including cattle, sheep, goats, pigs (domestic and wild), and deer. Researchers sought to describe prevalence rates in a wide range of wildlife species (Atwill et al. 2012; Jay-Russell 2013; Langholz and Jay-Russell 2013). In 2012 the produce industry, led by Western Growers and the CA LGMA Technical Committee, determined that it was more effective to focus on specific events (e.g., fecal contamination, feeding damage, animal intrusion into produce fields) rather than continuing attempts to catalogue pathogen prevalence in an ever-widening list of wildlife species (Western Growers 2012). In part, this reflected a pragmatic management reality. Episodes of animal intrusion or discovered fecal matter allow for concrete management decisions, while predicting the likelihood of a particular animal depositing pathogen contaminated fecal matter does not. As noted above, co-management strategies have increasingly focused attention on fecal matter as the risk factor, rather than animal presence or habitat in proximity of the crop. This targeted risk-based approach reduces the likelihood of adverse impacts on conservation resulting from a focus on wildlife habitat.

Concerns regarding potential impacts of food safety programs on conservation initiatives around the nation have led to cautionary language in amendments to the Proposed Produce Safety Rule for the FDA Food Safety Modernization Act (FSMA 2014). Recent additions state: "Nothing in this regulation authorizes the "taking" of threatened or endangered species as that term is defined by the Endangered Species Act (16 U.S.C. 1531–1544) (i.e., to harass, harm, pursue, hunt, shoot, wound, kill, trap, capture, or collect, or to attempt to engage in any such conduct), in violation of the Endangered Species Act. This regulation does not require covered farms to take measures to exclude animals from outdoor growing areas, or to destroy animal habitat or otherwise clear farm borders around outdoor growing areas or drainages." The language reflects public comment in which stakeholders expressed concern that growers might fail to effectively use co-management processes unless they were specifically reminded that conservation objectives, including how management actions might influence threatened and endangered species, must also be considered.

Co-management Process in Action

California's Central Coast region is a major produce growing region, with production values nearing $4 billion dollars (Langholz and DePaolis 2011a, b, c). Following the 2006 foodborne illness outbreak linked to spinach sourced from the region, many produce fields became a laboratory in which co-management processes evolved.

Co-management strategies are designed by a diverse range of professionals working collaboratively to identify risks, management objectives, and strategies to address these objectives. Where necessary, practices may be adapted to more effectively meet these management objectives. Many factors influence the potential for pathogen contamination, transport, and survival in the production environment (Atwill et al. 2012; Pachepsky et al. 2011). Co-management considers factors that influence pathogen contamination, transport and survival; in turn, growers design co-management strategies that fit unique farm settings. The discussion that follows demonstrates co-management processes using experiences from a range of operations in a major produce growing region in California. Examples come from small, diverse farming operations, large-scale operations with a single crop or multiple crops, and both organic and conventional operations.

The examples described below reflect information shared with the authors in private conversations as well as during the 2014 Forum of the Farm Food Safety Conservation Network (Kan-Rice 2014). The preference to share strategies without revealing identifying information about the grower or farming operation makes it challenging to attribute information to a specific source (e.g., by citation of Personal Communication). The caution with which growers share their food safety-related management decisions reflects the complex regulatory and market environment in which produce growers operate.

Frequently considered adaptations to incorporate co-management strategies often use spatial and temporal changes that address potential food safety concerns with common conservation practices. The analysis that guides management begins even before the crop is planted. For example, a pre-plant risk assessment considers proximity to potential contamination hazards in the production environment. This may include evidence of animal movement corridors, areas of dense wildlife populations, and food or shelter that may attract wildlife to the crop environment or influence movement of wildlife through the crop or adjacent water bodies, etc.

Broader considerations include noting any land use or activity that may introduce fecal matter (e.g., livestock operations, grazing operations, stockpiled manure, compost operations) or contaminated water (e.g., sloped rangeland draining to the crop field, flood waters). Where a grower has the ability to plant crops in a variety of locations, crops that are consumed raw (without a kill step for any pathogen contamination that may be present), may be planted further from identified hazards. In some instances, there may be specific guidance about recommended distance between crops and identified risks. For example, for both composting operations and Concentrated Animal Feeding Operations (CAFOs), the LGMA language proposes 400 ft distance from the edge of a crop field to these land uses. This guidance may be revised as additional research helps better understand risk in these situations (CA LGMA 2013). Flexibility of planting location is often dependent on the size of the operation, and many growers, particularly small, limited resource farmers, have few options about planting location. In addition, considerations such as crop rotation to prevent plant and soil-borne diseases may impact a grower's flexibility in crop selection and planting date.

Another strategy might be to make a temporal shift, for example planting produce crops at times of the year when wildlife intrusion is less likely. The long growing season on California's Central Coast allows for such strategies in a landscape with a diverse tapestry of crops. For example, growers report that feral pigs are attracted to ripening grapes. Knowing grapes ripen in the fall, a grower in an adjacent production block may choose not to plant a lettuce crop for harvest at the time grapes are ripening. Choosing to plant a crop destined for processing, with a validated kill step, in the field near the vineyard represents a temporal co-management strategy. It is important to note that not all processing procedures are kill steps. For example, blanching procedures commonly used on frozen vegetables, unless validated, are not considered a reliably effective kill step (Personal Communication, Laura Giudici Mills, LGM Consulting, Spreckels, CA). This scenario would rely upon both a good deal of operational flexibility and historical knowledge of feral pig activity in the area.

Irrigation water from open sources (e.g., streams, rivers, irrigation reservoirs, canals) may also have temporary risks associated with high wildlife populations. It is common to have large flocks of visiting waterfowl during migratory periods or concentrated use of the water source during seasonal dry periods. Standard co-management strategies may include robust monitoring for fecal contamination and evidence of animal intrusion in the crop, with defined "no-harvest" zones where contamination is determined to be a medium-high hazard/probable risk (CA LGMA 2013). Additional water testing may also be added to standard testing protocols and, in some cases, irrigation water may be treated to prevent pathogen contamination of the crop. Alternately, growers may opt to use a different water source during periods of high water fowl use, plant a crop destined for processing (with a kill step), or forgo planting altogether during the high animal activity period. Opting not to plant is an extreme option. Most growers need to maximize use of their farmland and won't idle their ground unless it is part of their normal crop rotation schedule, or they miss a planting due to weather event or later than normal harvest of previous crop (Personal Communication, Laura Giudici Mills, LGM Consulting, Spreckels, CA).

Co-management strategies require flexible responses to changing conditions. For example, in the 2014 production season in California's Central Coast, some vegetable crop growers reported more feral pigs moving down from adjacent rangeland, perhaps as a result of drought conditions and limited water in the rangeland. In some cases, growers have opted not to plant certain farms as the likelihood of wildlife intrusion is high. A high risk of contamination, and the potential inability to sell the crop, is economically unacceptable. For large-scale producers, such a decision may be possible as other locations can remain productive. For many small-scale producers, limited in their choice of fields to plant, the decision to forego planting may not be economically viable. A grower practicing co-management must necessarily consider all relevant factors, including risk associated with activities on neighboring properties.

Temporal considerations also apply to intervals between possible contamination and harvest of the crop. For example, many growers plant cover crops to build soil health and to protect the soil during fallow periods between crops (Smith et al. 2011). Cover crops may provide habitat for a range of wildlife species during their growth, and fecal material may be present at the time they are incorporated in the soil prior to planting a cash crop. An example of this is seen in the rotation of strawberries, vegetable crops, and cover crops. A planting sequence might be to plant cover crops in the fall, incorporate the cover crop in early spring, plant lettuce a few weeks after incorporation of the cover crop, harvest lettuce after approximately 3 months, replant lettuce (a double crop), and then plant strawberries in the following fall. Since the lettuce crop is harvested 3 months after the cover crops are incorporated, pathogens that might have been present have had an opportunity to die off. Co-management strategies consider the benefits of the cover crops in the cropping system and the potential contamination risk from wildlife fecal matter. Research provides some insight into how long pathogens survive on contaminated crop crops in various settings (Koike et al. 2010). Future research may provide more detailed understanding of timing for pathogen die-off, which will allow management to reduce the potential for food safety risk while meeting management objectives for soil health and protection.

Other factors may also influence how farmers can respond to wildlife activity near their crops. For example, small birds often build nests in farm structures (e.g., pump houses, equipment and packing sheds.) and food safety professionals may be concerned about their feces contaminating well water or equipment that may enter the crop fields. In some cases these birds are protected species (e.g., Migratory Bird Act, or if endangered, by the Endangered Species Act), and disturbing them may be illegal. If bird activity is noted, growers may install features (e.g., wires on perching or nesting sites in the structures) to discourage nesting, perching or roosting. Ideally this work is done outside of the nesting season. However, if the birds have nested and laid eggs, growers may not disturb them or their nests. In this case, growers may document the activity and guide workers to carefully inspect any materials stored in or near the structure to avoid introducing fecal matter to the crop. Food safety plans may note that, after young birds have fledged, the nest should be removed and features installed to discourage future bird use of the site.

In the case of birds, food safety professionals acknowledge that it is not possible to monitor all bird activity over a crop field. Deterrent actions such as noise cannons may be used to discourage birds, but in many cases bird activity in the crop is unavoidable. In this instance, guidance focuses on careful inspection of the crop for fecal contamination or other evidence of bird activity (e.g., feeding damage, feathers, tracks). General guidance for all animals is covered in Considerations in Assessing Potential Hazards and Risks Associated with Animal Activity in the Field (both domesticated and wild) in the LGMA Guidance Documents (CA LGMA 2013). Guidance includes the following: volume and concentration of fecal material in the field and production area; frequency of animal sightings and sign (e.g., tracks, scat, rubbing, animal damage to crop); animal species likely to aggregate (e.g., flocks and herds) and produce concentrated areas of fecal material and incidental contact with the crop; potential for animals to transport pathogens from a high-risk source (e.g., CAFO, garbage dump, sewage treatment facility) to the field; species with seasonal migrations that result in increased population density and potential for activity in the field. A low-hazard/negligible risk would likely result in the grower or handler following their company's Standard Operating Procedure (SOP), whereas a medium-high hazard would likely result in the grower or handler establishing a "no-harvest" buffer or, if the area cannot be effectively buffered, not harvesting the crop and, if necessary, consulting with state and regional experts (see CA LGMA's Appendix Z) to develop co-management strategies to prevent recurrence (Personal Communication, Mills, Laura Giudici, LGM Consulting, Spreckels, CA; CA LGMA 2013). See additional discussion in Chap. 7 of this volume by Giclas and Wetherington.

Co-management processes can also be built around practices intended to protect food safety through monitoring or controlling animal movement, such as fencing and bare ground buffers. These practices may be particularly important when addressing concerns regarding conservation practices that cannot simply be relocated. Riparian vegetation and wetlands are landscape features that require co-management strategies that recognize their importance in the production environment. While fencing and bare ground buffers may raise concerns within the conservation community, these strategies can be managed to balance both food safety and wildlife protections.

Fencing may provide an effective method of discouraging animal movement into crop fields. Co-management strategies typically emphasize fencing that allows maximum mobility of wildlife along habitat corridors, but discourages movement into crop fields. In one example, a grower has allowed entry and exit points along the fence line for wildlife. Strategic positioning of these entry and exit points allows wildlife to move to an adjacent park with abundant habitat for the animals, but effectively reduces traffic that transects planted fields.

Managing ground squirrel activity around a small wetland area in the center of a large produce farm provides another example of building potential co-management strategies. In this case the grower has erected low (approximately 3 ft tall) silt fencing around the wetland to discourage movement of the ground squirrels out of the vegetation surrounding the wetland while maintaining the water quality benefits provided by the wetlands. A grower visiting the wetland noted that at her operation, applying food-grade oil to the silt fence effectively reduced the ability of frogs to climb up and over the silt fencing. While some animals still move into the cropped area surrounding the wetland, numbers are reduced, thereby minimizing the risk of contamination. Such use of silt fencing is common along crop field edges that border riparian areas, wetlands, tail water basins, or other land features likely to support wildlife.

Fencing may not always be effective in keeping animals out of production fields. It is important to know what animals are likely to be a problem. For instance, the types of fencing recommended to exclude deer will not effectively exclude feral pigs. Where wildlife is not protected, other management options may be available and may be recommended in addition to fencing (Jay and Wiscomb 2008). For example, in California feral pigs are exotic and invasive, and are not protected. Depredation permits may be obtained from the California Department of Fish and

Wildlife, and feral pigs may be hunted or trapped. Growers routinely set traps for pigs and rodents that may enter crop fields.

Bare ground buffers that maintain an area free of vegetation immediately adjacent to the crop field are a management strategy that allows for direct observation of animal tracks into fields. Keeping the soil covered is a basic tenet of soil conservation. Co-management strategies focus on guiding the placement of bare ground buffers to minimize potential adverse impacts of bare ground on soil and water resources. For example, a minimal bare ground buffer placed immediately adjacent to a crop field, with a vegetated strip, or dense riparian vegetation between the bare ground buffer and a water body is less concerning than a broad bare ground buffer directly adjacent to a stream. In many cases, dirt roads around production blocks serve as bare ground buffers. One grower who farms along a riparian area routinely drives the roads, stopping to inspect and document animal activity evidence (e.g., tracks, fecal matter). These detailed records serve as a foundation for risk-based management decisions. At times, such documentation leads to a decision not to harvest large production blocks. The economic implications of such decisions will be discussed later in this chapter.

Co-management in a Broader Context

This chapter has focused on co-management processes as they relate to wildlife management strategies. It is important to note that co-management applies to all aspects of management. Consider the earlier example of a cropping system that includes cover crops, lettuce, and strawberries. A grower who manages such a system approached the USDA Natural Resources Conservation Service (NRCS) to explore the possibility of capturing water from subsurface drains for application on the farm. Through an astute evaluation of her farming operation, she recognized that her drainage water with elevated nitrate levels presented an environmental quality concern for the receiving waters of the Monterey Bay National Marine Sanctuary. Further, she noted the historic drought compelled her to find ways to make maximum use of water resources. Working with NRCS engineers she determined it was possible to capture the drainage water and store it for use on the farm. Her ultimate decision of whether to proceed with the project must consider volume of water likely to be captured, costs to develop the storage and re-use capacity, and suitability of the water for the intended uses. The latter must consider not only microbiological factors pertinent to food safety concerns, but also parameters that may influence agronomic suitability of the water (e.g., salinity, nutrients, plant pathogens, pesticides, heavy metals). In her case, application of the water to the lettuce and strawberries presents an unacceptable food safety risk, unless treated, which introduces another cost factor. Use of the water for dust abatement on farm roads or to the cover crop (not consumed) may be acceptable if there is no possibility of contaminating either the lettuce or the strawberry crops.

Impacts to the Bottom Line

Co-management seeks to balance food safety and sustainability goals in the context of maintaining a sound bottom line. The economic loss incurred from fields that must be abandoned just before harvest due to fecal contamination can be significant. In addition, compliance with both food safety and conservation goals may generate additional operational costs.

University of California Cost and Return Studies provide estimates of costs for many commodities, and can be used here to demonstrate the challenges producers face. As an example, using the most recent data available for growers of romaine lettuce hearts in California's Central Coast, growers may have total costs of $4109 per acre to grow the crop. Costs to harvest are an additional estimated $4739 per acre, assuming a 2009-based industry average yield of 700 cartons. Net returns above the total costs to grow and harvest vary according to the current prices per carton. At $12–14 per carton, net returns can range from a loss to $952 per acre (Smith et al. 2009). If animal intrusion evidence is extensive, a grower may not harvest large production blocks for fear that fecal contamination will not be detected in a preharvest inspection and contaminated product will be harvested and sold. While harvest costs are not incurred, there will be costs associated with destroying the un-harvested crop, typically by incorporating it back into the soil. In this case, total loss to the grower will include growing costs plus the cost to deal with the un-harvested crop. There are also hard to monetize costs associated with impact on future contracts with a buyer for whom the grower was unable to meet a contracted harvest.

Conservation practices also incur costs for the grower. For example, cover crops, previously discussed, can benefit soil health and help capture residual soil nitrogen and reduce runoff and leaching of excess nutrients, most notably nitrogen (Dabney et al. 2001), while vegetated buffers that reduce soil erosion may also reduce sediment and phosphorus movement to surface water (Abu-Zreig et al. 2003; Zhang et al. 2010). Both nitrogen and phosphorus have been linked to significant water quality concerns in numerous regions around the country (Carpenter et al. 1998). In California's Central Coast region, cover crops are commonly planted in the fall, grown during winter, and worked into the soil in spring. Tourte et al. (2003a, b) estimated representative costs for fall-planted cover crops to be $147 per acre. For grassed filter strips, Tourte et al. (2003a, b) estimated representative costs for a 1300 linear foot, 16-ft wide buffer strip (total area 0.5 acre) to be $229. This does not include lost productivity from this half acre area. These costs would be in addition to crop production costs noted above, but provide multiple benefits, including improved soil health and reduced nutrient runoff. Conservation practices may also support attainment of water quality-related regulatory requirements, though often there is a lag time in demonstrated efficacy of conservation practices in mitigating water quality concerns (Meals et al. 2010) so farmers may not find their investment of time and money immediately rewarded.

Management modifications for compliance with food safety protocols also come with additional costs. Hardesty and Kusunose (2009) surveyed producers on California's Central Coast on their costs for modifications made specifically for compliance with the CA LGMA food safety metrics. Responses from producers revealed that costs averaged $54.63 per acre in 2007 in addition to $13.60 per acre for average modifications costs. Smith et al. (2009) estimated food safety costs at $50 per acre. The current market demand to meet Global Food Safety Initiative standards (GFSI) and monthly water system testing has increased costs that have yet to be fully accounted. Also, the increased sharing of documentation (e.g., ranch assessments, ranch audits, pest controls used, copies of water tests, statements of ranch history) between produce buyers and growers adds a clerical cost to food safety programs that has yet to be calculated on a per acre basis (Personal Communication, Ken Stearns, Food Safety Director, D'Arrigo Brothers, Salinas, CA).

The previous example of lettuce illustrates the financial impacts of food safety management. The sum of the average modification costs and 2007 seasonal food safety costs—$68.23—represents almost 1 % (0.93 %) of the growers' average lettuce revenues. Additional costs not included in the Hardesty and Kusunose (2009) work include ongoing expenses related to crew training and equipment, and product abandoned due to food safety concerns (Personal Communication, Ken Stearns, Food Safety Director, D'Arrigo Brothers, Salinas, CA). Since it appears that growers may have excluded some costs when reporting their seasonal food safety costs, a combined per acre cost of approximately $100 could be a more accurate average per acre estimate (Hardesty and Kusunose 2009).

Market-driven requirements for produce growers generally include liability insurance as an additional cost. In the 2009 romaine lettuce example, liability coverage is included at $1 per acre, with a total cost of $1140 for the 1140 acre operation for which costs were analyzed (Smith et al. 2009). Liability insurance covers bodily injury and property damage claims made against the grower, including claims made by third parties for contaminated product. Cost for this insurance currently runs $1–2 per acre. More recently, larger produce buyers have required higher levels of coverage with significantly higher premiums. This liability insurance does not cover Workman's Compensation coverage for employee injury, which is based on payroll for an operation and it typically a much higher cost to the grower (Personal Communication, Don Winn, Arthur J. Gallagher Co., Hollister, CA, 2014).

In addition to various types of liability insurance, an increasing number of produce growers in California's Central Coast region are also opting to carry recall insurance. This insurance covers the costs a grower incurs in the event of accidental or malicious contamination of the crop. Recall insurance covers costs a grower may incur to take product off shelves, test and destroy product, as well as the cost to replace the product and the lost profits of the growers' customers (e.g., handlers/shippers/buyers). Claims against this coverage are only possible if the policy holder's own crop is contaminated; the coverage does not apply to losses that may result to an entire sector as the result of contamination of a given crop. For example, a grower with recall insurance for a spinach crop in New York cannot make a claim based on lost revenues related to reduced demand for spinach in the New York market following a contamination event in California. Also, of critical importance, this insurance does *not* cover losses incurred if product is contaminated in the field and has to be destroyed instead of sold. Recall insurance premiums are typically much higher and may average $10 per acre (Personal Communication, Don Winn, Arthur J. Gallagher and Co. Hollister, CA, 2014).

Supporting Co-management

Support for stakeholders with diverse skills and knowledge who are charged with finding a balance between food safety and conservation has proven a critical factor in the evolution of co-management.

As food safety management shifted toward increasing focus on practices at the field level, both food safety and conservation professionals found themselves in unfamiliar territory. In April 2007, shortly after the 2006 *E. coli* O157:H7 outbreak linked to leafy greens, more than 100 invited food safety and water quality leaders met in San Luis Obispo, California. The purpose of the meeting was to discuss research priorities needed to assure food safety while conserving water quality. The attendees represented government, industry, and academia and were active at national, state, and regional levels. Organizers hypothesized that the food safety and water quality communities were largely uneducated as to each other's concerns, constraints, and motivating interests (Crohn and Bianchi 2008). Small groups representing both constituencies visited area farms. During these farm visits, water quality leaders were asked to audit the farms from a food safety perspective and food safety professionals were expected to assess water quality concerns. The co-management process experienced in 2007, and that which has evolved over the past decade, relies on deliberate efforts by all involved decision makers to consider the potential effects of a management decision in multiple dimensions.

For this reason, a keystone for supporting co-management is providing well-documented language in terms that all stakeholders can understand. Many producers, food safety professionals, and conservation professionals now recognize the broader context in which management practices on produce farms must be considered. For example, producers are learning to describe the purpose of conservation practices, and to articulate their concerns about how food safety management objectives may undermine conservation objectives. Food safety professionals are learning to recognize individual conservation practices, their purpose and the language growers may use to describe their importance in the production environment. Conservation professionals are learning to understand food safety concerns and to respect the need for producers to consider food safety impacts of conservation activities on and around their produce farms.

With regard to wildlife, much of the effort to support understanding has been focused on helping food safety professionals and production personnel recognize the benefits of vegetation and water bodies in conservation stewardship and helping conservation professionals recognize the importance of designing conservation practices that both support conservation objectives and acknowledge food safety risks. The shift in focus to evidence of contamination (e.g., fecal matter, tracks, feeding damage) rather than the possibility that wildlife may frequent non-crop areas near the crop fields has allowed more opportunities to use co-management processes for best outcomes.

The following section describes resources for those seeking co-management support and training opportunities.

The resources to develop co-management strategies lie within the diverse communities of practice apparent in the definition. Agricultural producers, food safety and wildlife professionals, conservation professionals, and academics with a primary focus on all of the above are critical partners in the inter- and multidisciplinary dialogue required to develop effective strategies. There are both regional and national examples of ongoing efforts to support multidisciplinary teams: Farm Food Safety Conservation Network[1]; Produce Safety Alliance[2]; National Sustainable Agriculture Coalition.[3] The Center for Produce Safety[4] supports co-management research. The Western Institute for Food Safety and Security[5] and the Western Center for Food Safety,[6] are active in both research and extension surrounding co-management. Additionally, the USDA Natural Resource Conservation Service has supported development of co-management strategies through its Conservation Innovation Grants Program.[7]

Research and extension programs throughout the USA are actively pursuing information around the key production questions surrounding wildlife, water, and soil amendments. The integration of information continues to occur at the field level, in the dialogue between the producer and food safety and conservation professionals. In support of this dialogue, University of California academics in collaboration with USDA Natural Resources Conservation Service conservation professionals have created publications, video, and online training materials. These resources are found on the University of California's Food Safety website's Co-management of Food Safety and Sustainability page.[8]

On-farm food safety assessments are generally organized around food safety hazards in the production field, those associated with adjacent land, and potential

[1] http://www.awqa.org/water-quality/farm-food-safety-and-conservation-network/. Accessed 16 Jan 2015.

[2] http://producesafetyalliance.cornell.edu/. Accessed 16 Jan 2015.

[3] http://sustainableagriculture.net/. Accessed 16 Jan 2015.

[4] http://www.centerforproducesafety.org/. Accessed 16 Jan 2015.

[5] http://www.wifss.ucdavis.edu/. Accessed 16 Jan 2015.

[6] http://wcfs.ucdavis.edu/. Accessed 16 Jan 2015.

[7] http://www.nrcs.usda.gov/wps/portal/nrcs/main/national/programs/financial/cig/. Accessed 16 Jan 2015.

[8] http://ucfoodsafety.ucdavis.edu/Preharvest/Co-management_of_Food_Safety_and_Sustainability/. Accessed 16 Jan 2015.

hazards in the production environment. Co-management resource sheets have been developed that address individual conservation practices, their purpose and importance in the production environment, and when audit standards might consider the practices as addressing farming impacts on the environment and/or as potential contributors to food safety risk. Co-management resource sheets are organized into conservation practices that are found:

- Within the production field (cover crops, vegetative barriers, and soil amendments).
- Adjacent to the production field (critical area plantings, filter strips and grassed waterways; conservation cover and wetland wildlife habitat management; hedgerows, windbreaks, and herbaceous wind barriers).
- Near streams and water bodies (riparian forest buffer and riparian herbaceous cover).

Additional resource sheets cover practices related to water (sprinkler and microirrigation; irrigation field ditch, irrigation system tailwater recovery, and surface drainage ditches; irrigation reservoir and structure for water control; constructed wetlands) and sediments (sediment basin and water and sediment basin).

Within each resource sheet are practice descriptions, the potential advantages and disadvantages of these practices in the agricultural environment, and areas within some audit standards that may trigger concerns for assessing impacts on the environment as well as food safety concerns. Scenarios provide examples of how food safety concerns regarding these practices might be addressed. For example, co-management strategies for managing food safety risks from cover crops include chemically mowing the cover crop at 4–6 in. to reduce the potential for habitat, incorporating cover crops prior to planting adjacent fields, and routine monitoring for animal activity (Bianchi 2013a). In the case of riparian vegetation, monitoring of the riparian edge to understand animal movement patterns, with use of temporary fencing on an as needed basis as a control strategy where necessary may be a co-management strategy (Bianchi 2013b).

Future Directions

Key research questions remain, many defined during the critical conversations surrounding on-farm decisions regarding co-management. Additionally, responsibility lies with the research community for creating an open and integrated approach to interpretation, extension, and implementation of research results surrounding contamination, transport and survival of pathogens in the production environment. Agricultural, food safety, and conservation scientists must keep pace with co-management questions supporting food safety, the environment, and the bottom line.

References

Abu-Zreig M, Rudra RP, Whitely HR et al (2003) Phosphorus removal in vegetated filter strips. J Environ Qual 32:613–619

Atwill ER, Li X, Bond R et al (2012) Introduction to waterborne pathogens in agricultural watersheds. 2012. USDA Natural Resources Conservation Services, nutrient management technical note no. 9. http://directives.sc.egov.usda.gov/OpenNonWebContent.aspx?content=32935.wba. Accessed 12 Jan 2015

Bianchi M (2013a) Co-managing food safety and sustainability opportunities for co-management – cover crops and vegetative barriers. http://ucfoodsafety.ucdavis.edu/files/198471.pdf. Accessed 10 Nov 2014

Bianchi M (2013b) Co-managing food safety and sustainability opportunities for co-management. Vegetated practices near streams. http://ucfoodsafety.ucdavis.edu/files/198485.pdf. Accessed 10 Nov 2014

Bianchi M, K Lowell (2012) Balancing food safety and sustainability – opportunities for co-management. University of California Agriculture and Natural Resources. http://ucfoodsafety.ucdavis.edu/files/157154.pdf . Accessed 12 Jan 2015

Carpenter SR, Caraco NF, Correll DL et al (1998) Nonpoint pollution of surface waters with phosphorus and nitrogen. Ecol Appl 8(3):559–568

Crohn D, Bianchi M (2008) Research priorities for coordinating management of food safety and water quality. J Environ Qual 37:1411–1418

Dabney SM, Delgado JA, Reeves DW (2001) Using winter cover crops to improve soil and water quality. Commun Soil Sci Plant Anal 32(7–8):1221–1250

Díaz FJ, Dahlgren RA (2012) Agricultural pollutant removal by constructed wetlands: implications for water management and design. Agric Water Manage 104:171–183

Food Safety Modernization Act (FSMA) (2014) http://www.fda.gov/downloads/Food/GuidanceRegulation/FSMA/UCM417136.pdf. Accessed 12 Jan 2015

Hardesty SD, Kusunose Y (2009) Growers' compliance costs for the leafy greens marketing agreement and other food safety programs. UC small farm program brief. Accessed 20 Jan 2010

Hilty JA, Merenlender AM (2004) Use of riparian corridors and vineyards by mammalian predators in northern California. Conserv Biol 18(1):126–135

Jay MT, Wiscomb GW (2008) Food safety risks and mitigation strategies for feral swine (*Sus scrofa*) near agriculture fields. In: Timm RM, Madon MB (eds) Proceedings of the 23rd vertebrate. Pest conference. Published at University of California, Davis, CA, pp 21–25

Jay-Russell MT (2013) What is the risk from wild animals in food-borne pathogen contamination of plants? CAB Rev 8(040):1–16

Johnston LM, Wiedmann M, Orta-Ramirez A et al (2014) Identification of core competencies for an undergraduate food safety curriculum using a modified Delphi approach. J Food Sci Edu 13:12–21. doi:10.1111/1541-4329.12024

Kan-Rice P (2014) Balancing food safety and water quality not cheap, but it can be done. University of California Agriculture and Natural Resources Green Blog. http://ucanr.edu/?blogpost=15361&blogasset=52096 Accessed 20 Jan 2015

Kocher SD, Harris R (2007) Riparian vegetation. Forest stewardship series 10, University of California Agriculture and Natural Resources Publications, no. 8240. http://anrcatalog.ucdavis.edu/pdf/8240.pdf. Accessed 12 Jan 2015

Koike S, Suslow T, Cahn M (2010) Investigation of *E. coli* survival on contaminated crop residue. Final report center for produce safety. http://www.centerforproducesafety.org/amass/documents/researchproject/307/2010-Koike_Investigation%20of%20E.%20coli%20survival%20on%20contaminated%20crop. Accessed 12 Jan 2015

Langholz J, DePaolis PD (2011a) Economic contributions of Monterey County Agriculture. http://ag.co.monterey.ca.us/download_resource/222. Accessed 10 Nov 2014

Langholz J, DePaolis FD (2011b) Economic contributions of San Luis Obispo county agriculture. http://www.slocounty.ca.gov/Assets/AG/croprep/econ_study/Economic_Study_2013.pdf. Accessed 10 Nov 2014

Langholz J, DePaolis FD (2011c) Economic contributions of Santa Barbara county agriculture. http://www.countyofsb.org/uploadedFiles/agcomm/outreach/SB-Ag-Econ-vDec31-5pm.pdf. Accessed Oct 2014

Langholz J, Jay-Russell M (2013) Potential role of wildlife in pathogenic contamination of fresh produce. Hum Wildl Interact 7(1):140–157

California Leafy Greens Handlers Marketing Agreement (CA LGMA) (2008) Commodity specific food safety guidelines for the production and harvest of lettuce and leafy greens

California Leafy Greens Handlers Marketing Agreement (CA LGMA) (2013) Commodity specific food safety guidelines for the production and harvest of lettuce and leafy greens. http://www.lgma.ca.gov/wp-content/uploads/2014/09/California-LGMA-metrics-08-26-13-Final.pdf. Accessed 12 Jan 2015

Lowell K, Bianchi M (2011) Food safety and surface water quality. In: Goh KS, Bret BL, Potter TL, Gan J (eds) Pesticide mitigation strategies for surface water quality. American Chemical Society, Washington, DC, pp 351–372

Lowell K, Langholz J, Stuart D (2010) Safe and sustainable: co-managing for food safety and ecological health in California's central coast region. An initiative of The Pew Charitable Trusts at Georgetown University. The Nature Conservancy of California and the Georgetown University Produce Safety Project, San Francisco, CA

Meals DW, Dressing SA, Davenport TE (2010) Lag time in water quality response to best management practices: a review. J Environ Qual 39:85–96

Naiman RJ, Decamps H, Pollock M (1993) The role of riparian corridors in maintaining regional biodiversity. Ecol Appl 3(2):209–212

National Sustainable Agriculture Coalition (2014) http://sustainableagriculture.net/. Accessed 12 Jan 2015

Pachepsky Y, Shelton DR, McLain JE et al (2011) Irrigation waters as a source of pathogenic microorganisms in produce: a review. Adv Agron 113:73, http://afrsweb.usda.gov/SP2UserFiles/person/12091/2011PachepskyEtAl_AdvancesAgronomyV113pp73-138IrrigationWaters.pdf. Accessed 12 Jan 2015

Resource Conservation District (RCD) (2007) A grower survey: reconciling food safety and environmental protection. Resource Conservation District of Monterey County, Salinas, CA

Resource Conservation District (RCD) (2009) Challenges to co-management for food safety and environmental protection: a grower survey. Resource Conservation District of Monterey County, Salinas, CA

Smith R, Klonsky K, de Moura R (2009) Sample costs to produce romaine hearts leaf lettuce http://coststudies.ucdavis.edu/files/lettuceromcc09.pdf. Accessed 12 Jan 2015

Smith R, Bugg RL, Gaskell M (2011) Cover cropping for vegetable production: a grower's handbook. Publication 3517. University of California, Oakland, CA, p 90

Tourte L, Buchanan M, Klonsky K et al (2003a) Central coast conservation practices. Estimated costs and potential benefits for an annually planted cover crop. University of California Cooperative Extension p 4. http://coststudies.ucdavis.edu/conservation_practices/. Accessed 12 Jan 2015

Tourte L, Buchanan M, Klonsky K et al (2003b) Central coast conservation practices. Estimated costs and potential benefits for annually planted grassed filter strip strips. University of California Cooperative Extension p 4. http://coststudies.ucdavis.edu/conservation_practices/. Accessed 12 Jan 2015

Western Growers (2012) Proposed changes as a result of a CDFA-funded project. "Determining the potential impact of vegetable food safety regulations on wildlife and the environment" California Technical Subcommittee meeting, 10 Apr 2012. http://www.growershipper.com/uploads/expertpanel.pdf. Accessed 12 Jan 2014

Zhang X, Liu X, Zhang M et al (2010) A review of vegetated buffers and a meta-analysis of their mitigation efficacy in reducing nonpoint source pollution. J Environ Qual 39:76–84

Chapter 10
Recommendations to Regulations: Managing Wildlife and Produce Safety on the Farm

Gretchen L. Wall and Elizabeth A. Bihn

Abstract Successful fruit and vegetable production requires produce growers to not only have keen business acumen, but also a vast knowledge of science and agriculture, adaptability to changing farm and environmental conditions, an understanding of produce safety practices, and often times, sheer determination and dedication to rigorous labor. One long-standing and frustrating challenge for produce growers is managing wildlife on fruit and vegetable farms in an effort to protect crops from damage and preserve a full harvest to take to the market. In the last 15 years, however, focus regarding wildlife on farms has shifted to produce safety risks that may result from the presence of wildlife fecal material in produce fields and packinghouses. With buyer requirements for produce safety practices and the first-ever federal regulation of fruits and vegetables on the horizon, growers need to understand and implement food safety practices on the farm, including managing wildlife concerns to reduce risks, and make critical decisions to ensure their farm's long-term viability.

Keywords Food regulation • Food Safety Modernization Act • Good Agricultural Practices (GAPs) • Produce safety • Produce Safety Alliance • U.S. Food and Drug Administration • Wildlife • Domesticated animals

Introduction

Growing fruits and vegetables for human consumption and farm profit involves a delicate and complex process of balancing resources, including land, water, labor, equipment, capital, and time, among many other factors. Successful produce growers understand the high level of management that it takes to grow and harvest fruits

G.L. Wall
Department of Food Science, Cornell University, Ithaca, NY, USA

E.A. Bihn (✉)
Department of Food Science, Cornell University, Geneva, NY, USA
e-mail: eab38@cornell.edu

M. Jay-Russell, M.P. Doyle (eds.), *Food Safety Risks from Wildlife*,
Food Microbiology and Food Safety, DOI 10.1007/978-3-319-24442-6_10

and vegetables but, over the past 15 years, food safety requirements have evolved and in some cases, changed how resources are allocated. Many growers, regardless of size and scale, have trouble believing produce safety is an issue because they think they are already doing everything they can to grow safe produce (after all, they feed this food to their families). In addition, many food safety concerns involve factors which are largely out of the grower's control, including but not limited to, weather events (e.g., flooding, rain, wind), upstream point-source and non-point-source contamination, and wildlife habitat on or near the farm. Simply stated, nature cannot be taken out of the equation and the growing environment cannot be fully controlled.

Microbial contamination of fresh fruits and vegetables can occur during any stage of the production, harvest, and handling (Beuchat 2002; FDA 1998; Gorny 2006; Bihn and Gravini 2006). In addition, there are approximately 189,000 fruit and vegetable farms operating in the United States, the District of Columbia, and the Commonwealth of Puerto Rico, each of which have their own set of unique production practices, agricultural commodities, wildlife habitat, climates, natural resources, and marketing venues. Despite these wide ranging differences, one thing all farms have in common is wildlife, a natural, valuable, and unavoidable part of the farm environment.

This commonality also presents a unique risk to produce because a number of studies have revealed that a variety of animals, both domesticated and wild, can harbor human pathogens in their feces and can contaminate produce. Even though domesticated animals are much easier than wildlife to control in terms of their movement and access to produce fields and water sources, they still pose a risk since it is well known that cattle, sheep, pigs, chickens, cats, and dogs can harbor human pathogens and can shed them in their feces (Jay et al. 2007; Wang et al. 1996; Beuchat 2006). Although less likely than domestic animals, wildlife can also carry human pathogens such as *Salmonella* and *E. coli* O157:H7 (Langholz and Jay-Russell 2013). As wildlife interact more with domesticated animals and humans, such as through shared water sources and habitat, they are more likely to become carriers of human pathogens (Nielsen et al. 2004). Reptiles, amphibians, several bird species, flies, and larger warm-blooded animals such as elk, deer, coyotes, and feral pigs have been identified as carriers of human pathogens (Nielsen et al. 2004; Sanderson et al. 2006). Posing additional challenges to produce growers, the prevalence of human pathogens in most wildlife species is low and contamination events are often sporadic or go unnoticed, making prevention strategies extremely difficult (FDA 2011).

Using the scientific information that is currently available, a critical consideration is how researchers, educators, policy-makers, and industry members can help produce growers navigate the decision-making process that requires the evaluation of complex interactions between the environment, wildlife, and pathogen reservoirs. Recognizing that each farm is different, with a variety of wildlife present, crops grown, and market pressures, it seems likely that wildlife management decisions are best made by each individual farm owner or farm operator. This means fruit and vegetable growers are expected to have the expertise to make difficult decisions about wildlife management that may affect produce safety, market access, regulatory compliance, and conservation initiatives. This chapter will present issues related to wildlife management on the farm, including the ambiguity that growers are forced to navigate as they strive to meet regulatory, buyer, and consumer demands for safe produce.

Attributes That Contribute to the Complexity of On-Farm Food Safety Decisions

Mitigation of Fruit and Vegetable Crop–Wildlife Conflicts

Growers are acutely aware of how wildlife may affect their farm operations since harvest and yield depend on the mitigation of crop consumption, damage, or destruction by wildlife or other pests (Anderson and Lindell 2013; Baldwin et al. 2012). A survey conducted in 1998 with 2000 farmers and ranchers revealed that 80 % of fields suffered crop damage from wildlife in the previous year, and 53 % reported that the damage exceeded their level of tolerance (Conover 2002). Mitigating these negative crop–wildlife interactions can often bring up additional conflicts related to conservation efforts, including preservation of wildlife habitat, interaction with endangered species, soil erosion prevention, and protecting water ways from run-off (Dickman 2010). One way to address these conflicts is through co-management, which is an approach designed "to minimize microbiological hazards associated with food production while simultaneously conserving soil, water, air, wildlife, and other natural resources" (Lowell et al. 2010) (see also Chap. 9).

Co-management is a relatively new area of on-farm food safety programs; therefore, many growers may be unfamiliar with the topic. It is also likely that growers, along with Extension educators and other organizations who work closely with growers, will need to modify their approaches, perspectives, and expectations on the way they interact with wildlife on fruit and vegetable farms. This shift in focus is not only to support safe and sustainable stewardship of the land and its natural resources, but also to ensure that growers are aware of the contamination risks that wildlife may pose to the produce. There are many educational programs that have been developed to assist growers with understanding and implementing farm food safety practices, such as Good Agricultural Practices (GAPs), which will be discussed later in this chapter. However, implementing effective food safety practices on the farm requires that growers develop a working knowledge of produce safety risk assessment since no two farms are the same. Additionally, growers must tailor guidance to fit their own farm's needs and incorporate strategies that foster a more holistic approach to food safety and wildlife management.

Perceptions of Wildlife as a Valuable Part of the Farming Landscape

The concern about wildlife goes beyond the farm gate since buyers, regulatory personnel, third-party auditors, and consumers may also have opinions and demands about management strategies that farms employ for food safety and conservation efforts. These opinions and demands are not necessarily consistent or uniform, causing growers to balance multiple, and sometimes conflicting, priorities. Consumers, especially in U-pick and Pick-Your-Own environments, enjoy being on a farm because it is integrated into the natural environment. Farm visitors and customers often delight in seeing wildlife in the farm environment, unaware of the food safety concerns, and may even view food safety management practices that deter or eliminate wildlife as unnatural and unnecessary. Growers have even expressed concern about the high level of scrutiny being applied to the presence of wildlife in rural environments where is it expected that wildlife would naturally be present (see also Chap. 9).

In effort to align with the Natural Resource Conservation Service (NRCS) and other voluntary conservation program goals, the current proposed regulations outlined in the FSMA's Produce Safety Rule attempt to balance food safety practices with environmental and conservation practices. Although the proposed Produce Safety Rule will be discussed in detail later in the chapter, it is valuable to mention here because the rule relies on each farm owner or manager making the final management decisions to comply with the regulation since no single regulatory solution would satisfy the diversity of agricultural production needs across the country. This regulatory approach requires growers to have enough knowledge to make the right decisions about farm management in order to balance food safety and conservation. Some federal agencies, such as USDA-NRCS, support growers through additional initiatives, including an estimated $15 million 2014 fiscal year budget commitment, to stimulate the development and adoption of innovative conservation practices, including those which demonstrate and quantify the effects of conservation practices in reducing pathogen transport from manure or wildlife to produce crops (NRCS 2014).

The impact that buyers have on how food safety practices are implemented on the farm should not be overlooked. As buyers raise expectations and requirements for food safety, they have also required that farms participate in third-party audits to verify that practices have been implemented. There are many different audit schemes, audit companies, and individual auditors, often resulting in inconsistent food safety expectations imposed on growers. Some auditors have interpreted audit questions to demand that growers take steps toward food safety that conflict with their conservation practices (Langholz and Jay-Russell 2013; Lowell et al. 2010; Baumgartner and Runsten 2013). Though it is hoped that the greater focus on co-management will limit this from happening in the future, it is important to acknowledge past challenges and inconsistent demands in attempting to balance wildlife, conservation, and food safety issues on the farm.

Consumer Confidence and Health Initiatives

Food safety decisions can also impact the confidence consumers have in a product as well as the demand and market for fruit and vegetable commodities. When produce is implicated in foodborne illness outbreaks, the consumer and media response is often dramatic (Chapman et al. 2012; Terry 2011). After the 2006 *E. coli* O157:H7 outbreak associated with spinach, the spinach industry incurred a $60 million loss as sales plummeted due to consumer aversion and dwindling confidence in the safety of the commodity, regardless of growing region or association with the outbreak (Arnade et al. 2010). The outbreak investigation identified several risk factors including feral pig presence in and around spinach fields, a nearby cow-calf operation, and irrigation wells used for crop production that were subject to potential contamination by surface water. Unfortunately, no definitive link to the outbreak was determined (Jay et al. 2007; Gelting et al. 2011). Despite the lack of a conclusive contamination source, many growers in California's Central Coast region made significant changes in an attempt to reduce food safety risks associated with wild animals, in some cases resulting in negative impacts to the environment, farm landscapes, and wildlife management practices. These examples highlight the drastic ripple effect in the actions of both consumers and growers, as evidenced in the history of documented consumption pattern changes as well as food safety and ecological health conflicts that occurred in California's Central Coast region post-2006 *E. coli* outbreak in spinach (Lowell et al. 2010).

It is in the best interest of everyone that safe and abundant production of fruits and vegetables continue. This is critical because growing fruits and vegetables is not only important to farm viability and food availability, but essential to the health of consumers. Consuming fruits and vegetables reduces the risk of developing certain types of cancers, promotes maintenance of proper body weight, and provides necessary vitamins and nutrients that support overall health (Bhupathiraju et al. 2013; American Cancer Society 2014). Reduction in the volume of produce grown may limit its availability and increase the price consumers pay, resulting in decreased consumption and the loss of health benefits associated with produce consumption. Ensuring growers have the tools to make safe, sustainable, and economical food safety and wildlife management decisions on the farm will aid in keeping fresh produce available and affordable in local communities as well as the food system at large. This is much easier said than done, as evidenced by the many factors growers must contend with in the farm environment and the marketplace.

Grower Challenges to Balancing Food Safety Goals and Wildlife Management Strategies

Wildlife and pest management within the production environment is not new to fruit and vegetable growers. However, part of the focus has now shifted from protecting the crop yield to protecting the crop from microbial contamination. Understanding and identifying risks associated with pests and wildlife is much more obvious in the case of crop damage or loss than it is in identifying microbial contamination risks which, in many cases, are not visible to the human eye. It is also important to acknowledge that food safety is just a small portion of overall farm management goals. Every farm has limited resources available to meet all of its production, food safety, and marketing goals. Farming is inherently a high-risk venture, from fluctuating prices and yields, variable weather patterns, and inconsistent demands for everything from the types of commodities and varieties of produce a farm grows for the market to the food safety practices they implement. The value of a crop, overall revenue of the farm, farm scale (large farm vs. small farm), and type of farming operation (organic vs. conventional) may dictate what management strategies the grower has available to use and may affect their willingness and ability to adopt certain strategies.

Critical food safety decisions often must be made at harvest and can affect a large portion of, or even an entire seasons' worth of production and income. For example, a grower who adheres to the California Leafy Green Products Handler Marketing Agreement (LGMA) may be required to create a minimum five foot radius buffer around a contamination event (i.e., fecal deposit) and not harvest that product within the buffer (LGMA 2011). Larger farms may be able to absorb the financial loss resulting from the lower yield because they can spread the cost over more acres. A small farm with only a few acres may end up creating buffer zones that cover a large percentage of the crop acreage, resulting in loss of substantial farm income. Research has revealed that the five foot radius buffer successfully reduces microbial contamination risks, but there is not enough scientific evidence to determine if a smaller buffer would be sufficient or what the impact this standard would have on overall fruit and vegetable production in the United States (Jeamsripong et al. 2013).

A survey of 43 California leafy green growers participating in the LGMA program found that one-third of the growers reported removing produce acreage to meet the buffer zone requirements (Hardesty and Kusunose 2009). These growers reported losing an average of 21 acres of leafy greens (values ranged from 1 to 150 acres) due to this buffer requirement alone. While the size distribution of growers (80 % with gross revenues over $1 million) within the study may not be representative of growers nationwide, the numbers show the potential impact of commodity-specific standards. Whether these management practices are employed for food safety purposes or to minimize overall damage to the crop, or perhaps for both, the economic and food safety benefit must be worth the cost incurred, otherwise the practice may not be sustainable for the farm long-term. This again highlights the complexity of decision-making on the farm and the need for additional research to elucidate the impact of implementing numeric standards for produce safety as well as how risks to produce crops impact production decisions.

In the case of produce safety, more than one factor may influence what action a grower should take to mitigate food safety risks associated with wildlife. Corrective actions can depend on the extensiveness of the problem (e.g., widespread contamination or localized), the amount and consistency of fecal material, weather events (e.g., rain which can cause splash or run-off), current production practices (e.g., overhead or drip irrigation), history of previous events, commodities affected, and target market. For example, growers of crops such as cherries or apples that are destined for processing (e.g., canning or sauce-making) may have a higher tolerance for bird damage to the crop. Imperfect fruit may still be saleable since it will receive a processing treatment that would adequately mitigate the presence of harmful microorganisms that may have been deposited by birds. In this case, both food safety and minimum quality standards are achieved with reduced financial liability to the grower. However, fresh market growers will likely have a lower (or no) tolerance for pest damage and microbial contamination of the crop because buyer demand requires undamaged fruit produced using food safety practices that meet a specific standard (Anderson and Lindell 2013). This scenario requires a different set of risk management strategies that the grower must decide upon based on his or her farming operation and target market.

The efficacy and practicality of the strategy used to reduce the problem are primary factors as to whether growers are willing to invest in the solution. Exclusionary control methods such as fencing can be very effective but are often cost-prohibitive. Wildlife deterrents such as propane cannons, reflective tape, or decoys may be affordable, but they require management such as moving or modulating delivery to remain effective because wildlife can become accustomed to these methods. Dogs are an effective pest control management option, resulting in minimized damage to crops; however, the benefit must be weighed against the risk since domesticated animals, including dogs, can harbor and spread human pathogens in their feces (VerCauteren et al. 2005; Jay-Russell et al. 2014). Some growers advocate that one or two well-trained dogs present in the field are better than a herd of deer or flock of geese, and may limit other risks such as groundhog holes that represent a physical risk (e.g., broken ankles) experienced by farm workers.

The location of the farm may also determine what type of strategies growers are able to use. Rural farms may have the option of wildlife control through hunting or nuisance (depredation) permits, but farms in residential and urban areas may have limitations on gun discharge or noise ordinances that must be obeyed. The U.S. Fish and Wildlife Service Endangered Species Act must also be considered when attempting to control or minimize wildlife access to farms. Additionally, local influences such as neighbors may play into management decisions made on the farm (Anderson and Lindell 2013; Baldwin et al. 2012; Conover 2002). As one produce grower in a focus group conducted by researchers at Cornell University in 2013 stated regarding his neighbor's wildlife deterrent methods, "Wildlife may become accustomed to the noise, but the one thing that will not are the neighbors!" then he added "...every time that cannon goes off, my dog becomes incontinent." This very concisely (and humorously) highlights the complexity of situations many growers must navigate. Ultimately, growers must choose which food safety management strategies are the most appropriate for their farm, recognizing that the growing environment can never be considered zero risk.

Research Needs for Better Decision Making

The complex nature of the farming environment not only presents hurdles to the grower decision-making process, but also challenges to researchers who are tasked with generating data to identify risks on the farm and effective methods to reduce them. Significant scientific advancements have allowed researchers to delve deeper into the interactions between human pathogens associated with fresh produce, their hosts and vectors, as well as identify methods for mitigating risks that primarily emphasize the prevention of microbial contamination of fresh fruits and vegetables (Beuchat 2006; Lynch et al. 2009; Sivapalasingam et al. 2004; Harris et al. 2003). Unfortunately, mitigation steps are not always practical or affordable from a grower's perspective. The scope of produce safety scientific research will always be challenged by the unpredictable and uncontrollable nature of the growing environment as well as the need to develop functional mitigation strategies. For example, if routine irrigation water testing indicates a high level of generic E. coli in the water that directly contacts the harvestable portion of the crop, the grower must choose an action that will reduce the risk of crop contamination, be financially feasible, and meet the crops water demands. In a region experiencing drought, a decision not to irrigate because of an identified food safety risk in the water might mean the grower will have no crop to sell since the plants may not survive without water.

Contamination of produce involves complex interactions within the "epidemiologic triad" representing animal reservoirs, pathogens, and the local environment (Park et al. 2012). Applied research that defines prevalence of pathogens in the environment, the most likely animal reservoirs, and how pathogens move within the farm environment will help guide the development of management strategies that minimize microbial contamination of crops by wildlife. Designing and conducting research projects is difficult given the open and highly variable farm landscape, as well as the proclivity of wildlife to move at will. Research-oriented organizations, such as the Center for Produce Safety Campaign for Research, have been bridging these knowledge gaps by funding applied research projects so that the science can inform the development of on-farm best practices. Although significant progress has been made in building the body of science to support recommendations, providing sound metrics for the management of food safety risks on the farm continues to be a challenge and focus of the scientific community.

Establishing science-based standards for food safety on the farm has been difficult, if not impossible in many cases (FDA 2011). Scenarios such as fecal contamination in the field from wildlife or gross contamination of water bodies due to large migratory fly-ways must be evaluated on a case-by-case basis since it is extremely difficult to replicate real wildlife events in the laboratory or research setting. The risks posed may differ depending on the commodity, type of animal involved (if known), climate, weather events, topography, production practices, and time within the production season that the contamination event occurred (e.g., close to harvest) (Beuchat 2002; Gorny 2006). Furthermore, the study of actual contamination events is rare since the production environment is constantly changing, contamination events are fleeting, and some growers may not want to be involved in studies which implicate that their fields or produce have been exposed to human pathogens, regardless of the protection and coding of data collected. Much of the research to support scientific decision-making processes within these complex environments is still in the early stages; however, examination of these variables may help develop more sophisticated models that growers can use for predicting when and where contamination might occur (Strawn et al. 2013).

The nature of produce safety is also such that most of outbreaks cannot be traced back to the original source due to a number of factors. The relatively short shelf life of produce and infrequent, sporadic nature of contamination events make traceback investigations extremely difficult since the product has either been consumed or thrown away, or fields have been plowed or replanted by the time illnesses are reported. Even if animal fecal contamination does occur, it is likely that the origin of the outbreak will remain unidentified, leaving growers to question where the risks on their farms exist and whether the strategies they employ to reduce risks are actually working. Furthermore, there are many opportunities for produce to become contaminated after it is harvested and leaves the farm. These risks include subsequent handling steps which can spread or amplify a contamination event such as washing, packaging, fresh-cut preparation, or consumer handling in the home. The complexity of the food system can lead state health department and federal investigators down multiple paths, and leave holes in the attempt to trace contamination, identify the origin, and determine steps to prevent future contamination events from occurring.

To address such an extensive network of interactions in the growing environment, research regarding produce safety is scattered across multiple disciplines, including horticulture, microbiology, ecology, epidemiology, food science, soil science, veterinary medicine, wildlife biology, hydrology, and general production agriculture. Bringing together this research from diverse disciplines to synthesize meaningful and useful recommendations for the farming community is critical to establishing and implementing proven methods of risk reduction. Unfortunately, this represents another research challenge to initiate and sustain communication between such diverse groups of researchers in various fields of study. Researchers studying these complex aspects of produce safety have varied expertise and knowledge bases. This complicates communication among scientists in different disciplines because they may not use the same research protocols or technical language, or even be aware of the foundational literature that is critical to understanding their colleagues' viewpoints and research approaches. In an effort to unite multi-disciplinary research scientists toward reaching produce safety goals, in 2008, the FDA provided funding to the Western Center for Food Safety to outline frameworks for developing research protocols in the areas of agricultural water and manure management. The resulting efforts were published in the Journal of Food Protection and will aid in communication and collaboration of future produce safety research proposals across disciplines (Harris et al. 2012, 2013).

The goal of research across multiple disciplines is to create actionable items that growers can incorporate into their expanding farm management tool kit and enable them to make science-based decisions for both food safety and the sustainability of their businesses. As the climate of produce safety shifts from what were once recommended guidelines to mandatory regulations, the agricultural community will need resources that emphasize principles of risk assessment that enable growers to evaluate food safety risks on their farm and implement practices which utilize limited farm resources in the most efficient and effective way.

From Recommendations to Regulations: A Sweeping Reform of Produce Safety

Voluntary Compliance

In response to a significant number of foodborne illness outbreaks associated with produce, in 1998, the United States Food and Drug Administration (FDA) published the *Guidance for Industry: Guide to Minimize Microbial Food Safety Hazards for Fresh Fruits and Vegetables* (referred to as "The Guide"). The Guide summarized produce-associated outbreak data and outlined practices that growers were encouraged to adopt through voluntary compliance on farms and in packinghouses to prevent the microbial contamination of fruits and vegetables. Through collaboration and internal projects, the FDA has supported increased awareness about produce safety issues and the implementation of food safety practices since the release of the Guide, but the agency has not been alone in its efforts.

Pressure on farms from the industry and buyers to implement food safety practices has continued to evolve along with government-sponsored programs. Some buyers sent letters to their suppliers (growers) requesting they follow the practices outlined in the Guide. Initially, most buyers were not willing to pay more for crops grown using GAPs or provide any incentives to growers for implementing GAPs. In addition, crop supply often determined how committed buyers were to purchasing produce that was grown using the outlined practices. When crops are abundant, buyers can be more demanding, and growers that have implemented GAPs are often able to secure contracts that other growers could not. When crops are in short supply and high demand, the need to have the commodity available outweigh the desire to have it grown using GAPs. This inconsistency has made many growers uncertain of the value of adopting GAPs given the increased cost and management effort required.

The third-party audit industry saw an opportunity to expand into the farm environment as a way to help buyers ensure the commodities they were purchasing were meeting the GAPs standards. Growers, accustomed to having United States Department of Agriculture (USDA) personnel on their farms to grade commodities, began requesting that the USDA develop a GAPs audit to streamline the process at the farm level. In 1999, the USDA Agricultural Marketing Service's Fruit and Vegetable Program and Specialty Crops Inspection (SCI) Division began developing a voluntary audit-based verification program, and today offers two audit options. The USDA is not the only group conducting on-farm third-party audits; GlobalGAP, Scientific Certification Systems, Inc. (SCS), and PrimusLabs, are examples of other companies that offer GAPs audits, though they may not use the same auditing standards. Some buyers tend to prefer one audit over another so the hope that USDA involvement would streamline the audit process never came to fruition. Unfortunately, farms that sell to multiple buyers often have to participate in multiple third-party audits from different auditing companies. This audit fatigue is more prevalent on larger farms, but the cost, time, effort, and frustration was significant enough to lead United Fresh Produce Association (2010) to convene the Produce GAPs Harmonization Initiative aimed at developing one harmonized audit (Gombas 2013). Though this process was successful and resulted in a harmonized audit that is openly available for use by auditing firms, variation within audits is still prevalent because not every company is using the harmonized audit metrics or they are adding additional questions through addendums that are not included in the original harmonized standard.

Even if there was only one audit, the experience and expertise of individual auditors may result in variability of how audit schemes are administered. Auditors have various educational and professional backgrounds that may not include knowledge of agricultural production environments. Auditor trainings vary by auditing agency, so each company is free to decide the level of training required for the auditors they hire. Navigating the world of audits is challenging for all growers and adds to the cost of production. It is not clear if audits reduce risks, but with continued buyer requirements for third-party audits, it is apparent audits will continue to be a part of many farm business operations in the future.

Commodity groups have played a major part in the evolution of the produce safety landscape. Not only do commodity groups support growers, several produce commodity groups have developed their own commodity-specific food safety guidelines such as those for melons (2005), strawberries (2005), lettuce and leafy greens (2006), tomatoes (2006 1st edn, 2008 2nd edn), almonds (2009), green onions (2010), mushrooms (2010), citrus (2011), and avocado (2011) (Fleming et al. 2005; Gorny et al. 2006; NATTWG 2008; Western Growers Association and Intertox Incorporated 2010). In the case of tomatoes, the release of the commodity-specific guidelines was followed by the Florida legislature passing a Tomato Good Agricultural Practices Program (TGAPs) law in 2007 that required tomato producers in Florida to follow these standards and be subject to enforcement by the Florida Department of Agriculture and Consumer Services. This was the first instance of voluntary guidance becoming a mandatory, statewide government inspection and audit program for tomato production, handling, and packing, effective on July 1, 2008 (Florida Administrative Code 2007).

Despite the widespread, decade-long effort to promote and facilitate the adoption of GAPs outlined within the voluntary guidelines, a continued history of produce-associated outbreaks in the United States drove the FDA to publish its intention to develop a produce regulation in the federal docket in December 2009. In preparation for the development of mandatory standards, the FDA collaborated with the

Pew Charitable Trust Produce Safety Project (PSP) to gather information on the implications of federal standards for produce and to identify key challenges and concerns that growers had about implementation of food safety practices on the farm. Launched in 2008, the PSP held a summit in 2009 followed by six stakeholder discussion sessions throughout the U.S. in 2010 (Produce Safety Project 2009; Produce Safety Project 2010). Common themes emerged regarding the goals and expectations that a new regulation should try to achieve, as well as challenges to meeting new regulatory requirements:

- Universal standards should ensure a "level playing field."
- Standards must take into account differences between commodities and growing regions (i.e., one size does not fit all).
- The regulation should not be prescriptive on how to implement practices, so each farm has the flexibility to meet the standards that are established.
- A scarcity of resources may complicate the effective implementation and enforcement of mandatory standards.
- The issue of produce safety requires a science- and risk-based approach.
- Standards must be flexible, continuously updated, and implemented in a phased approach.
- The development process must be transparent and engage all stakeholders.
- Education is paramount prior to the regulation.

During these meetings, farmers and stakeholders also expressed concern that aggressive food safety practices lacked scientific basis, were unrealistic, and may contradict laws designed to protect wildlife. Some of the comments heard during the stakeholder sessions were foreshadowing of public feedback yet to come on the newly proposed federal regulations.

Food Safety Modernization Act and the FDA Produce Safety Rule

As the first and most comprehensive reform of U.S. food safety policy in over 70 years, the Food Safety Modernization Act (FSMA) was signed into law by President Obama on January 4, 2011 to address the public health burden associated with food-borne illness outbreaks in human food, animal feed, and imported food products (FDA 2011). The FDA was required to establish science-based, minimum standards for the safe production, harvest, and handling of fruits and vegetables as part of FSMA's directive, also called the Produce Safety Rule (FDA 2013a). The rule proposes standards to minimize microbial hazards in the following areas: worker health, hygiene, and training; agricultural water; soil amendments; domesticated and wild animals; equipment, tools, and buildings; and sprout production.

Although the rule does outline specific commodities that are considered "covered produce," the regulation is not commodity-specific, per se. Based on historical produce-related outbreak data and evaluation of contamination sources, the FDA has chosen to focus on production practices rather than to take a commodity-specific approach since research indicates that all types of fruits and vegetables have the potential to become contaminated during the production, harvest, and handling process. In general, agricultural commodities that are consumed raw or in an unprocessed state would be subject to the regulation, including peanuts, tree nuts, mushrooms, sprouts, herbs, and fresh fruits and vegetables.

The proposed rule does provide provisions for exemption to farms and covered produce, provided certain conditions are met. First, produce farms are not covered by the regulation if they have an average annual monetary value of all produce sales of $25,000 or less. This accounts for approximately 34,000 farms in the United States. An exemption may also be established under any of the following three criteria: (1) the produce is destined for commercial processing which includes a kill step (e.g., canning) that adequately reduces the presence of microorganisms of concern; (2) the produce is only grown for personal consumption or on-farm consumption; or (3) the monetary value of all food sales must average less than $500,000 per year for the last 3 years *and* more than 50 % of all food sales[1] are direct to a "qualified end user".[2] These exemptions originated from the Tester-Hagan Amendment in the FSMA legislation. Senators Jon Tester (D-MT) and Kay Hagan (D-NC) sponsored the amendment in an effort to remove smaller scale, local producers from the burdensome costs of federal oversight, while leaving them still subject to the current regulatory framework of state and local health entities. It is projected that the overall public foodborne illness health burden posed by these farms that primarily direct market or sell produce locally is relatively low; however, the risk to individuals is not necessarily lower. Even small market venues present the possibility of consumers experiencing significant health consequences. In 2011, an outbreak of *E. coli* O157:H7 caused by deer fecal contamination in strawberries sold at roadside stands and a farmer's market was identified in Oregon. Out of the 15 cases that were identified, six were hospitalized, including four with hemolytic uremic syndrome (HUS). Two of the individuals who had HUS later died (Laidler et al. 2013).

After accounting for farms that qualify for one or more of the exemptions mentioned above, including farms that are not covered because they have produce sales of less than $25,000 and farms that grow commodities which are destined for further processing, approximately 40,000 farms of the total 189,000 remain subject to the produce regulation (Fig. 10.1).

Many have questioned whether the impacts of these exemptions will create weaknesses within the food safety system, allowing smaller growers to produce higher risk commodities because they are not subject to the same regulatory oversight as larger growers. While some growers may breathe a sigh of relief after

[1] The FDA defines "food sales" to include all products grown or raised for animal or human consumption or to be used as ingredients for any such item.

[2] A qualified end user is defined as consumers of the food, regardless of location, or a restaurant or retail establishment in the same state or not more than 275 miles from the farm.

Fig. 10.1 Breakdown of farms eligible for qualified exemptions and farms not covered by the produce safety rule

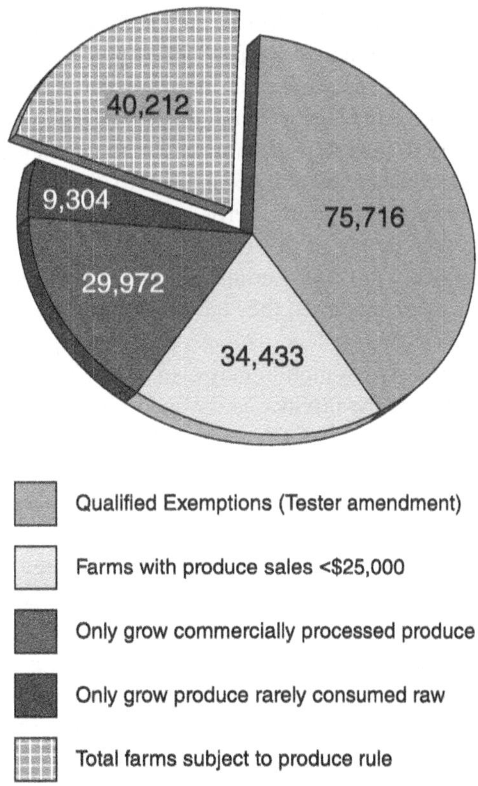

confirming their exemption from the final regulations, many are already noticing that exemption from the regulation does not mean exemption from buyer demand and market pressure to implement food safety practices. Even small-scale marketing venues such as local farm-to-school programs, farmers' markets, auction houses, and grower cooperatives are beginning to ask for verification of basic food safety practices through specified procurement standards, food safety plans, or even third-party audits. It is anticipated that the federal regulations will set the foundation for food safety practices, while buyers may continue to build their expectations and requirements. For example, the proposed Produce Safety Rule does not require a food safety plan or third-party audit, but many buyers and nearly all audit schemes do.

Drafting a Regulation to Fit a Diverse Industry and Complex Topic

Crafting a regulation to fit a diversity of produce commodities, growing environments, production practices, and farm sizes calls for flexibility in the established standards and must accommodate variation in the agricultural environment. The

FDA and other federal agencies such as the USDA have taken a number of actions to address produce safety over the last two decades, leading up to the development of the Food Safety Modernization Act. In addition to commodity-specific guidance documents developed for the fresh produce industry, the FDA has relied on data from inspections and outbreak investigations, public hearings to gather information from stakeholders, and surveys of current industry practices to target and develop goals to reduce foodborne illnesses in fresh produce through the development of regulatory standards. The agency has provided an evaluation of produce safety risks and an analysis of economic impacts of the proposed Produce Safety Rule through two documents, the *Qualitative Assessment of Risk to Public Health from On-Farm Contamination of Produce* (QAR) and the *Analysis of Economic Impacts—Standards for the Growing, Harvesting, Packing and Holding of Produce for Human Consumption* (AEI) (FDA 2013b, c).

Pursuant to Section 419 of the Food, Drug, and Cosmetic Act, the FDA must develop minimum science-based standards for the safe production and harvest of fruits and vegetables. The statute directs the FDA to base these standards on "known safety risks and to include procedures, processes, and practices that are reasonably necessary to prevent the introduction of known or reasonably foreseeable biological, chemical, or physical hazards into fruit and vegetable raw agricultural commodities…" As a result of the QAR, several key focus areas were identified for inclusion and expansion of the regulatory language for the proposed Produce Safety Rule including:

(a) Emphasis should be placed on biological hazards because they account for the vast majority of produce-related foodborne illnesses (not chemical or physical food safety hazards).
(b) There are many routes of contamination including farm worker health and hygiene, soil amendments (e.g., manure, compost), agricultural water (both for field production and post-harvest uses), post-harvest handling and storage, and wildlife and domestic animals.
(c) All types of produce commodities have the potential to become contaminated.
(d) Specific practices associated with a particular produce commodity may affect the potential routes and likelihood of contamination.
(e) Postharvest practices such as cooking before consumption will have an impact on the likelihood of illness occurring.

The prioritization of biological hazards above chemical and physical hazards is supported by data that indicates microbial hazards result in the greatest health consequences to consumers. At the farm level, these risks are present throughout the agricultural environment, including biological contamination risks presented by agricultural water, soil amendments, wildlife, and storage areas as well as the workers who harvest and pack fruits and vegetables. As mentioned earlier, no commodities are excluded from this risk of contamination, but practices during both production and postharvest handling can impact food safety risks.

Because the proposed Produce Safety Rule was designated an "economically significant rule," the FDA had executive orders to assess all costs and benefits of regulatory actions and alternatives, all the way down to the cost of paper towels and soap for worker health and hygiene practices. This process encourages selecting regulatory approaches that maximize net benefits, including economic, environmental, and public health and safety effects. Growers, especially those who consider themselves “small-scale,” expressed particular concern about the overall cost of implementing the regulatory standards, if they must comply. Some grassroots consumer and agricultural interest groups focusing on small-scale and organic farm viability, advocated for the Tester-Hagan Amendment, believing that the regulation may put many small farms out of business or limit their opportunities to diversify their farm operations. On the other hand, major industry players claimed that the amendment fell short of providing adequate protection to consumers by exempting small farms (United Fresh 2010).

Although the estimates of economic impact may indicate the need for significant investment by produce growers to comply with the regulation, it relies on one major assumption—that no grower is currently doing any of these practices on their farm. For those familiar with most farm operations in the United States, this simply is not the case. Many basic GAPs are being implemented on farms, although they may not be formally documented. One standard in the proposed regulation that accounts for a significant proportion of the estimated costs is the worker health and hygiene standard. A large proportion of the proposed rule’s total cost is attributed to the requirement that farm workers must wash hands before handling produce, which accounts for the loss of productivity in time spent hand washing (47 cents worth of labor for 2 min of time). Supplying toilet and hand washing facilities, training workers, and accounting for the time required for them to use these facilities is already being done on most farms, so these are costs they already expend. Another significant area of cost was allocated for testing agricultural water. This may be a new cost for many farms, but the overall cost will depend on how often the water is tested, the type of water source (well or surface), and how many water source(s) a grower uses. The supplemental to the proposed Produce Safety Rule which was released in September 2014 decreased the number of water tests that may be required and there is some possibility that lower cost testing options will be available in the future as more laboratories open to provide water testing services (FDA 2014a, b).

In the proposed wildlife and domesticated animal standards, the FDA estimates that monitoring for wild animal intrusion in production fields will occur at least three times per production season on the average farm; once at the beginning of the season, once “as needed,” and immediately prior to harvest. According to the 2007 NASS Census of Agriculture, it is estimated that farms have, on average, two production seasons per year for a total of six monitoring events (USDA NASS 2007). With an estimated cost of \$3.36 for monitoring animal intrusion per acre, the average cost per affected farm will vary depending on total acreage, but could be anywhere between \$126 for a very small farm to \$840 for a large farm.[3] Some feel these

[3] FDA estimates the costs for each farm size category defined earlier in the analysis by multiplying the estimated costs per acre by the midpoint of the acreage that defines each farm size category (112.5, 375, and 750 acres for very small, small' and large, respectively).

estimates are inflated, since many of the tasks related to wildlife, including monitoring, are done while completing other tasks on the farm. Whether these estimates are excessive or not, growers' perceptions regarding the cost of time and resources spent mitigating wildlife intrusion will likely vary depending on how they perceive the human–wildlife conflict and the severity of damage or risk to their crops (Dickman 2010).

The QAR and the AEI were essential to the development of the proposed Produce Safety Rule, but it is worth noting that assessing the impacts to the environment as a result of implementing this portion of the regulation was not initially addressed. With the release of the proposed rule in January 2013, the agency initially prepared a categorical exclusion for the need to draft an Environmental Impact Statement (EIS), as directed by the National Environmental Policy Act of 1969 (NEPA). Upon further analysis and feedback received during the open comment period, it was determined that the preparation of an EIS was necessary to evaluate significant environmental impacts that could result from certain provisions, based on the currently proposed standards included within the Produce Safety Rule. Further analysis concluded that provisions outlined in the wildlife and domesticated animal management, biological soil amendments, and water quality standards may have direct, indirect, and cumulative impacts along with other FSMA proposed standards (i.e., Preventive Controls for Human Food and Animal Feed) on produce farms. Specific concerns were raised about excessive regulation of surface water sources that may result in increases in ground water pumping and in application of surface water chemical treatments to reduce microbial hazards in production water, which in turn, affect wildlife habitat and their water sources. As part of the EIS, which is anticipated to be released in January 2015, the agency will propose alternatives within a range, incorporating extremes on either end of the spectrum, from taking no action to actions that have the potential to impact the environment significantly (FDA 2014a, b). For advocates of environmental health, conservation, and co-management, the preparation of the EIS has been a welcome process since many of the issues documented during listening sessions and through the open comment period reflected concern about the proposed Produce Safety Rule's impact on wild life and the greater natural environment.

The proposed FSMA's Produce Safety Rule was released for comment in the Federal Register on January 4, 2013, and received over 15,000 comments to the docket during the first open comment period which ended on November 22, 2013. The themes and recommendations developed during the Produce Safety Project stakeholder sessions in 2008–2010 were echoed in listening sessions held throughout the country in 2013 after the release of the proposed rule (Taylor 2013). As a result of the overwhelming feedback received, on December 19, 2013, the FDA announced that it planned to revise the language in certain provisions to make them more flexible and less burdensome, and to publish the supplemental provisions in the Federal Register for public comment. The supplemental document was published in the Federal Register for a 90-day comment period on September 29, 2014 (FDA 2014a, b). Revisions were made to the water quality standards and testing requirements, manure and compost requirements, the definition of a "covered" farm, withdrawal of certain qualified exemptions, and provisions related to wildlife.

The changes to the wildlife management standards released in the supplemental to the proposed Produce Safety Rule were heavily influenced by information gathered during the EIS scoping process. In response to concerns that the regulation may inadvertently promote practices that adversely affect wildlife habitat, habitat connectivity, and impact threatened or endangered species, the FDA consulted with the USDA's Natural Resource and Conservation Service (NRCS) and the U.S. Fish and Wildlife Service. These revisions emphasized that the regulation does not authorize farms to take actions that would require the "taking[4]" of threatened or endangered species and encourages the co-management of food safety, conservation, and environmental protection.

Overall, the FDA and associated agencies participated in an unprecedented amount of outreach, inviting feedback prior to the development of the draft regulation, after the proposed regulation's initial release, and again after the supplemental release. The task of developing an effective regulation that increases the safety of produce, while not overburdening growers, is certainly a challenging one. The FDA must respond to the needs of a variety of farms—small, large, organic, conventional, domestic, foreign, and everything in-between. Most importantly, the FDA is charged with bringing all growers up to a basic standard of safety. The crux of regulatory action is clarity about what is expected from farms, identifying sufficient resources to help growers understand produce safety risks, and helping growers implement the required produce safety practices.

Education and Extension to Assist Produce Growers

One anticipated result of the proposed Produce Safety Rule will be the continued need for growers to have access to education and extension opportunities to increase their produce safety knowledge, their ability to assess risks and implement GAPs to reduce food safety risks, and meet regulatory requirements. Even growers who are exempt from the final regulation will need access to this information and training because it is likely that they will be subject to food safety specifications in the marketplace. Expectations from regulators, buyers, and consumers will continue to drive the need to understand and implement food safety practices on the farm, so it is in the best interest of all growers to be aware of practices that enhance the safety of produce.

There have been many successful produce safety programs developed by Land-Grant Universities, produce industry groups, non-profit organizations, federal agencies, and for-profit organizations to help inform growers of produce safety risks that can reduce produce safety risks. The value of Land-Grant Universities cannot be

[4] "Taking" is defined by the Endangered Species Act to include any actions that harass, harm, pursue, hunt, shoot, wound, kill, trap, capture, or collect threatened or endangered species.

overstated in the continued effort to educate growers. In a series of focus groups conducted by the Produce Safety Alliance in 2012 with growers around the country, University Extension programs were rated by growers as the most reliable source of information (PSA 2012). Growers trust information provided by Land-Grant University Extension programs and Extension professionals are uniquely positioned to provide educational opportunities for growers because they are already working with the growers in the area of produce safety as well as in fruit and vegetable production.

Historically, Land-Grant Universities have been involved in produce safety education and extension since the term was coined through both research and extension programs. The University of California, Davis, North Carolina State University, University of Florida, Rutgers University, University of Minnesota, and Texas A & M University are just a few of the universities providing well-established GAPs programs available to assist growers. Cornell University has been home to the National GAPs Program since 1999, and includes collaborators in all of the universities listed above, plus an additional 25 Land-Grant Universities, resulting in 32 collaborative states nationwide. This extensive collaborative network and experience in GAPs education and training resulted in the Produce Safety Alliance being established at Cornell University in 2010.

The Produce Safety Alliance (PSA, http://producesafetyalliance.cornell.edu/), Sprout Safety Alliance (SSA, http://www.iit.edu/ifsh/sprout_safety/), and the Food Safety Preventive Controls Alliance (FSPCA, http://www.iit.edu/ifsh/alliance/) are three federally funded alliances tasked with developing core curriculum, training, and outreach programs to assist with implementation of the proposed FSMA regulations to farms and companies producing human and animal food. The PSA was formed through a cooperative agreement between the FDA, USDA, and Cornell University for the purpose of providing fundamental, science-based on-farm food safety knowledge to fresh fruit and vegetable farmers, packers, regulatory personnel and others interested in the safety of fresh produce. Prior to beginning development of the curriculum, the PSA convened an educational materials conference with 28 state collaborators to review existing programs and resources including key aspects of successful GAPs training from collaborators across the United States. Ten national working committees were formed, engaging 178 individuals from 36 states, including growers, researchers, extension educators, government personnel, retailers, and produce industry representatives to identify challenges to implementing food safety practices on small farms. This working committee effort included 72 calls over a period of a year and a half, resulting in key curriculum content recommendations. In 2012, eight produce grower focus groups were conducted in the Southeast (3), West Coast (3), Midwest (1), and Northeast (1) to assess grower expectations and preferences for produce safety training and educational materials (PSA 2012). These actions enabled the PSA to prioritize grower educational needs, establish learning objectives for the core curriculum, and assemble content to meet the learning objectives. The resulting educational program for growers includes seven curriculum modules designed to be delivered in approximately 7 h of dedicated instruction time (Fig. 10.2). The curriculum is one way growers can satisfy the

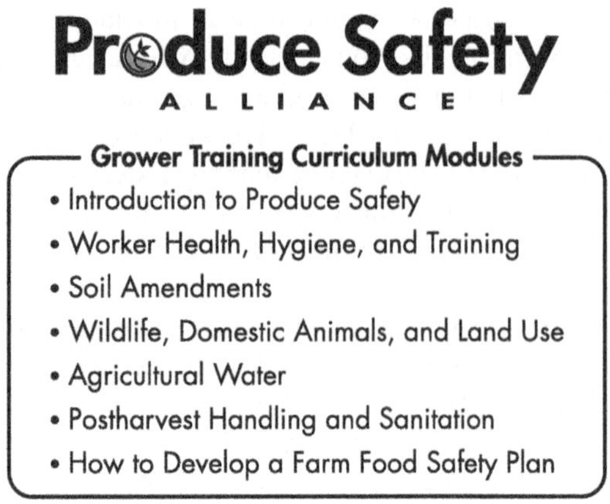

Fig. 10.2 Produce safety alliance grower training curriculum modules

requirements of proposed Produce Safety Rule §112.22(c), which state that "at least one supervisor or responsible party for the farm to successfully complete food safety training at least equivalent to that received under the standardized curriculum recognized as adequate by the Food and Drug Administration."

To address the critical issues of wildlife and food safety management on the farm, the curriculum outlines six key learning objectives in the PSA Wildlife, Domesticated Animals, and Land Use Module, including:

1. Identifying potential routes of contamination associated with wildlife, domesticated animals, and land use.
2. Describing practices to reduce risks associated with wildlife, domesticated animals, and land use.
3. Describing co-management strategies that address both conservation and food safety goals.
4. Describing the importance of conducting a pre-harvest assessment of fields.
5. Describing corrective actions that could be used if significant risks are present in production fields.
6. Identifying records that should be kept to document any management, monitoring, or corrective actions taken to reduce produce safety risks.

These objectives outline key information that is necessary to identify produce safety risks and implement practices that reduce the risks posed by both wildlife and domesticated animals. The PSA curriculum content was designed to increase grower understanding and provide guidance for how to implement effective produce safety practices on a diversity of farms, so all growers, whether they are subject to the regulation or not, will find the information relevant.

In addition to grower trainings, train-the-trainer workshops will be conducted throughout the country to increase the availability of qualified trainers to assist growers. The successful launch of the PSA trainings will require collaboration from personnel at Land-Grant Universities, the produce industry and grower organizations, federal agencies such as USDA, FDA, and NRCS, and other groups tasked with assisting growers in complying with the federal regulation. Reaching fruit and vegetable growers, typically located in remote and rural areas across all 50 states, will require a sustained effort from a broad coalition of stakeholders, but chief among them will be the commitment from growers themselves.

Conclusion

As the need to implement produce safety requirements continues to grow, through the requirements to meet federal regulations proposed in the FSMA's. Produce Safety Rule and ever-increasing buyer demands, produce growers will need to understand and adapt their farming practices to meet these new criteria. One commonality in the production of fresh fruits and vegetables is the complex growing environment and the natural presence of wildlife in these landscapes. Although wild animals can harbor human pathogens and spread contamination to produce through their feces, growers can identify where risks exist and prioritize the implementation of GAPs using principles of co-management on their farms to reduce risks. Decisions related to wildlife management and food safety can be challenging; therefore, growers must have access to the best science-based information available to help them make effective decisions that are financially feasible. Fortunately, there are many resources available to assist growers in making the best decisions for their farms with produce safety, environmental sustainability, market, and regulatory requirements in mind. Most importantly, everyone needs to recognize that the growing environment cannot be considered zero risk, but that produce safety risks on all farms and packinghouses can be reduced through risk assessment and implementation of GAPs.

References

American Cancer Society (2014) Cancer facts & figures 2014. American Cancer Society. Atlanta, GA. http://www.cancer.org/research/cancerfactsstatistics/cancerfactsfigures2014/. Accessed 30 October 2014

Anderson A, Lindell C (2013) Bird damage to select fruit crops: the cost of damage and benefits of control in five states. Crop Prot 52:103–106

Arnade C, Calvin L, Kuchler F (2010) Consumers' response to the 2006 foodborne illness outbreak linked to spinach. USDA Amber Waves. http://www.ers.usda.gov/amber-waves/2010-march/consumers%E2%80%99-response-to-the-2006-foodborne-illness-outbreak-linked-to-spinach.aspx#.VD6zk_ldV8G. Accessed 30 Oct 2014

Baldwin RA, Salmon TP, Schmidt RH et al (2012) Wildlife pests of California agriculture: regional variability and subsequent impacts on management. Crop Prot 46(2012):29–37

Baumgartner JA, Runsten D (2013) Farming with food safety and conservation in mind. Wild Farm Alliance (WFA) and Community Alliance with Family Farmers (CAFF) http://www.wildfarmalliance.org/resources/WFA-CAFF_Food_Safety-Conservation.pdf. Accessed 30 Oct 2014

Beuchat LR (2002) Ecological factors influencing survival and growth of human pathogens on raw fruits and vegetables. Microbes Infect 4:413–423

Beuchat LR (2006) Vectors and conditions for pre-harvest contamination of fruits and vegetables with pathogens capable of causing enteric diseases. Brit Food J 108:38–53

Bhupathiraju SN, Wedick NM, Pan A et al (2013) Quantity and variety in fruit and vegetable intake and risk of coronary heart disease. Am J Clin Nutr 98:1514–1523

Bihn EA, Gravini RB (2006) Role of good agricultural practices in fruit and vegetable safety. In: Matthews KR (ed) Microbiology of fresh produce. ASM Press, Washington, DC, pp 21–53

California Leafy Green Products Handler Marketing Agreement (LGMA) (2011) Commodity specific food safety guidelines for the production and harvest of lettuce and leafy greens. http://www.lgma.ca.gov/wp-content/uploads/2014/09/California-LGMA-metrics-08-26-13-Final.pdf. Accessed 9 Sept 2014

Chapman B, Kreske A, Powell D (2012) Crisis management: how to handle outbreak events. Food Safety Magazine, June/July 2012. http://www.foodsafetymagazine.com/magazine-archive1/junejuly-2012/crisis-management-how-to-handle-outbreak-events/. Accessed 9 Sept 2014

Conover M (2002) Resolving human–wildlife conflicts: the science of wildlife damage management. CRC Press, Boca Raton, FL

Dickman AJ (2010) Complexities of conflict: the importance of considering social factors for effectively resolving human–wildlife conflict. Anim Conserv 13:458–466

Fleming P, Pool W, Gorny J (eds) (2005) Commodity specific food safety guidelines for the melon supply chain, 1st edn. Produce Marketing Association and United Fresh Produce Association. http://www.fda.gov/Food/FoodSafety/Product-SpecificInformation/FruitsVegetablesJuices/GuidanceComplianceRegulatoryInformation/ucm168609.htm. Accessed 4 Sept 2014

Florida Administrative Code (2007) Rule 5G-6: tomato inspection. https://www.flrules.org/gateway/chapterhome.asp?chapter=5g-6. Accessed 9 Sept 2014

Gelting RJ, Baloch MA, Zarate-Bermudez MA et al (2011) Irrigation water issues potentially related to the 2006 multistate *E. coli* O157:H7 outbreak associated with spinach. Agric Wat Manag 98(9):1395–1402

Gombas DE (2013) Produce GAPs harmonization: the goal is in sight. Food Safety Magazine. http://www.foodsafetymagazine.com/magazine-archive1/junejuly-2013/produce-gaps-harmonization-the-goal-is-in-sight/. Accessed 29 Oct 2014

Gorny J (2006) Microbial contamination of fresh fruits and vegetables. In: Sapers GM, Gorny JR, Yousef AE (eds) Microbiology of fruits and vegetables. CRC, Taylor and Francis Group, Boca Raton, FL, pp 3–32

Gorny J, Giclas H, Gombas D, Means K (eds) (2006) Commodity specific food safety guidelines for the lettuce and leafy greens supply chain, 1st edn. International Fresh Cut Produce Association, Produce Marketing Association, United Fresh Fruit and Vegetable Association, and Western Growers. http://www.fda.gov/downloads/Food/FoodSafety/Product-SpecificInformation/FruitsVegetablesJuices/GuidanceComplianceRegulatoryInformation/UCM169008.pdf. Accessed 4 Sept 2014

Hardesty SD, Kusunose Y (2009) Growers' compliance costs for the leafy greens marketing agreement and other food safety programs. UC small farm program research brief. http://sfp.ucdavis.edu/files/143911.pdf. Accessed 9 Sept 2014

Harris LJ, Farber JN, Beuchat LR et al (2003) Outbreaks associated with fresh produce: incidence, growth, and survival of pathogens in fresh and fresh-cut produce. Comp Rev Food Sci Food Saf 2:78–141

Harris LJ, Bender J, Bihn EA et al (2012) A framework for developing research protocols for evaluation of microbial hazards and controls during production that pertain to the quality of agricultural water contacting fresh produce that may be consumed raw. J Food Prot 75(12):2251–2273

Harris LJ, Berry ED, Blessington T et al (2013) A framework for developing research protocols for evaluation of microbial hazards and controls during production that pertain to the application of untreated soil amendments of animal origin on land used to grow produce that may be consumed raw. J Food Prot 76(6):1062–1084

Western Growers Association and Intertox Incorporated (2010) Commodity specific food safety guidelines for the production, harvest, post-harvest, and valued-added unit operations of green onions. http://www.fda.gov/Food/GuidanceRegulation/GuidanceDocumentsRegulatoryInformation/ProducePlantProducts/ucm203094.htm. Accessed 4 Sept 2014

Jay MT, Cooley M, Carychao D et al (2007) Escherichia coli O157:H7 in Feral Swine near Spinach Fields and Cattle, Central California Coast. Emerg Infect Dis 13:1908–1911

Jay-Russell MT, Hake AF, Bengson Y et al (2014) Prevalence and characterization of *Escherichia coli* and *Salmonella* strains isolated from stray dog and coyote feces in a major leafy greens production region at the United States-Mexico border. PLoS One 9(11), e113433

Jeamsripong S, Jay-Russell M, Carabez JA et al (2013) Simulation of wildlife fecal contamination of Romaine lettuce by indicator *Escherichia coli*. Poster presented at the IAFP annual conference, Charlotte, NC, 31 July 2013

Laidler MR, Tourdjman M, Buser GL et al (2013) *Escherichia coli* O157:H7 infections associated with consumption of locally grown strawberries contaminated by deer. Clin Infect Dis 57:1129–1134

Langholz J, Jay-Russell M (2013) Potential role of wildlife in pathogenic contamination of fresh produce. Hum Wild Interact 7:140–157

Lowell K, Langholz J, Stuart D (2010) Safe and sustainable: co-managing for food safety and ecological health in California's Central Coast Region. The Nature Conservancy of California and the Georgetown University Produce Safety Project, San Francisco, CA. http://ucfoodsafety.ucdavis.edu/files/198568.pdf. Accessed 1 Oct 2014

Lynch MF, Tauxe RV, Hedberg CW (2009) The growing burden of foodborne outbreaks due to contaminated fresh produce: risks and opportunities. Epidemiol Infect 137(3):307–315

Natural Resources Conservation Service (NRCS) (2014) Conservation innovation grants fiscal year (FY) 2014 Announcement for program funding. http://www.nrcs.usda.gov/Internet/FSE_DOCUMENTS/stelprdb1250984.pdf. Accessed 31 Oct 2014

Nielsen EM, Skov MN, Madsen JJ et al (2004) Verocytotoxin-producing Escherichia coli in wild birds and rodents in close proximity to farms. Appl Environ Microbiol 70(11):6944–6947

North American Tomato Trade Work Group (NATTWG) (2008) Commodity specific food safety guidelines for the fresh tomato supply chain, 2nd edn. http://www.fda.gov/downloads/Food/GuidanceRegulation/UCM171708.pdf. Accessed 4 Sept 2014

Park S, Szonyi B, Gautam R et al (2012) Risk factors for microbial contamination in fruits and vegetables at the pre-harvest level: a systematic review. J Food Prot 75:2055–2081

Produce Safety Project (2009) Executive summary: implications of mandatory safety standards, 5 Mar 2009. http://www.pewtrusts.org/~/media/legacy/uploadedfiles/phg/content_level_pages/reports/PSPRPTSummitMandatoryStandardspdf.pdf. Accessed 10 Oct 2014

Produce Safety Project (2010) Executive summary: stakeholders' discussion series. February 19–April 27, 2010. http://www.pewtrusts.org/~/media/legacy/uploadedfiles/phg/content_level_pages/issue_briefs/PSPProduceSafetyProjectFinalSummariespdf.pdf. Accessed 10 Oct 2014

Produce Safety Alliance (PSA) (2012) Farm focus group summary. http://producesafetyalliance.cornell.edu/newsletters/FocusGroupSummary.pdf. Accessed 10 Oct 2010

Sanderson MW, Sargeant JM, Shi X et al (2006) Longitudinal emergence and distribution of Escherichia coli O157 genotypes in a beef feedlot. Appl Environ Microbiol 72:7614–7619

Sivapalasingam S, Friedman CR, Cohen L, Tauxe RV (2004) Fresh produce: a growing cause of outbreaks of foodborne illness in the United States, 1973 through 1997. J Food Prot 67:2342–2353

Strawn LK, Fortes ED, Bihn EA et al (2013) Landscape and meteorological factors affecting prevalence of three food-borne pathogens in fruit and vegetable farms. Appl Environ Microbiol 79(2):588–600

Taylor MR (2013) Let's keep talking and listening about food safety. FDA voice blog – May 6, 2013. http://blogs.fda.gov/fdavoice/index.php/2013/05/lets-keep-talking-and-listening-about-food-safety/. Accessed 9 Sept 2014

Terry L (2011) Oregon confirms deer droppings caused *E. coli* outbreak tied to strawberries. The Oregonian. http://www.oregonlive.com/washingtoncounty/index.ssf/2011/08/oregon_confirms_deer_droppings.html. Accessed 9 Oct 2014

U.S. Department of Health and Human Services and U.S. Food and Drug Administration (2011) FSMA facts: background of the FDA Food Safety Modernization Act. http://www.fda.gov/downloads/Food/GuidanceRegulation/UCM263773.pdf. Accessed 20 Sept 2014

U.S. Department of Health and Human Services and U.S. Food and Drug Administration (2013a) Proposed standards for the growing, harvesting, packing, and holding of produce for human consumption. http://www.fda.gov/Food/GuidanceRegulation/FSMA/ucm334114.htm. Accessed 9 Sept 2014

U.S. Department of Health and Human Services and U.S. Food and Drug Administration (2013b) Analysis of economic impacts – standards for the growing, harvesting, packing and holding of produce for human consumption. http://www.fda.gov/downloads/Food/GuidanceRegulation/FSMA/UCM334116.pdf. Accessed 9 Sept 2014

U.S. Department of Health and Human Services and U.S. Food and Drug Administration (2013c) Draft qualitative assessment of risk (QAR) to public health from on-farm contamination of produce. http://www.regulations.gov/#!documentDetail;D=FDA-2011-N-0921-0001. Accessed 9 Sept 2014

U.S. Department of Health and Human Services and U.S. Food and Drug Administration (2014a) Supplemental for the proposed rule for produce safety. http://www.regulations.gov/#!documentDetail;D=FDA-2011-N-0921-0973. Accessed 19 Sept 2014

U.S. Department of Health and Human Services and U.S. Food and Drug Administration (2014b) Environmental impact statement (EIS) for the FSMA proposed rule for produce safety, 5 Aug 2014. http://www.fda.gov/Food/GuidanceRegulation/FSMA/ucm396564.htm. Accessed 9 Sept 2014

U.S. Food and Drug Administration (1998) Guidance for industry: guide to minimize microbial food safety hazards for fresh fruits and vegetables. http://www.fda.gov/downloads/Food/GuidanceComplianceRegulatoryInformation/GuidanceDocuments/ProduceandPlanProducts/UCM169112.pdf. Accessed 1 Sept 2014

United Fresh Produce Association (2010) Letter to the Senate: S.510 of the Food Safety Modernization Act. November 18, 2010. http://www.unitedfresh.org/assets/files/Letter%20on%20Passage%20of%20S%20%20510%20and%20Tester%20Amendment.pdf Accessed 30 Oct 2014

United States Department of Agriculture National Agricultural Statistics Service. Statistics by State – as reported by the Washington Field Office (2007) Census of agriculture. http://www.nass.usda.gov/Statistics_by_State/. Accessed 1 Sept 2014

VerCauteren KC, Seward NW, Hirchert DL et al (2005) Dogs for reducing wildlife damage to organic crops: a case study. In Nolte DL, Fagerstone KA (eds) Proceedings of the eleventh wildlife damage management conference. National Wildlife Research Center, Animal and Plant Health Inspection Service, U.S. Department of Agriculture, pp 286–293

Wang G, Zhao T, Doyle MP (1996) Fate of enterohemorrhagic *Escherichia coli* O157:H7 in bovine feces. Appl Environ Microbiol 62:2567

Chapter 11
A One Health Approach to Wildlife and Food Safety

Amanda Arens, Cheryl Scott, and Bennie Osburn

Abstract Global health problems including the assurance of safe and secure food are becoming more numerous and complex and require sensitive and transdisciplinary problem solving efforts. One Health provides the framework to approach food safety risks from the whole ecosystem of the food system by using a Web of Causation approach instead of an 'us vs. them' approach. This whole ecosystem, One Health approach focuses on prevention through the integration of wildlife, environmental, human, and domestic health sectors improving our ability to prevent rather than react to disease events. A true One Health viewpoint understands that all life is connected to its habitat, and the health of the whole sits squarely on a robust and sustainable environment. Safe food and water, thus ecological health, can be ensured using an evidence-based, transdisciplinary, collaborative based approach to the solution of food production and public health.

Keywords Agriculture • Climate change • Conservation • Ecosystem • Environmental health • Food safety • Global health • One Health • Public health • Wildlife

Introduction

There seems to be little doubt, the planet is changing. Climatic alterations, human population expansion, habitat alterations, ecosystem shifts, and hunger are profound. Global health problems including the assurance of safe and secure food are becoming even more numerous and complex, and require sensitive and transdisciplinary problem solving efforts. The globalization of our world now means that what is happening in one village in remote Africa or Asia will have repercussions

A. Arens (✉) • C. Scott • B. Osburn
Western Institute for Food Safety and Security, School of Veterinary Medicine, University of California, Davis, Davis, CA, USA
e-mail: alarens@ucdavis.edu

M. Jay-Russell, M.P. Doyle (eds.), *Food Safety Risks from Wildlife*, Food Microbiology and Food Safety, DOI 10.1007/978-3-319-24442-6_11

that affect the health and welfare of all communities across the globe. Expansive and rapid movement of people, pathogens, animal products, and produce around the world lends urgency to the common goal of health.

Food is only as safe and nutrient dense as the environment from which it comes. Food that is grown or processed in a contaminated environment becomes a food safety risk; food that is grown in nutrient-poor soil is less nutritious than food grown in a nutritionally richer environment; food from sick animals or animals carrying zoonotic pathogens becomes a food safety risk. Thus, healthy animals and a healthy environment are required to ensure a safe food supply.

The interdependency of human, animal, and environmental or ecosystem health in many aspects including food safety necessitates that problems in any of these sectors cannot be addressed in isolation, but rather need to be addressed by a larger, more systems-based approach in which all sectors are considered as part of the solution. One such approach that has come to the forefront is that of One Health—"the collaborative effort of multiple health science professions, together with their related disciplines and institutions – working locally, nationally, and globally – to attain optimal health for people, domestic animals, wildlife, plants, and our environment" (King et al. 2008). A One Health approach to food safety aims to have a safe food supply while at the same time ensuring the health and welfare of animals intended for food and preserving the health of the ecosystem in which the food lives or is grown.

One Health and Food Safety

One Health is an expanding area of professional global health advocacy arising from the recognition of the growing interconnections and overlap—economic, cultural, and physical—at the interface of human, animal, and ecosystem health (Fig. 11.1).

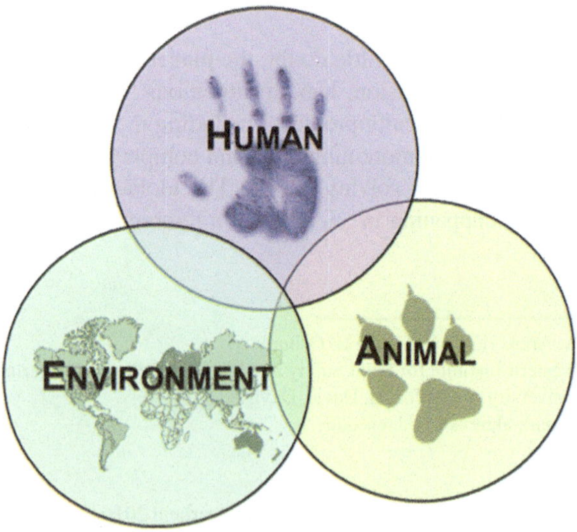

Fig. 11.1 One Health triad: Interconnection of humans, animals, and the environment

Although One Health is becoming an increasingly mainstream field of study today, the origins of One Health go back at least two centuries. In the nineteenth century, Rudolf Virchow, a German physician and pathologist, formally recognized the connection between human and animal health, stating, "Between animal and human medicine there is no dividing line, nor should there be. The object is different, but the experience obtained constitutes the basis of all medicine" (Kahn et al. 2007). Subsequently, the medical and veterinary professions noted the impact of animal diseases and ecological change on public health. Calvin Schwabe introduced the "One Medicine" concept in *Veterinary Medicine and Human Health* (Schwabe 1984) long after interest in the field had waned in the early 1900s. In recent years, the One Health concept has steadily gained recognition within the human and animal health sciences. In July 2008, the American Veterinary Medical Association (AVMA) released the report, *Executive Summary of the AVMA One Health Initiative Task Force*. In collaboration with the American Medical Association, the initiative provides groundbreaking recommendations and strategic action to support and expand the One Health concept across both veterinary and human health professions.

One Health seeks to shift the paradigm from the current "individual" and "disease-centered" approach that focuses on treatment to a "system-" or "community-based" approach that focuses on prevention. One Health is a creative way to view human, animal, and ecosystem health as a cooperative endeavor between health practitioners and environmental scientists in a collaborative and synergistic effort (Fig. 11.2). One Health provides the framework to address food safety issues in a transdisciplinary way in which solutions come from both within and beyond the various disciplines creating new perspectives to address these global, complex issues.

Safe Food, Wildlife Preservation, and Ecosystem Conservation through One Health

As our world population grows from 7 billion today to 9.1 billion by 2050 (United Nations) food security, food safety, and adequate nutrition will become increasingly more important. Everyone wants to trust that the food we eat and feed our families will not make us sick. In the USA, foodborne illness affects 48 million people, causes 128,000 hospitalizations, and results in over 3000 deaths annually (CDC 2011). While there are no current statistics of the global impact of foodborne diseases specifically, food and waterborne diseases together are estimated to kill 2.2 million people worldwide (WHO 2010). Foodborne illnesses arise from contamination from a number of pathogens including bacteria, viruses, parasites, and prions but can also be due to toxins, chemicals, metals, and allergens that are transmitted via food or water.

These microbial foodborne pathogens are part of the ecosystem where they live, survive, and find new hosts. These pathogens adapt to local conditions whether in

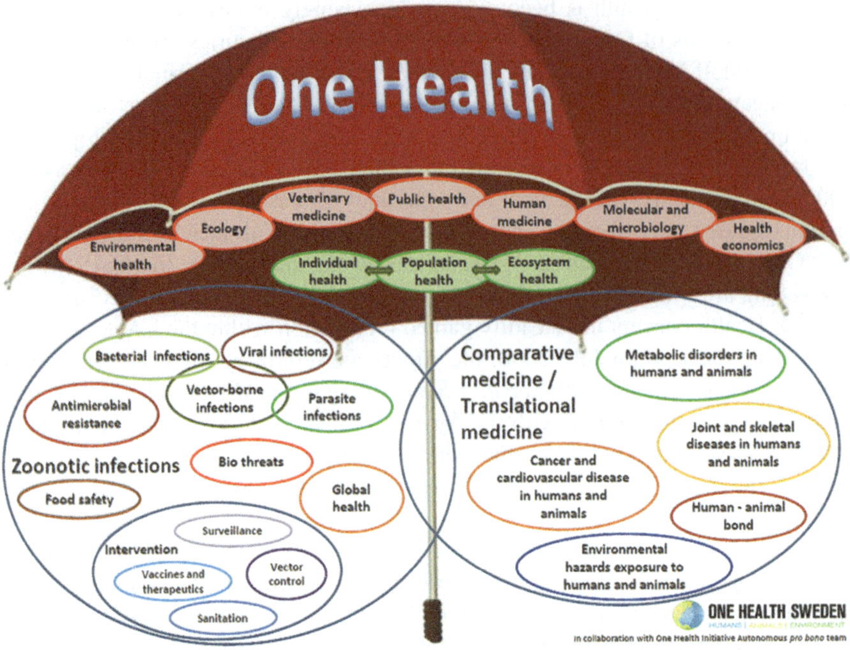

Fig. 11.2 One Health Umbrella—Broader vision of One Health demonstrating the relationships and interconnectedness of a variety of disciplines. Working together, these disciplines comprise a one health approach to solve global issues. "One Health Umbrella" graphic was developed under the auspices of One Health leader Dr. Olsen and the *One Health Sweden* team [two physicians and two veterinarians] in collaboration with the One Health Initiative Autonomous *pro bono* team [two physicians, two veterinarians, one PhD health research scientist] in December 2013

animals, plants, soil, or water. In some instances, these pathogens may replicate in the environment or find effective ways to propagate in more favorable conditions in animal or human hosts. The current hypothesis is that most of these pathogens are carried and multiply within the intestinal tracts of their animal hosts before they are eventually passed into the environment in the feces. Feces are often the rich and protective condition which allows the pathogens to remain viable and infective.

In the past few decades, 75 % of new human infections of all kinds are of zoonotic origin, meaning they can be spread from animals to people, and approximately 30 % of all globally emerging infections over the past 60 years have included pathogens that are commonly transmitted through food (Jones et al. 2008). Examples of zoonotic diseases that started as a foodborne disease and then became transmissible by human-to-human contact include HIV, Ebola, and SARS; examples of zoonotic pathogens that continue to be spread through food include *Salmonella*, *E. coli O157:H7*, *Listeria monocytogenes*, *Campylobacter*, and *Cryptosporidium*.

While foodborne illnesses have historically been associated with undercooked meat, the vehicles for human contamination have changed in the past decade. Between 1998 and 2008, 46 % of foodborne illnesses were associated with fresh produce, 22 % were associated with meat and poultry, 20 % were associated with dairy and eggs, and 6.1 % of illnesses were associated with fish and shellfish (Painter et al. 2013). Even though most cases of produce-associated illnesses are often attributed to contamination with *Norovirus*, a human pathogen, enteric zoonotic foodborne pathogens such as pathogenic *E. coli*, *Salmonella*, and *Campylobacter* cause a significant amount of produce-associated foodborne outbreaks. Raw produce is at risk because there is often no kill step to reduce or eliminate the pathogen(s) that may contaminate the products at any point along the food production continuum. Further, many fresh-cut fruits and vegetables are not amenable to treatments to kill pathogens and some are field-packed and thus not subject to a processing step (Jay-Russell 2013). Thus, contamination at any place along the production chain can cause foodborne illness. The change in dietary preferences in Western cultures to consume more raw agricultural products, thus failing to have this final kill step, whereby produce could be sterilized, is one of the reasons for the increase in produce-associated outbreaks. With this increase in occurrence of outbreaks in fresh produce, people have looked for the cause or source of contamination of these specific commodities.

There are 25 animal-derived foodborne pathogens which have been implicated as the causative agent of disease in people. Of these 25 pathogens, nine are considered of greatest importance by the Centers for Disease Control and Prevention. Of these nine highly important pathogens, eight may be of domestic and wild animal origin and include *Salmonella*, *Campylobacter*, *Cryptosporidium*, *E. coli* O157:H7, *Clostridium*, *Listeria*, *Toxoplasma*, and *Yersinia*. Routes of contamination of these pathogens onto fresh produce can be direct fecal contamination; through water, soil amendments such as manure or compost, or wind; or as secondary contamination from unclean equipment, clothes, or workers.

The current approach to a foodborne outbreak is to focus on the human illness and "trace back" the outbreak to find a "root cause." Once a plausible cause has been identified, recommendations are made, often solely focused on food safety to prevent the same type of contamination from occurring again. Wildlife may pose a risk to food safety as a probable source of contamination (California Department of Public Health 2007; Jay-Russell 2013; Rice 2014). However, there is often a lack of conclusive evidence implicating wildlife in foodborne illness outbreaks because they typically are not present at the time the traceback investigation is performed. An example of this is the 2006 outbreak of *E. coli* O157:H7 in California's Salinas Valley in bagged spinach. This was the first major outbreak involving fresh produce and sickened almost 200 people across 26 states and led to 3 deaths (CDC 2006). The "root cause" of the outbreak was not conclusively determined; however wildlife, especially feral swine, and grazing cattle were both implicated based on epidemiological and laboratory findings during the outbreak investigation (Jay et al. 2007, California Department of Public Health 2007). The leafy greens industry rapidly responded to this outbreak by creating the Leafy Greens Marketing Agreement (LGMA).

Because many of the known foodborne pathogens are zoonotic and may be found in wildlife and environmental reservoirs, addressing these sources is certainly a critical piece to the development of control measures aimed at the environmental level to reduce the incidence of human exposure. However, mitigation of wildlife contamination is a challenge. There are no economically feasible mechanisms to prevent direct contact. For example, barriers can be used to prevent access from some animals, but they are not all exclusive; poisons are toxic for many animals, not just the target animals, and have downstream effects such as decimating raptor populations; and habitat removal is detrimental to the environment and overall ecology. We know that intact ecosystems contribute to agricultural productivity by providing soil fertility, improved water quality, recharging of groundwater, and pollination of plants. So, how do we maintain the ecosystem and keep our food safe?

A One Health approach focuses on prevention. One Health shifts the "focus upstream to ecological, animal and environmental sources and influences responsible for these illnesses and helps identify the most effective points for the initiation of food safety actions" (King 2012). Coordination of wildlife, environmental, human, and domestic health sectors improves our ability to prevent disease events rather than simply reacting to them. Prevention is always preferable to control because it actively avoids the impacts of disease.

Balanced Solutions to the Food Safety and Wildlife Interface through One Health

Food safety has historically been recognized as, and measured by, the impact on people and the risk to human health (Rabinowitz et al. 2008). Animals and wildlife have been viewed as a direct threat to food safety. However, this "Us vs. Them" approach has led to policies for avoidance and vector/reservoir population control. Ultimately, risks are mitigated with barriers (Rabinowitz et al. 2008). This "Us vs. Them" approach focusing solely on the animals ignores other sources or routes of contamination.

Instead of a root cause approach, a more One Health approach is to view the whole ecosystem of the food system and analyze the Web of Causation (Fig. 11.3). Because of the intricate relationships between people, animals, and the environment, there is an intricate array of relationships similar to a spider's web that includes the commodity, agricultural practices, sources of contamination (e.g., domestic and wild animals, people, water, and soil), environmental conditions including weather, and routes of contamination (e.g., tools, farm equipment, people, water, soil amendments, wind, and direct contamination).

The Web of Causation is the first step in addressing food safety at the preharvest level where it provides a whole ecosystem perspective to use when faced with determining potential routes for pathogens to reach commodities. In this open environment, all the factors associated with the Web of Causation must be considered when assessing potential routes or pathways of contamination. In contrast, post-harvest processing and manufacturing of food products occur in a much more rigid and confined environment that can be controlled. Thus, preharvest food safety poses a

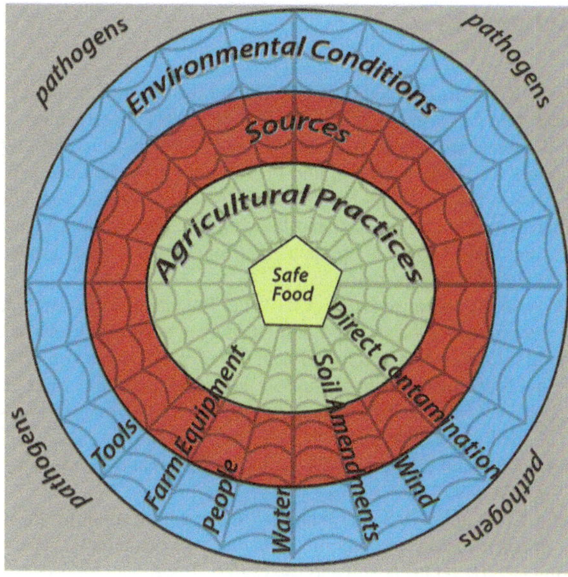

Fig. 11.3 Web of Causation for contamination of safe food by pathogens demonstrating that there is often no one point source, but rather a series of events throughout the ecosystem that are interconnected like the strands on a spider's web

more complex and greater challenge than post-harvest food safety and thus needs a more complex and dynamic approach to addressing food safety challenges.

The Web of Causation provides the opportunity to move beyond "Us vs. Them" and develop preventive strategies that are in the best interest of people, animals, and the environment. Thus, the Web of Causation is a One Health approach to food safety where the vision of One Health is to "optimize human-environmental interactions while minimizing health hazards to humans and animals and preserving a balanced ecosystem" (Zinstaag et al. 2009).

One example of a transdisciplinary, One Health approach is the concept of co-management which offers a comprehensive solution to the problem. "Co-Management is an approach to conserving soil, water, air, wildlife, and other natural resources while simultaneously minimizing microbial hazards associated with food production" (Leafy Green Marketing Association). The expectation of co-management is that safe food now becomes a collaborative priority for all stakeholders including landowners, farmers, conservation groups, buyers, industry, public health, ecosystem scientists, and wildlife agencies (see also Chap. 9).

A true One Health viewpoint understands that all life is connected to its habitat, and the health of the whole sits squarely on a robust and sustainable environment. Safe food and water, and thus ecological health, can be ensured using an evidence-based, transdisciplinary, collaborative based approach to the solution of food production and public health. It seems incumbent upon this generation of scientists and problem solvers to attempt to leave the world to our children in a more logical, balanced, and sustainable direction.

References

Bryce E. Wildlife forced out of California 'salad bowl' by food safety regulations. The Guardian

California Department of Public Health (2007) Investigation of an *Escherichia coli* O157:H7 Outbreak associated with dole pre-packaged spinach. http://www.cdph.ca.gov

Centers for Disease Control and Prevention (2011) Estimates of foodborne illness in the United States. www.cdc.gov/foodborneburden/. Accessed 25 Sept 2014

Centers for Disease Control and Prevention (2006) Multistate outbreak of E. coli O157:H7 infections linked to fresh spinach (final update). www.cdc.gov/ecoli/2006/spinach-10-2006.html. Accessed 25 Sept 2015

UC ANR Cooperative Extension (2012) Balancing food safety and sustainability – opportunities for co-management. http://ucfoodsafety.ucdavis.edu/files/157154.pdf

Jay MT, Cooley M, Carychao D et al (2007) *Escherichia coli* O157:H7 in feral swine near spinach fields and cattle, Central California coast. Emerg Infect Dis 13:1908–1911

Jay-Russell M (2013) What is the risk from wild animals in food-borne pathogen contamination? CAB Rev 8:no. 040

Jones KE, Patel GN, Levy MA et al (2008) Global trends in emerging infectious diseases. Nature 451:990–993

Kahn LH, Kaplan B, Steele JH (2007) Confronting zoonoses through closer collaboration between medicine and veterinary medicine (as 'one medicine'). Vet Ital 43(1):5–19

King LJ (2012) One health and food safety. Improving food safety through a One Health approach: workshop summary, pp 218–225

King LJ, Anderson LR, Blackmore CG et al (2008) One Health initiative task force report. J Am Vet Med Assoc 233(2):259–261

Painter JA, Hoekstra RM, Ayers T et al (2013) Attribution of foodborne illnesses and hospitalizations and deaths to food commodities by using outbreak data, United States 1998–2008. Emerg Infect Dis 19:407–415

Rabinowitz PM, Odofin L, Dein FJ (2008) From "us vs. them" to "shared risk": can animals help link environmental factors to human health. Ecohealth 5:224–229

Rice DH (2014) Produce contamination by other wildlife. The produce contamination problem (second edition) causes and solutions. Food Laboratory Division, New York State Department of Agriculture and Markets, Albany, NY, pp 167–183

Schwabe CW (1984) Veterinary medicine and human health. Williams and Wilkins, Baltimore, MD

World Health Organization (2010) Foodborne diseases. www.who.int/foodsafety/areas_work/foodborne-diseases/en/. Accessed 25 Sept 2015

Zinstaag J, Shelling E, Bonfoh B et al (2009) A 'One Health' research and application tool box. Vet Ital 45(1):121–133

Index

A

Adaptive management, 185–186, 188, 194
American Veterinary Medical Association (AVMA), 243
Arizona Leafy Green Marketing Association (AZLGMA) board, 179
Asian flu, 102
Avian influenza (AI) viruses. *See* Influenza A viruses (IAV)
Avian paramyxovirus (APMV). *See* Newcastle disease viruses (NDV)

B

Bacteria
- antimicrobial resistance, 70–71
- *Campylobacter*
  - antimicrobial resistance, 100–101
  - bird feeder, 91
  - *C. coli*, 67
  - *C. concisus*, 67
  - *C. jejuni*, 67, 99
  - *C. upsaliensis*, 67
  - in Europe, 98
  - gastrointestinal and extra-gastrointestinal infections, 67
  - genetic diversity, 68
  - genus, 98
  - in Germany, 68
  - *Listeria monocytogenes*, 69–70, 101
  - microbiological methodology, 67
  - in North-Western Italy, 68
  - in Norway and Finland, 68
  - public health concerns, 99–100
  - seasonality, 69
  - in Spain, 68–69
  - in Sweden, 68
  - transmission pathways, 67–68
- *Clostridium perfringens*, 101
- *E. coli*, 101
- non-typhoidal *Salmonella* spp., 63–65
- *Salmonella* (*see Salmonella*)
- STEC
  - in Austrian Alps, 66
  - game meat chain and human infection, 66–67
  - infection, sources, 65
  - prevalence surveys, 65
  - reservoirs of, 65
  - risk factors, 66
  - sporadic cases, 65
  - trend for, 65
- *Yersinia enterocolitica*, 101

Biodiversity, 32, 44, 52
Bovine tuberculosis, 77–78
Brucella, 3, 76
Brucellosis, 76–77
Bushmeat, 32, 35, 36, 48, 49, 52

C

CAFOs. *See* Concentrated animal feeding operations (CAFOs)
California Leafy Green Products Handler Marketing Agreement (CALGMA), 170, 173, 174, 179, 203, 207

M. Jay-Russell, M.P. Doyle (eds.), *Food Safety Risks from Wildlife*, Food Microbiology and Food Safety, DOI 10.1007/978-3-319-24442-6

Campylobacter, 13, 16, 20
  antimicrobial resistance, 100–101
  bird feeder, 91
  C. concisus, 67
  in Europe, 98
  gastrointestinal and extra-gastrointestinal infections, 67
  genetic diversity, 68
  genus, 98
  in Germany, 68
  Listeria monocytogenes, 69–70
  microbiological methodology, 67
  in North-Western Italy, 68
  public health concerns, 99–100
  seasonality, 69
  in Spain, 68–69
  in Sweden, 68
  transmission pathways, 67–68
Campylobacter jejuni, 4, 6, 11, 15, 27, 67, 98–100
Campylobacter upsaliensis, 67
Ceftriaxone, 97
Co-management processes
  adaptations, 205
  balancing food safety, 202
  bare ground buffers, 208, 209
  CAFOs, 205
  conservation practices, 212
  contamination risk, 204
  costs, 210–211
  cover crops, 206, 210
  cropping system, 209
  eliminate wildlife, 203
  Endangered Species Act, 204
  factors, pathogen transport and survival, 205
  FDA Food Safety Modernization Act, 204
  fecal contamination, careful inspection, 207
  fencing, 208–209
  food safety professionals, 202, 212
  grower practice, 206
  irrigation water, 206
  LGMA, 205
  liability insurance, 211–212
  low-hazard/negligible risk, 208
  management decisions, 203
  noise cannons, 207
  on-farm decisions, 214
  on-farm food safety programs, 219
  pre-plant risk assessment, 205
  resource conservation, 202
  set traps, 209
  stakeholders, 212
  support and training opportunities, 213–214
  temporal shift, 206
  vegetation, 203
  water, agronomic suitability, 209
  water bodies, 203
  water quality communities, 212
Commodity Specific Food Safety Guidelines for the Production and Harvest of Lettuce and Leafy Greens (CSGLLG), 170, 171
Concentrated animal feeding operations (CAFOs), 188, 205
Coxiella burnetii, 60, 78
Cryptococcus, 118
Cryptosporidium, 9, 13, 16, 20
Culex spp., 109, 115, 116

D

DNA-based serotyping. *See* Molecular serotyping

E

Eastern equine encephalitis virus (EEEV), 109, 110, 115–117
Echinococcus granulosus, 73–74
Echinococcus multilocularis, 73–74
Emerging infectious diseases (EIDs), 31, 36–37
Environmental impact statement (EIS), 233
Epidemiological cut-off values (ECOFF), 100
Escherichia coli (*E.coli*), 16, 17, 20, 21
Escherichia coli O157:H7, 3, 4, 7, 15, 17, 20, 65, 66, 175, 202, 212, 218, 221, 229

F

Food and Drug Administration (FDA)
  covered farm, 234
  draft regulation, 234
  Produce Safety Rule
    AEI, 231, 233
    biological hazards, 231
    commodity-specific guidance documents, 231
    cost estimation, 232
    economically significant rule, 231
    FSMA, 228–230
    EIS, 233
    on-farm food safety decisions, 220
    QAR, 231, 233
    small-scale, 232
Foodborne pathogens
  antibiotic resistance, reservoir, 23

contaminated meat consumption, 2–4
emerging pathogens, reservoir for, 23
mitigation and control strategies
adaptive management, 184–186
agricultural crops and livestock, 183
developing strategies, 188–190
fencing, 193
guard dogs, 194
habitat modification, 192
lethal control, 190–191
mitigation measures, 190
monitoring, 194
nonlethal techniques, 192, 193
physical barriers, 192
problem identification, 187
repellants, 192, 193
reproductive control, 191–192
risk assessment, 187–188
scare devices, 192, 193
prevalence, 4–5
climatic/seasonal impact, 20–21
cultivation bias, 5
detection method, specificity, 13
hosts behavioral features, 17–20
host's physiological state, 16–17
management, 21
sample size and number, 13
sample source, 5
produce field contamination, 4
water sources contamination, 22–23
wildlife–livestock–human interaction, in Europe
bacteria (*see* Bacteria)
behavioral changes, 62
bovine tuberculosis, 77–78
brucellosis, 76–77
extensive farming, 60
game meat, 62
game species, 60
HEV, 71–73
intensive farming, 60
management practice, 62
Mycobacterium bovis, 77–78
parasites (*see* Parasites)
proximity, 61
Q fever, 78
socioeconomic changes, 62
tularemia, 79
zoonotic diseases, 2, 60
Food regulation
assist produce growers, education and extension, 234–237
balancing food safety, 221–223
decision making, 218, 224–226
domestic animals, 218
food safety requirements, 218
on-farm food safety decisions
buyers, 220
consumer confidence, 220, 221
crop–wildlife conflicts, mitigation, 219
farm visitors and customers, 220
federal agencies, 220
health initiatives, 221
Produce Safety Rule, 220
wildlife management strategies, 221–223
produce safety (*see* Produce safety)
Food safety and wild life
co-management processes, 202, 212
global health problems, 241
healthy animals and environment, 242
One Health approach, 242
co-management approach, 247
ecosystem conservation, 243–246
human and animal health, 243
individual and disease-centered approach, 243
One Medicine concept, 243
public health, 243
system/community-based approach, 243
transdisciplinary way, 243
us *vs.* them approach, 246
Web of Causation, 246–247
wildlife preservation, 243–246
Food safety, leafy green
CALGMA, 169
California vegetable growers, 170
CSGLLG, 170
expert panel recommendations, 179
recommended changes
adjacent land, 177–178
animal intrusion, 177
animals of significant risk, 175–177
contamination, 178
crop damage, 178–179
spinach outbreak, 169
survey
California leafy green food safety guidelines, 171
co-management issues, 172
co-management practices, 171
conservation practices, 174
data collection and analysis, 172
distribution, 171
expert panel review, 172
food safety programs, 173–174
Western Growers, 170

Food Safety Modernization Act (FSMA), 132, 228–230
Food Safety Preventive Controls Alliance (FSPCA), 235
Francisella tularensis, 79
Fungi, 118

G
GenomTrakr WGS database, 143
Giardia, 10, 16, 23
Global Food Safety Initiative standards (GFSI), 211
Good Agricultural Practices (GAPs), 219, 226, 235
Guillian–Barré syndrome (GBS), 99

H
Hendra virus, 152
Hepatitis E virus (HEV), 71–73
Highlands J virus (HJV), 109, 117
Highly pathogenic avian influenza (HPAI)
- H5N1, 45, 102, 104
- H7N7, 105

Histoplasma capsulatum, 118
H1N1, 102
Hong Kong flu, 102

I
iDecisionSciences (IDS), 171
Influenza A viruses (IAV)
- *Anseriformes*, 103
- antigenic drift and shift, 102
- antiviral treatment, 103
- Asian flu, 106
- causes, 101
- epidemics, 102
- H17N10 and H18N11 viruses, 101
- H9N2, 105
- hemagglutinin, 101
- HPAI
  - H5N1, 102, 104
  - H7N7, 105
- LPAI viruses, 103, 104, 106
- neuraminidase, 101
- noneradicable zoonosis, 103
- pandemic, 102
- personnel protective equipment, 103
- swine influenza, 102
- vaccination, 103

Integrated pest management (IPM), 191
Irritable bowel syndrome (IBS), 99

J
Japanese encephalitis virus (JEV), 114

L
Listeria monocytogenes, 4
Live-attenuated influenza vaccines (LAIV), 103
Low pathogenicity avian influenza (LPAI) viruses, 103, 104

M
Molecular serotyping, 136–137
Most probable number (MPN) method, 133
Multilocus sequence typing (MLST), 100, 141–142
Multiple-locus variable-number tandem repeat analysis (MLVA), 141
Murray Valley encephalitis virus (MVEV), 115
Mycobacterium tuberculosis complex (MTC), 77

N
National Antimicrobial Resistance Monitoring system (NARMS), 97, 100
National Environmental Policy Act of 1969 (NEPA), 233
Natural Resource Conservation Service (NRCS), 220, 234
Newcastle disease viruses (NDV)
- acute disease, 107
- arboviruses, 111
  - *Culex* spp., 109, 110
  - disease manifestations, 108
  - EEEV, 115
  - families, 109
  - geographical range, 110
  - HJV, 117
  - hosts, 109
  - humans and domestic mammals, 108
  - JEV, 114
  - long-distance spread, 111
  - MVEV, 115
  - public health risk, 110
  - SLEV, 114
  - transmission cycles, 108, 109
  - USUV, 117
  - WEEV, 116
  - WNV (*see* West Nile virus (WNV))
- chickens, 107
- human infections, 108
- lentogenic/respiratory pathotype, 106
- mesogenic pathotype, 106
- natural reservoirs, 107

- neurotropic velogenic pathotype, 106
- outbreaks, 106
- peracute disease, 107
- species, 106
- strict biosecurity measures, 108
- transmission, 107
- vaccination, 108
- velogenic pathotype, 106
- virulent NDV, 106, 107
- viscerotropic velogenic pathotype, 106

Nipah virus
- foodborne risk, 163
- henipavirus, 152, 163
- human infection, 152
- mode of transmission
  - bat bitten fruit consumption, 157
  - foodborne, 155–156, 164
  - pandemic, 164
  - person-to-person, 152, 157, 164
- outbreak
  - India/Bangladesh, 153, 154, 163
  - Malaysia/Singapore, 152–153
- *Pteropus bats*, 152, 157
- stage III zoonotic disease, 152
- transmission, horsemeat, 163
- vaccination, 164

Non-typhoidal *Salmonella* spp., 63–65

O

One Health approach, 242
- co-management approach, 247
- ecosystem conservation, 243–246
- human and animal health, 243
- "individual" and "disease-centered" approach, 243
- "One Medicine" concept, 243
- public health, 243
- system/community-based approach, 243
- transdisciplinary way, 243
- Web of Causation, 246–247
- wildlife preservation, 243–246

P

Parasites
- *Echinococcus granulosus*, 73–74
- *Echinococcus multilocularis*, 73–74
- *Toxoplasma gondii*, 74–75
- *Trichinella* spp., 75–76

Phasianidae, 112

PrimusLabs.com GAP programs, 173

Produce safety
- FDA, 226 (*see also* Food and Drug Administration (FDA))
- FSMA, 228–230
- voluntary compliance
  - buyers, 226
  - commodity groups, 227
  - farmers and stakeholders, 228
  - PSP, 227–228
  - regulatory requirements, 228
  - third-party audit industry, 226–227

Produce Safety Alliance (PSA), 235, 236

Produce Safety Project (PSP), 228

Pulsed-field gel electrophoresis (PFGE), 138–140

Q

Q fever, 76, 78

Qualified end user, 229

R

Raw date palm sap
- discouraging consumption, 158–159
- safety improvement, 159–163
  - bamboo skirts, 160
  - only safe sap intervention, 162
  - sap stream and collection pot, 159–160

Red-legged partridges, 112

S

Salmonella
- antimicrobial resistance
  - backyard chickens, 97
  - genes, 97
  - geographical distribution, 97
  - isolation, 97
  - NARMS, 97
  - wild birds, 98
- bird feeder, 91
- detection, molecular tools
  - DNA microarrays, 134–135
  - PCR, 134, 135
  - qPCR, 134
  - sample preparation, 135
- enteritidis, 93
- and foodborne disease, 132
- genus, 93
- molecular serotyping, 136–137
- molecular typing methods
  - features, 138
  - gram-negative bacterial foodborne pathogens, 137
  - MLST, 141–142
  - PCR-based typing, 140–141

Salmonella (*cont.*)
PFGE, 138–140
WGS, 142–143
non-typhoidal salmonellosis, 92
normal gut microflora, 92
organic food, 91
public health concerns
outbreak, 94
pigeons, 96
potential health risks and mitigation strategies, 96
poultry, 95
S. Typhimurium, 94, 97
transmission, mode, 92, 96
S. Cholerasuis, 64
virulence potential, evaluation, 144
Shiga toxin-producing *Escherichia coli* (STEC)
game meat chain and human infection, 66–67
in Austrian Alps, 66
infection, sources of, 65
prevalence surveys, 65
reservoirs of, 65
risk factors, 66
sporadic cases, 65
trend for, 65
Spanish flu, 102
Spoligotyping, 141
Sprout Safety Alliance (SSA), 235
Staphylococcus aureus, 13
St. Louis encephalitis virus (SLEV), 114
Supershedders, 66
Swine influenza (SI), 102

T
Toxoplasma gondii, 74–75
Trichinella spp., 3, 75–76
Tularemia, 79

U
United States Department of Agriculture (USDA), 90, 209, 213, 226, 227, 231
Usutu virus (USUV), 117

W
Western equine encephalitis virus (WEEV), 116
West Nile virus (WNV)
age-associated differences, 112
Corvidae family, 111
Galliformes, 112
infection route, 113
isolation, 112
lineage 1 viruses, 111
lineage 2 viruses, 111
mature chickens, 112
prevention, 113
rate and incidence, 111
seroconversion, 112
vaccination, 114
Whole Genome Sequence Analysis (WGS), 142–143
Wild birds and free-range chickens
IAV (*see* Influenza A viruses (IAV))
one health issues
bird feeders, 91
human food, 91
meat, 90
organic food, 90
pathogen testing, 90
risk, 90, 91
zoonotic pathogens, 92–101
bacteria (*see* Bacteria)
fungi, 118

Z
Zoonotic pathogens
bacteria (*see* Bacteria)
fungi, 118
IAV (*see* Influenza A viruses (IAV))
Zoonotic viruses
animal-based food systems, 37–42
contact, 42
distal factors, 43–44
transmission routes, 42
animal production system
cost of, 32
global trends, 45–46
risk, 32
scale of, 33–35
current surveillance systems, 51
EID, 31, 36–37
intervention and risk mitigation options
conservation and health, 47–48
farmed wildlife, 50–51
policy landscape, 48
proximal and distal risk factors, 46
realistic interventions, 49–50
risk analysis, 48–49
wild harvest system, 32
costs, 32
global trends, 44–45
risk, 32
scale of, 35–36
sophisticated sale and distribution systems, 32

The manufacturer's authorised representative in the EU is Springer Nature Customer Service Centre GmbH, Europaplatz 3, 69115 Heidelberg, Germany. If you have any concerns regarding our products, please contact ProductSafety@springernature.com

Printed and bound by CPI Group (UK) Ltd, Croydon, CR0 4YY

22/07/2026

02174142-0004